LONDON MATHEMATICAL SOCIETY STUDENT TEXTS

Managing Editor: Ian J. Leary,
Mathematical Sciences, University of Southampton, UK

68 Hyperbolic geometry from a local viewpoint, LINDA KEEN & NIKOLA LAKIC
69 Lectures on Kähler geometry, ANDREI MOROIANU
70 Dependence logic, JOUKU VÄÄNÄNEN
71 Elements of the representation theory of associative algebras II, DANIEL SIMSON & ANDRZEJ SKOWROŃSKI
72 Elements of the representation theory of associative algebras III, DANIEL SIMSON & ANDRZEJ SKOWROŃSKI
73 Groups, graphs and trees, JOHN MEIER
74 Representation theorems in Hardy spaces, JAVAD MASHREGHI
75 An introduction to the theory of graph spectra, DRAGOŇ CVETKOVIĆ, PETER ROWLINSON & SLOBODAN SIMIĆ
76 Number theory in the spirit of Liouville, KENNETH S. WILLIAMS
77 Lectures on profinite topics in group theory, BENJAMIN KLOPSCH, NIKOLAY NIKOLOV & CHRISTOPHER VOLL
78 Clifford algebras: An introduction, D. J. H. GARLING
79 Introduction to compact Riemann surfaces and dessins d'enfants, ERNESTO GIRONDO & GABINO GONZÁLEZ-DIEZ
80 The Riemann hypothesis for function fields, MACHIEL VAN FRANKENHUIJSEN
81 Number theory, Fourier analysis and geometric discrepancy, GIANCARLO TRAVAGLINI
82 Finite geometry and combinatorial applications, SIMEON BALL
83 The geometry of celestial mechanics, HANSJÖRG GEIGES
84 Random graphs, geometry and asymptotic structure, MICHAEL KRIVELEVICH et al
85 Fourier analysis: Part I – Theory, ADRIAN CONSTANTIN
86 Dispersive partial differential equations, M. BURAK ERDOĞAN & NIKOLAOS TZIRAKIS
87 Riemann surfaces and algebraic curves, R. CAVALIERI & E. MILES
88 Groups, languages and automata, DEREK F. HOLT, SARAH REES & CLAAS E. RÖVER
89 Analysis on Polish spaces and an introduction to optimal transportation, D. J. H. GARLING
90 The homotopy theory of $(\infty, 1)$-categories, JULIA E. BERGNER
91 The block theory of finite group algebras I, MARKUS LINCKELMANN
92 The block theory of finite group algebras II, MARKUS LINCKELMANN
93 Semigroups of linear operators, DAVID APPLEBAUM
94 Introduction to approximate groups, MATTHEW C. H. TOINTON
95 Representations of finite groups of Lie type (2nd Edition), FRANÇOIS DIGNE & JEAN MICHEL
96 Tensor products of C^*-algebras and operator spaces, GILLES PISIER
97 Topics in cyclic theory, DANIEL G. QUILLEN & GORDON BLOWER
98 Fast track to forcing, MIRNA DŽAMONJA
99 A gentle introduction to homological mirror symmetry, RAF BOCKLANDT
100 The calculus of braids, PATRICK DEHORNOY
101 Classical and discrete functional analysis with measure theory, MARTIN BUNTINAS
102 Notes on Hamiltonian dynamical systems, ANTONIO GIORGILLI
103 A course in stochastic game theory, EILON SOLAN
104 Differential and low-dimensional topology, ANDRÁS JUHÁSZ
105 Lectures on Lagrangian torus fibrations, JONNY EVANS
106 Compact matrix quantum groups and their combinatorics, AMAURY FRESLON
107 Inverse problems and data assimilation, DANIEL SANZ-ALONSO, ANDREW STUART & ARMEEN TAEB
108 Künneth geometry, M.J.D. HAMILTON & D. KOTSCHICK
109 ADE: Patterns in mathematics, PETER J. CAMERON et al

London Mathematical Society Student Texts 110

Sets and Transfinite Algebra

THOMAS MÜLLER
University of Vienna

CAMBRIDGE
UNIVERSITY PRESS

CAMBRIDGE
UNIVERSITY PRESS

Shaftesbury Road, Cambridge CB2 8EA, United Kingdom

One Liberty Plaza, 20th Floor, New York, NY 10006, USA

477 Williamstown Road, Port Melbourne, VIC 3207, Australia

314–321, 3rd Floor, Plot 3, Splendor Forum, Jasola District Centre,
New Delhi – 110025, India

Cambridge University Press is part of Cambridge University Press & Assessment,
a department of the University of Cambridge.

We share the University's mission to contribute to society through the pursuit of
education, learning and research at the highest international levels of excellence.

www.cambridge.org
Information on this title: www.cambridge.org/9781009737876
DOI: 10.1017/9781009737883

When citing this work, please include a reference to the DOI 10.1017/9781009737883

First published 2026

A catalogue record for this publication is available from the British Library

A Cataloging-in-Publication data record for this book is available from the Library of Congress

ISBN 978-1-009-73784-5 Hardback
ISBN 978-1-009-73787-6 Paperback

CONTENTS

§0. Preface

This book comes in two parts; the first, consisting of §§1–7, offers an informal axiomatic introduction to the basics of set theory, including a thorough discussion of the axiom of choice and some of its equivalents. The second part, consisting of §§8–14, is written at a somewhat more advanced level, and treats selected topics in transfinite algebra; that is, algebraic themes where the axiom of choice, in one form or another, is useful or even indispensable.

The book contains a number of topics not usually found in introductory texts: we discuss, among other things, Tarski's fixed point theorem for complete lattices, as well as his generalisation of Weierstraß' intermediate value theorem to densely ordered sets, Hamel's general solution of Cauchy's functional equation

$$f(x+y) = f(x)+f(y),$$

Novotný's apparently little known construction of all morphisms between given mono-unary algebras, Hodges' surprising result to the effect that (a special case of) Krull's maximal ideal theorem is equivalent to the axiom of choice, a thorough exposition of Birkhoff's famous subdirect decomposition theorem, a discussion of (the basics of) matroid theory in arbitrary cardinality, various characterisations of the Jacobson radical, and, finally, Artin's solution of Hilbert's 17th problem concerning the decomposition of definite rational functions over $\mathbb{Q}$ as a sum of squares of rational functions. To the best of the author's knowledge, ours is the first treatment of this beautiful and important argument in English (Artin's original publication having been written in German).

Part II as a whole may be seen as one long advertisement for the usefulness of transfinite induction arguments in algebraic contexts; in that sense Parts I and II are meaningfully connected. Nevertheless, a considerable amount of effort has gone into allowing flexibility in perusing the material presented: Part I is designed to be readable and well motivated on its own, and to provide the basis of a short introduction to some of the main ideas and techniques of set theory, which should already be useful at an (advanced) undergraduate level; a lecturer might just pick topics from Part I as the basis of an interesting (one-semester) course. On the other hand, a reader or lecturer using this book might also, after a short introduction to the axiom of choice (and those of its equivalents as are needed), with or without proofs, delve straight into some of the topics in Part II.

The main groups of readers this text aims at are advanced undergraduates and graduate students of mathematics, as well as professional mathematicians, in particular those preparing a course of lectures in this general area. Several devices are employed to try to engage and keep the reader's attention. For instance, the book offers more than 140 carefully planned exercises, many (but by far not all) being rather routine, inviting the reader to actively engage with the material, test understanding, and, occasionally, develop further certain aspects of the theory presented. The appendix provides full solutions to all problems. Moreover, for standard concepts from algebra, definitions are not always included in the text; for instance, the reader will look in vain for the definition of a group, that of a subgroup, or of a ring. In this case, the student is expected to actively explore alternative sources like books, lecture notes, or the internet, if necessary. Usually, some suitable reference is given in the text, at the beginning of the pertinent section.

While composing this text, the author allowed himself the pleasure of reading, respectively re-reading, some of the relevant research literature having given rise to the theory presented. This made the task of writing a pleasure, and if some of the joy and enthusiasm I felt passes on to my readers the effort will have been a worthwhile one. Various parts of the material presented here have been used successfully in second-, third-, and fourth-year courses, respectively, at Queen Mary & Westfield College. I wish the reader fun in exploring some of the most fundamental and beautiful mathematics created during the twentieth century, and hope that lecturers, when preparing courses overlapping the material offered here, will find the large spread of attractive topics to choose from enjoyable and useful.

University of Vienna, T. Müller

Part I: Set Theory

Set theory, while being an important and ongoing research topic in its own right, may also be viewed as the basis of all mathematical theory. For instance, any algebraic object like a group, ring, or lattice, or a topological structure like the Euclidean plane $\mathbb{E}^2$, is, first of all, a set, endowed with extra structure (binary operations, order relations, metrics, etc.). As originally conceived during the last quarter of the 19th century, set theory is the creation of one single man: Georg Cantor (1845–1918). It was developed in a number of research papers during the period 1874–1891, and summarised in his "Beiträge zur Begründung der transfiniten Mengenlehre", published in two parts in 1895 and 1897; cf. [21].

Having studied in Zürich, Berlin, and Göttingen, Cantor obtained his doctorate in 1867 under the supervision of the number theorist E. Kummer (Berlin) with a thesis on integral solutions of equations of degree two in finitely many variables, entitled "De aequationibus secundi gradus indeterminatis". Subsequently, in 1869, he submitted a habilitation thesis to the university in Halle on equivalence of ternary forms in two variables entitled "De transformatione formarum ternariarum quadraticarum", after which he stayed in Halle for the rest of his life, from 1877 on as a full professor.

After his early work in number theory Cantor, following a suggestion of Heine, turned his attention to an important problem concerning the representation of a real function by a convergent trigonometric series of the form

$$\frac{b_0}{2} + \sum_{\nu=1}^{\infty} (a_\nu \sin x + b_\nu \cos x),$$

in particular proving uniqueness of such a representation; cf. [16], [17]. The question whether this uniqueness theorem also holds true in the presence of countably many exceptional points, led him to a new construction of real numbers as (equivalence classes of) rational sequences; cf. [18]. In this context, Cantor also showed that the set $\mathbb{A}$ of real algebraic numbers (which properly contains all rationals) is countable, while the set $\mathbb{R}$ of real numbers is uncountable; see [19] and [20].[1] Thus, the theory of Fourier series led Cantor into the new and fascinating realm of sets, which became his main research topic; some of the more elementary facets of this theory we are going to explore in what follows. The reader who would like to know more about Cantor's biography, mathematics, and philosophy, is referred to Dauben's excellent monograph [26].

[1] As a consequence of these two results, Cantor obtains a new proof for the existence of real transcendental numbers. Unlike Liouville's earlier proof, Cantor's is not constructive, but yields extra information: since $\mathbb{R} \setminus \mathbb{A}$ is uncountable, almost all real numbers are in fact transcendental.

We now turn to some particulars concerning this first part. Technically speaking, Part I provides an informal axiomatic introduction to set theory up to cardinal and ordinal arithmetic, thereby setting out the minimum of set theory which, according to the author's prejudices, any mathematician ought to know. The word "informal" refers to the use of ordinary language, instead of the more technical language of first order logic; see, however, §6 for some comments in this direction.

A number of topics, not usually found in books of this kind and introductory level, are included: for instance, we discuss Tarski's fixed point theorem for complete lattices, including his student's rather difficult proof of the converse, Tarski's generalisation of Weierstraß' intermediate value theorem to densely ordered sets, the existence of algebraic closures due to Steinitz, Hamel's general solution of Cauchy's functional equation

$$f(x+y) = f(x) + f(y),$$

the invariance of dimension for arbitrary vector spaces over skew fields, and Novotný's set-theoretic construction of all morphisms between two given mono-unary algebras. Both Ann Davis' and Novotný's arguments (which are presented, among other things, in §7), require more or less all the set theory learned in §1–§5.

In §4, we provide a careful discussion of the axiom of choice and some of its equivalents, describing in detail the rather sophisticated and sometimes quite deep proof methods, while §7, among several other things, also presents Wilfrid Hodges' surprising result to the effect that (a special case of) Krull's maximal ideal theorem implies the axiom of choice, giving full details of the beautiful and rather non-trivial proof.

§ 1. The axioms of set theory

Set theory considers objects called *classes*, between which a binary relation may hold:

$$A \in B. \tag{1.1}$$

The relationship (1.1) is expressed by saying that A *is a member (or element) of* B, or that A *belongs to* B. We define a *set* to be a class which is a member of some class. Two classes A and B are said to be *equal*, denoted by $A = B$, if they have the same members. Negation of the assertions (1.1) and $A = B$ is written as $A \notin B$, respectively $A \neq B$. From a more philosophical point of view, two objects are usually said to be equal, if they have the same properties; that is, if the same assertions are true of both. In order to obtain this usual interpretation of equality, we introduce the following axiom.

A1. *If $A = B$ and $\mathfrak{P}(X)$ is any sentence about classes, then $\mathfrak{P}(A)$ holds if, and only if, $\mathfrak{P}(B)$ holds.*

In the light of the above definition of equality, Axiom A1 may be taken as limiting the kind of statement concerning classes we are willing to discuss. Without wanting to enter into logical technicalities at this stage, we simply note that some care needs to be taken in defining what exactly we mean by a sentence here. Suffice it for the moment to say that a sentence is a *sequence of symbols involving set variables, class variables, the logical signs (connectives), and quantifiers acting on set variables, formed according to certain syntactical rules;* see §6 for more information concerning this aspect.

Our next axiom requires, for every meaningful statement involving a class variable X, a class to exist whose members are precisely the sets X for which the statement holds.

A2. *If $\mathfrak{P}(X)$ is any sentence about classes, then there exists a class whose members are precisely the sets X for which $\mathfrak{P}(X)$ holds.*

The class alluded to in Axiom A2 will be denoted by

$$\{X : \mathfrak{P}(X)\}.$$

In general, this class will not be a set; when it is a set, the sentence $\mathfrak{P}(X)$ is said to be *collectivising* in X. Thus, for instance, if A is a given set, the sentence '$X \in A$' is collectivising in X, since the class of all sets X such that $X \in A$ is simply the set A itself. On the other hand, the sentence '$X \notin X$' cannot be collectivising. For, if the class

$$\mathfrak{X} = \{X : X \notin X\},$$

the class of all sets X which do not contain themselves as an element, were a set, then the question *whether or not $\mathfrak{X}$ contains itself* leads to a contradiction (Russell's Paradox). In the presence of Axiom A2, the argument leading to Russell's Paradox merely shows that the above class is not a set; in particular, *there exist classes which are not sets.* Such classes are called *proper classes.*

By means of Axiom A2 we can define certain new classes as well as the familiar operations on sets, although we cannot at this stage assert that the resulting classes are sets; we shall have to postulate special axioms to this effect.

(i) The *empty class* $\emptyset$ is defined via the equation

$$\emptyset := \{X : X \neq X\}.$$

(ii) The *total class* $\mathfrak{T}$ is defined as

$$\mathfrak{T} := \{X : X = X\}.$$

(iii) If A and B are classes, then the *singleton* consisting of A, and the *pair* consisting of A and B, are defined as

$$\{A\} := \{X : X = A\},$$

respectively

$$\{A,B\} := \{X : X = A \text{ or } X = B\},$$

when these are sets, and are not defined otherwise. (Later we shall see that these constructions are defined whenever A and B are sets.)

(iv) If A is any class, then the *union* $\bigcup A$ of (or over) A is defined via

$$\bigcup A := \{X : X \in Y \text{ and } Y \in A, \text{ for some } Y \in A\}.$$

(v) If A is a class, then the *intersection* $\bigcap A$ of (or over) A is defined as

$$\bigcap A := \{X : X \in Y \text{ for all } Y \text{ such that } Y \in A\}.$$

(vi) If A and B are sets, then the *ordered pair* (or *couple*) consisting of A and B (in that order) is

$$(A,B) := \{\{A\},\{A,B\}\}.$$

We note that, according to this last definition,

$$(A,B) = (C,D) \iff A = C \text{ and } B = D; \tag{1.2}$$

cf. Exercise 3 below. When $C = \{A,B\}$, we shall write $A \cup B$, $A \cap B$ instead of $\bigcup C$ and $\bigcap C$, respectively. Given classes A and B, A is said to be *disjoint* from B, if $A \cap B = \emptyset$. Given classes A and B, A is said to be a *subclass* of B, denoted by $A \subseteq B$, if

$$X \in B \textit{ for all } X \textit{ such that } X \in A.$$

A subclass which is also a set is called a *subset*. If $A \subseteq B$ and $A \neq B$, then A is said to be a *proper* subclass, in symbols: $A \subset B$. We also write '$B \supseteq A$' or '$B \supset A$' in place of '$A \subseteq B$' or '$A \subset B$', respectively. Negation of any of these relations is indicated by the same symbol with a line drawn through it.

(vii) If A and B are classes, then the class

$$A - B := \{X : X \in A \text{ and } X \notin B\}$$

is called the *difference* of A and B, sometimes also the *complement* of B in A (we tend to reserve the latter expression for the case when $B \subseteq A$, in which case we also write $A \setminus B$ for the complement of B in A).

(viii) If A is any class, then

$$\mathscr{B}(A) := \{X : X \subseteq A\}$$

is called the *Boolean* of A (after George Boole, 1815–1864). Instead of $\mathscr{B}(A)$ one also finds other notation like $\mathscr{P}(A)$ ($\mathscr{P}$ for power class), or the more combinatorially inspired notation 2^A in the literature.

(ix) If A and B are classes, then

$$A \times B := \{Z : Z = (X, Y) \text{ where } X \in A \text{ and } Y \in B\}$$

is called the *Cartesian product* of A and B.

A *function* from a class A to a class B is a subclass F of $A \times B$ such that for each $X \in A$ there exists precisely one $Y \in B$ for which $(X, Y) \in F$. The class A is called the *domain* of F, and the class of *values* of F, that is,

$$\{Y : Y \in B \text{ and } (X, Y) \in F \text{ for some } X \in A\}$$

is called the *range* of F. A somewhat different notation is often used when focussing attention on the range. If A is a class and I is any set, then the values of a function from I to A are called a *family of elements* of A, *indexed* or *coordinated* by I. If x_i is the element of A corresponding to $i \in I$, then the family is denoted by $\{x_i\}_{i \in I}$, and x_i is called its *i-coordinate*, i the *index*, and I the *index set*. Every set can be indexed, e.g., by itself; this means that we describe the set A by a function from A onto itself, say by the identity function, in which every element of A corresponds to itself: $A = \{a\}_{a \in A}$. Thus, in dealing with a set, there is no loss of generality in taking it to be indexed.

If $F \subseteq A \times B$ is a function from a class A to a class B, and if $A' \subseteq A$ is a subclass, then we call $F' = F \cap (A' \times B)$ the *restriction* of F to A', denoted by $F' = F|_{A'}$; unless stated otherwise, F' is interpreted as a function from A' to B.

We now come to the main group of axioms. These state essentially that the empty class is a set, and that all reasonable constructions, when applied to sets, again yield sets.

A3. $\emptyset$ *is a set.*

A4. *Any subclass of a set is a set.*

A5. *If A and B are sets, then so is the pair $\{A,B\}$.*

A6. *If A is a set, then so is its Boolean $\mathscr{B}(A)$.*

A7. *If A is a set, then so is its union $\bigcup A$.*

A8. *If F is a function whose domain is a set, then its range is a set.*

We note some immediate consequences of these axioms.

(i) If A and B are sets, then so is the couple (A,B), and if (A',B') is another couple of sets, then

$$(A,B) = (A',B') \Longleftrightarrow A = A' \text{ and } B = B'.$$

For, $\{A,B\}$ and $\{A\} = \{A,A\}$ are sets by Axiom A5, hence so is $(A,B) = \{\{A\},\{A,B\}\}$. Moreover, given (A,B), we can reconstruct (and distinguish) A and B by examining first the members of (A,B), then their members.

(ii) If $\mathfrak{P}(X)$ is any sentence about classes, and A is a set, then

$$\{X \in A : \mathfrak{P}(X)\} := \{X : X \in A \text{ and } \mathfrak{P}(X)\}$$

is a subclass of A, and hence a subset by A4. In particular, the complement $A - B$ of any class B in a set A is a subset of A.

(iii) If A, B are sets, then so is their Cartesian product $A \times B$. For, if $X \in A$ and $Y \in B$, then $\{X\},\{X,Y\} \subseteq A \cup B$, hence $\{X\},\{X,Y\} \in \mathscr{B}(A \cup B)$, and so $(X,Y) \in \mathscr{B}\mathscr{B}(A \cup B)$. Thus, $A \times B$ is a subclass of $\mathscr{B}\mathscr{B}(A \cup B)$, the latter being a set.

(iv) Any function whose domain is a set is itself a set. For, let F be a function whose domain is a set A, say. Then, by A8, its range is a set B, and since $F \subseteq A \times B$, it follows from (iii) and A4 that F is a set. If I is a set and A is a class, then any function from I to A is a set; hence, we may form a class whose members are all functions from I to A. This class is denoted by A^I and is called a *Cartesian power*. Since each function from I to A is an element of $\mathscr{B}(I \times A)$, it follows that $A^I \subseteq \mathscr{B}(I \times A)$. In particular, if both I and A are sets, then so is A^I.

(v) If $A \neq \emptyset$, then $\bigcap A$ is a set. For, if $a \in A$, then $\bigcap A$ is a subclass of a, and a is a set (being a member of a class), so $\bigcap A$ is a set by Axiom A4.

The axioms given so far allow us to construct arbitrarily large finite sets, but no infinite ones; their existence has to be postulated separately. This is usually done by means of an *axiom of infinity*, asserting existence of an infinite set (once this concept has been defined). We shall use an alternative approach, essentially due to Gabriel [32], via universal sets. This axiom will turn out to be considerably stronger than the axiom of infinity; it will in fact eliminate the need, in most problems, to consider classes which are not sets.

Definition 1.1. A set U is said to be *universal*, or a *universe*, if it satisfies the following:

(i) If $X \in U$, then $X \subseteq U$.

(ii) If $X \in U$, then $\mathscr{B}(X) \in U$.

(iii) If $X, Y \in U$, then $\{X, Y\} \in U$.

(iv) If $F = (F_i)_{i \in I}$, where $F_i \in U$ for all $i \in I$ and $I \in U$, then $\bigcup F \in U$.

We now add as axiom of infinity:

A9. *Every set is a member of some universe.*

This axiom effectively does away with our need for the consideration of classes; for instance, instead of the class of all sets we can work with the class of all sets in a given universe, which is again a set. In what follows, we shall therefore reserve the term 'class' to refer to a set which is not necessarily a member of the universe under consideration.

In order to illustrate the use of Axiom A9, we shall define the natural numbers and show that they form an infinite set. Let U be a universe such that $\emptyset \in U$. Call a subset V of U *numeral*, if $\emptyset \in V$ and if $X \cup \{X\} \in V$ whenever $X \in V$. For example, U itself is numeral: if $X \in U$, then $\{X\} = \{X, X\} \in U$ by Property (iii) in Definition 1.1, hence $\{X, \{X\}\} \in U$ by applying (iii) again; and indexing the set $\{X, \{X\}\}$ by itself, we see that $X \cup \{X\} \in U$. Let $\mathbb{N}$ be the intersection of all numeral subsets of U; in particular, $\mathbb{N} \subseteq U$. Then, by definition, $\mathbb{N}$ itself is numeral. Writing x' for $x \cup \{x\}$, we observe the following properties of $\mathbb{N}$:

N1. $\emptyset \in \mathbb{N}$.

N2. *If* $x \in \mathbb{N}$, *then* $x' \in \mathbb{N}$.

N3. *Any subset of* $\mathbb{N}$ *satisfying* $N1$ *and* $N2$ *coincides with* $\mathbb{N}$ (this, of course, is just the principle of induction).

The natural numbers are now defined as the elements of $\mathbb{N}$, the first few being

$$0 = \emptyset,\ 1 = \{0\},\ 2 = \{0,1\},\ 3 = \{0,1,2\}, \ldots$$

(an immediate induction shows that, in general, $n = \{0, 1, \ldots, n-1\}$).

By a *positive integer* we mean a natural number $\neq 0$. A class is said to be *finite*, if it can be indexed by a natural number; otherwise it is *infinite*. With this definition, we can show that the set $\mathbb{N}$ of natural numbers is infinite. In preparation for this argument, we first note the following:

(a) *If* $n \in \mathbb{N}$, *then* $n' \not\subseteq 0$. For, $n' = n \cup \{n\} \neq \emptyset$, while $0 = \emptyset$.

(b) *If* $m, n \in \mathbb{N}$ *and* $m \in n$, *then* $m \subseteq n$. This clearly holds for $n = 0$, since then $m \notin n$ for all $m \in \mathbb{N}$, so that the hypothesis is always false. Denote by $\mathscr{M}$ the set of all $n \in \mathbb{N}$ such that $m \subseteq n$ for all $m \in \mathbb{N}$ satisfying $m \in n$; so, by the above, $0 \in \mathscr{M}$. If $n \in \mathscr{M}$ and $m \in n'$, then either $m = n$, in particular $m \subseteq n$, or $m \in n$, which implies $m \subseteq n$ since $n \in \mathscr{M}$ by hypothesis. Thus, $m \subseteq n$ in either case, and $n \subseteq n'$, so $m \subseteq n'$. Thus $n' \in \mathscr{M}$, so $\mathscr{M}$ satisfies $N1$ and $N2$, hence $\mathscr{M} = \mathbb{N}$ by $N3$.

(c) *If* $m' \subseteq n'$, *then* $m \subseteq n$. For, if $m \cup \{m\} \subseteq n \cup \{n\}$, then $m \in n \cup \{n\}$; hence, either $m \in n$, so $m \subseteq n$ by (b), or $m = n$, again implying $m \subseteq n$.

By applying (c) twice, we obtain:

(d) *If* $m' = n'$, *then* $m = n$.

Properties (a) and (d) together with N1–N3 constitute what is known as the *Peano axioms* for the natural numbers. Henceforth, we shall write $n+1$ for n'.

Proposition 1.2. *The set* $\mathbb{N}$ *of natural numbers is infinite.*

Proof. We have to show that, if n is any natural number, then there exists no function from n to $\mathbb{N}$ with $\mathbb{N}$ as range. This certainly holds for $n = 0$, since a function with domain $0 = \emptyset$ must have empty range, whereas $\mathbb{N} \neq \emptyset$. Let $\mathscr{M}$ be the set of all numbers $n \in \mathbb{N}$ such that no function from n to $\mathbb{N}$ with range $\mathbb{N}$ exists. By the above, $0 \in \mathscr{M}$. Let $n \in \mathscr{M}$, and suppose that $n+1 \notin \mathscr{M}$; thus, there exists a function f from $n+1$ to $\mathbb{N}$ with range $\mathbb{N}$. Since every element $\neq 0$ of $\mathbb{N}$ is of the form $x+1$ for some $x \in \mathbb{N}$, and since $x+1$ determines x uniquely (by (d)), we may, for any $y \in \mathbb{N} \setminus \{0\}$, denote the unique element $x \in \mathbb{N}$ satisfying $x+1 = y$ by $y-1$. Moreover, denote by $f(i)$ the unique element j such that $(i, j) \in f$. With these notations, we define a function g from n to $\mathbb{N}$ via

$$g(k) := \begin{cases} f(k), & f(k) \subseteq f(n) \\ f(k)-1, & \text{otherwise} \end{cases} \quad (k \in n).$$

Since $0 \subseteq f(n)$, this indeed defines a function from n to $\mathbb{N}$; and its range is seen to be $\mathbb{N}$, contradicting our hypothesis that $n \in \mathscr{M}$; cf. Exercise 15 below. Therefore, $\mathscr{M}$ satisfies N2, so $\mathscr{M} = \mathbb{N}$ by N3. □

In the construction which led to the set $\mathbb{N}$, the set U was any universe containing the empty set; however, every non-empty universe must contain $\emptyset$ as a member by parts (i) and (ii) of the definition of a universe. Hence, we have the following.

Corollary 1.3. *Every non-empty universe is infinite.*

There are certain set constructions which, although they may not lead to contradictions, nevertheless give rise to pathological situations having no counterpart in the intuitive interpretation. For example, we cannot, on the basis of the above axioms, decide whether or not a set can be a member of itself, $X \in X$; thus, we cannot in general tell whether the sets X and $\{X\}$ are equal or not. Since a situation where $X \in X$ never arises in practice, it is best to exclude it explicitly. This is done by the *axiom of foundation.*

A10. *Every non-empty class A has an element which is disjoint from A.*

This axiom clearly rules out the possibility that $X \in X$ for some set X (consider $\{X\}$). More generally, there are no infinite chains $X_0, X_1, X_2, \ldots$ of sets (not necessarily distinct) such that

$$X_{n+1} \in X_n \quad (n \in \mathbb{N}). \tag{1.3}$$

For, if we had a family of sets satisfying (1.3), then the class whose members are $X_0, X_1, \ldots,$ would contradict A10.

The following extension of the notation is sometimes useful. Let $\mathscr{A}$ be a class of sets, indexed in some way, say $\mathscr{A} = (A_i)_{i \in I}$. Then in place of '$\bigcup \mathscr{A}$', '$\bigcap \mathscr{A}$' we shall also write '$\bigcup_i A_i$', respectively '$\bigcap_i A_i$'. In particular, if $\mathscr{A}$ is finite, say $\mathscr{A} = \{A_0, A_1, \ldots, A_n\}$, then we write

$$\bigcup \mathscr{A} = \bigcup_i A_i = A_0 \cup \cdots \cup A_n$$

and

$$\bigcap \mathscr{A} = \bigcap_i A_i = A_0 \cap \cdots \cap A_n.$$

Further, we define the *Cartesian product* of the family of sets $\mathscr{A} = (A_i)_{i\in I}$ as

$$\prod_i A_i := \left\{x \in \Big(\bigcup_i A_i\Big)^I : x = (x_i)_{i\in I} \text{ where } x_i \in A_i \text{ for all } i \in I\right\}.$$

In the case of two factors, this definition does not immediately coincide with our earlier definition, but the difference is immaterial; we are free to interpret Cartesian products in either way, when there are two (or any finite number of) factors.

Clearly, $\prod_i A_i$ is again a set, but even if each A_i is non-empty, there is in general no argument available at this stage of development to show that the product is non-empty. This will follow from the axiom of choice, which will be introduced in §3, and further discussed in §4.

Unlike the union and intersection, where the indexing of $\mathscr{A}$ was used only for convenience, the product depends in an essential way both on the indexing as well as on the class of coordinates; for instance, the product is affected if a coordinate appears more than once. To take an extreme case, if $A_i = A$ for all i, then $\bigcup \mathscr{A} = \bigcap \mathscr{A} = A$, whereas $\prod_i A_i = A^I$. If I is finite, $I = n+1$ say, we write again

$$\prod_i A_i = A_0 \times A_1 \times \cdots \times A_n.$$

Exercises for §1.

1. Show that the total class $\mathfrak{T}$ is not a set.
2. Show, without using Axiom A10, that $n \notin n$ holds for each natural number n.
3. Making use of our definition of ordered pairs, show that $(x_1,x_2) = (y_1,y_2)$ holds if, and only if, $x_1 = y_1$ and $x_2 = y_2$.
4. Show that $\{x,y\}$ cannot serve as definition of an ordered pair.
5. For a natural number $n \geq 1$, define n-tuples $(x_1,x_2,\ldots,x_n)$ either (i) as families indexed by the set $[n] := \{1,2,\ldots,n\}$, or (ii) inductively via $(x_1) = \{x_1\}$ and $(x_1,\ldots,x_n) = ((x_1,\ldots,x_{n-1}),x_n)$ for $n \geq 2$, making use of our previous definition of ordered couples. For each of these definitions, show that

   $$(x_1,x_2,\ldots,x_n) = (y_1,y_2,\ldots,y_n)$$

   holds if, and only if, $x_1 = y_1$, $x_2 = y_2$, …, $x_n = y_n$.
6. Express the following sets in the form $\{X : \mathfrak{P}(X)\}$.
   (i) $\{2,4,6,8,\ldots\}$,
   (ii) $\{1,-1,i,-i\}$,
   (iii) $\{-4,-\frac{3}{2},-\frac{2}{3},-\frac{1}{4},\frac{1}{4},\frac{2}{3},\frac{3}{2},4\}$.

7. Prove that the subclass relation $\subseteq$ is an order relation on the class of all sets; that is, show the following:
(i) $X \subseteq X$ for all sets X (reflexivity);
(ii) $X \subseteq Y$ and $Y \subseteq X$ imply $X = Y$ for all sets X, Y (anti-symmetry);
(iii) For all sets X, Y, Z, if $X \subseteq Y$ and $Y \subseteq Z$, then $X \subseteq Z$ (transitivity).

8. Suppose that three sets A, B, C satisfy $A \subseteq B$, $B \subseteq C$, and $C \subseteq A$. Show that $A = B = C$.

9. Prove the *distributive laws* for union and intersection:

$$(D_\cap) \qquad A \cap (B \cup C) = (A \cap B) \cup (A \cap C),$$

$$(D_\cup) \qquad A \cup (B \cap C) = (A \cup B) \cap (A \cup C).$$

10. Show that the following assertions for sets A, B are pairwise equivalent:
(i) $A \subseteq B$,
(ii) $A \cap B = A$,
(iii) $A \cup B = B$.

11. Define the *symmetric difference* $A \Delta B$ of two sets A, B via

$$A \Delta B = (A - B) \cup (B - A).$$

Prove the following:
(i) $A \Delta \emptyset = A$,
(ii) $A \Delta B = B \Delta A$,
(iii) $(A \Delta B) \Delta C = A \Delta (B \Delta C)$,
(iv) $A \cap (B \Delta C) = (A \cap B) \Delta (A \cap C)$,
(v) $A - B \subseteq A \Delta B$,
(vi) $A = B$ if, and only if, $A \Delta B = \emptyset$,
(vii) $A \Delta C = B \Delta C$ implies $A = B$.

12. Show that the Cartesian product is distributive with respect to union; that is, we have

$$X \times (Y \cup Z) = (X \times Y) \cup (X \times Z)$$

for all X, Y, Z.

13. Show that $X \times X = Y \times Y$ for any X, Y implies $X = Y$.

14. Suppose that, for classes X, Y, Z, we have $X \times Y = X \times Z$ and $X \neq \emptyset$. Show that necessarily $Y = Z$.

15. (a) How should one define the order relation $\leq$ on the set $\mathbb{N}$ of natural numbers?

(b) Complete the proof of Proposition 1.2, by showing that the function $g : n \to \mathbb{N}$ constructed there has range $\mathbb{N}$.

16. For a natural number $q \geq 2$, let

$$M_q = \{n \in \mathbb{N} : n \text{ is divisible by } q\}.$$

Describe the set $M_q \cap M_r$.

17. Show that the system of equations $A \cup X = A \cup B$, $A \cap X = \emptyset$ has exactly one solution X for given sets A and B.

18. Show that

$$\bigcap_{n=1}^{\infty} \left(-\frac{1}{n}, 1+\frac{1}{n}\right) = \bigcap_{n=1}^{\infty} \left[-\frac{1}{n}, 1+\frac{1}{n}\right] = [0,1].$$

§2. Correspondences, mappings, and quotient sets

Let A and B be sets. A *correspondence* from A to B (or between A and B) is a subset of the Cartesian product $A \times B$. Thus, a correspondence is simply a set of pairs (x,y), where $x \in A$ and $y \in B$. We shall usually denote correspondences by capital Greek letters. If Φ is a correspondence from A to B, and $A' \subseteq A$, then we define

$$\Phi(A') := \{y \in B : (x,y) \in \Phi \text{ for some } x \in A'\}.$$

In case $B = A$, Φ is called a correspondence *in* A, or a (binary) *relation on* A. Every set A has the correspondence

$$\Delta_A := \{(x,x) : x \in A\},$$

called the *diagonal* in A. We introduce two operations on correspondences, inversion and composition. The *inverse* Φ^{-1} of a correspondence Φ is given by

$$\Phi^{-1} := \{(x,y) : (y,x) \in \Phi\}.$$

Clearly, if $\Phi \subseteq A \times B$, then $\Phi^{-1} \subseteq B \times A$; in particular, Φ^{-1} is again a correspondence. A correspondence Φ in A is called *symmetric* if $\Phi^{-1} = \Phi$, *anti-symmetric* if $\Phi \cap \Phi^{-1} \subseteq \Delta_A$, and *reflexive* if $\Phi \supseteq \Delta_A$. Thus, a correspondence $\Phi \subseteq A \times A$ in A is symmetric, if $(a,b) \in \Phi$ implies $(b,a) \in \Phi$. Similarly, $\Phi \subseteq A \times A$ is anti-symmetric, if $(a,b) \in \Phi$ and $(b,a) \in \Phi$ implies $a = b$.

Given two correspondences Φ and Ψ, we define their *composition* $\Psi \circ \Phi$ via

$$\Psi \circ \Phi := \{(x,y) : (x,z) \in \Phi \text{ and } (z,y) \in \Psi \text{ for some } z\}.$$

Clearly, if $\Phi \subseteq A \times B$ and $\Psi \subseteq C \times D$, then $\Psi \circ \Phi \subseteq A \times D$; in particular, $\Psi \circ \Phi$ is again a correspondence (and thus a set). We note that $\Psi \circ \Phi = \emptyset$ if $B \cap C = \emptyset$. The following laws are easily verified:

$$\Theta \circ (\Psi \circ \Phi) = (\Theta \circ \Psi) \circ \Phi \tag{2.1}$$

$$(\Psi \circ \Phi)^{-1} = \Phi^{-1} \circ \Psi^{-1} \tag{2.2}$$

$$(\Phi^{-1})^{-1} = \Phi. \tag{2.3}$$

They are valid for any correspondences Φ, Ψ, Θ. Furthermore, if Φ is a correspondence from A to B, then

$$\Delta_B \circ \Phi = \Phi = \Phi \circ \Delta_A. \tag{2.4}$$

With the help of these operations we can define several important types of correspondences, which will occur frequently in what follows. A correspondence Φ in A is said to be *transitive* if $\Phi \circ \Phi \subseteq \Phi$; that is, if $(a,b) \in \Phi$ and $(b,c) \in \Phi$ implies $(a,c) \in \Phi$. A transitive and reflexive correspondence in a set A is called a *pre-ordering* of A (or a *pre-order* on A). Clearly, if Φ is a pre-ordering of A, then so is its inverse Φ^{-1}; in this case, Φ^{-1} is said to be *opposite* or *dual* to Φ. An anti-symmetric pre-ordering of A is called an *ordering* of A (or an *order relation* on A), sometimes also a *partial ordering*, to distinguish it from a *total* ordering, which is required, in addition, to satisfy $\Phi \cup \Phi^{-1} = A^2$. By an *ordered set* we shall mean a set together with an ordering defined on it.

If Φ is a correspondence from A to B, then $\Phi^{-1} \circ \Phi$ is a correspondence in A, and $\Phi \circ \Phi^{-1}$ is a correspondence in B. Moreover, Φ is a function from A to B if, and only if,

$$\Phi^{-1} \circ \Phi \supseteq \Delta_A \tag{2.5}$$

$$\Phi \circ \Phi^{-1} \subseteq \Delta_B; \tag{2.6}$$

cf. Exercise 2 below. Using this characterisation, one proves the following.

Proposition 2.1. *If f and g are functions, then so is their composition $g \circ f$.*

Note that $g \circ f$ may well be the empty set, namely if the range of f is disjoint from the domain of g.

Proof of Proposition 2.1. Let f be a function from A to B, and let g be a function from C to D. We claim that $h := g \circ f$ is a function from

$$f^{-1}(B \cap C) = \{x \in A : f(x) \in B \cap C\}$$

to D. Indeed,

$$\begin{aligned} h^{-1} \circ h &= (g \circ f)^{-1} \circ (g \circ f) \\ &= f^{-1} \circ g^{-1} \circ g \circ f \\ &\supseteq f^{-1} \circ \Delta_C \circ f \\ &= f^{-1} \circ f|_{f^{-1}(B \cap C)} \\ &\supseteq \left(f|_{f^{-1}(B \cap C)}\right)^{-1} \circ f|_{f^{-1}(B \cap C)} \\ &\supseteq \Delta_{f^{-1}(B \cap C)}. \end{aligned}$$

Here, we have used the fact that g is a function in step 3 and the fact that $f|_{f^{-1}(B \cap C)}$ is a function in the last step. Also,

$$\begin{aligned} h \circ h^{-1} &= (g \circ f) \circ (g \circ f)^{-1} \\ &= g \circ f \circ f^{-1} \circ g^{-1} \\ &\subseteq g \circ \Delta_B \circ g^{-1} \\ &\subseteq g \circ g^{-1} \\ &\subseteq \Delta_D, \end{aligned}$$

completing the proof. □

For a given x in the domain of a function f, we denote the unique element y such that $(x, y) \in f$ by $f(x)$; and for functions f and g we often write 'gf' instead of $g \circ f$.

A correspondence Φ from A to B is said to be a *bijection*, if Φ is a function from A to B and Φ^{-1} is a function from B to A. Two sets A and B are said to be *equipotent* (or *of the same cardinality*), denoted by A eq B or $|A| = |B|$, if there exists a bijection between them.

By an *n-ary relation* or *n-place relation* on a set A (where n is a positive integer) we mean a subset Φ of A^n; we frequently refer to Φ as a relation in A. Thus, for example, a binary relation (i.e., $n = 2$) is merely a correspondence in a specified set, while a unary relation ($n = 1$) is a subset of a specified set.

Definition 2.2. By a *mapping* (or *map*) we mean a triple (A,B,f) consisting of a set A, a second set B, and a function f from A to B. The set A is called the *source,* and B the *target* of the mapping.

The mapping (A,B,f) is usually denoted by $f : A \to B$, or sometimes by $A \xrightarrow{f} B$. The latter notation is used especially in diagrams, to illustrate the composition of maps. Unlike functions, which can always be composed, two mappings $f : A \to B$ and $g : C \to D$ are *composable* if, and only if, $B = C$, in which case the composition is defined to be $gf : A \to D$.

Given maps $f : A \to B$, $g : B \to C$, and $h : A \to C$, if the equation $gf = h$ holds, this may be expressed by saying that the diagram

$$\begin{array}{ccc} A & \xrightarrow{f} & B \\ {\scriptstyle 1_A}\downarrow & & \downarrow{\scriptstyle g} \\ A & \xrightarrow[h]{} & C \end{array}$$

is commutative (here 1_A is the identity map on A, see below). More generally, by a *commutative diagram* we mean a network of arrows between sets, representing mappings, such that any two paths (going along arrows) from one set to another define the same mapping between these sets. In practice, most diagrams are made up of triangles or squares.

Given a mapping $f : A \to B$, then for each $x \in A$, the unique element $f(x)$ of B such that $(x, f(x)) \in f$ is called the *image* of x under f. The set $f(A)$ consisting of all images of elements of A is called the image of A; it is simply the range of f, viewed as a function. If $f(A) = B$, that is, $f \circ f^{-1} \supseteq \Delta_B$, f is said to be *surjective*, or *onto* B, or a *surjection.* If $f^{-1} \circ f \subseteq \Delta_A$, f is said to be *injective*, or *one-to-one*, or an *injection.* A mapping which is both injective and surjective is called *bijective*, or a *bijection*; this is merely a bijection between two specified sets as defined earlier. We note that a correspondence Φ from A to B defines a bijection between A and B if, and only if, $\Phi^{-1} \circ \Phi = \Delta_A$ and $\Phi \circ \Phi^{-1} = \Delta_B$. A bijection of a set A onto itself is also called a *permutation* of A. The set of all permutations on A forms a group under composition, denoted $\mathrm{Sym}(A)$, and called the *symmetric group on* A.

Next, we give important examples of mappings which will occur frequently in what follows.

(i) For every set A the diagonal Δ_A defines a bijective map $1_A : A \to A$, called the *identity mapping* or *identity* on A.

(ii) If A is any set and B is a set containing A as a subset, then the diagonal on A defines an injective map $\iota_A : A \to B$, called the *inclusion map* from A to B. We sometimes also write 1_B^A for this map.

(iii) Given a Cartesian product $P = \prod_{i \in I} A_i$ then, for each fixed index $i \in I$, the function assigning to $x \in P$ its i-th coordinate $x(i)$ defines a mapping $\pi_i : P \to A_i$, called the *projection* of P onto the factor A_i. (Projections are surjective, provided we accept the axiom of choice or one of its equivalents; cf. §3–§4.)

(iv) Given a map $f : A \to B$ and a subset A' of A, the mapping $f \circ \iota_{A'} : A' \to B$ is called the *restriction* of f to A', denoted by $f|_{A'}$. Similarly, if $B' \subseteq B$ and $f(A) \subseteq B'$, then f may be cut down to a mapping $f' : A \to B'$ by restricting the target (no special notation will be used for this). Such a map f' is sometimes called a *co-restriction* of f.

(v) Given two mappings $f : A \to B$ and $f' : A' \to B$, where $A' \subseteq A$, then f is called an *extension* of f', provided $f|_{A'} = f'$. Thus, for instance, if $f : A \to B$ is any mapping and $A' \subseteq A$, then f is an extension of $f|_{A'}$, but of course, in general, f is not uniquely determined by $f|_{A'}$.

(vi) Given a set A and a natural number n, a mapping $\alpha : A^n \to A$ is called an *n-ary operation* on A. For $n = 0$, we have a noughtary operation, essentially specifying an element of A, while for $n = 1$ we just have mappings of A into itself. Using the natural bijection between $A^n \times A$ and A^{n+1} given by

$$((x_1, \dots, x_n), y) \mapsto (x_1, \dots, x_n, y),$$

we see that an n-ary operation is nothing but a special case of an $(n+1)$-ary relation. The mappings from A^n to A for arbitrary finite n are sometimes called *finitary operations*, to distinguish them from infinitary operations, that is, mappings $A^I \to A$, where I is infinite.

Let α be an n-ary operation on A, and let $B \subseteq A$. Then $B^n \subseteq A^n$, so α defines a mapping $B^n \to A$, namely the restriction $\alpha|_{B^n}$ of α to B^n. If the image of B^n under this mapping is contained in B, then we say that B is *closed with respect to* α, or that B *admits the operation* α. The restriction can then be cut down to a mapping $B^n \to B$, that is, to an n-ary operation on B. This operation is denoted by $\alpha|_B$, and is called the *restriction* of α to B.

(vii) Let $f : I \to A$ be a mapping, and write 'a_i' instead of $f(i)$ for $i \in I$. Then $\{a_i\}_{i \in I}$ is just a family of elements of A, as defined previously. This point of view of regarding a mapping is adopted when we want to stress the image set.

(viii) If A is a given set, then every subset B of A determines a mapping $\chi_B : A \to 2 = \{0, 1\}$, defined by

$$\chi_B(x) := \begin{Bmatrix} 1, & x \in B \\ 0, & x \notin B \end{Bmatrix} \qquad (x \in A).$$

χ_B is termed the *characteristic function* of the subset B. The mapping $\mathscr{B}(A) \to 2^A$ given by $B \mapsto \chi_B$ is easily seen to be a bijection; cf. Exercise 3.

(ix) Given a map $f : A \to B$ we define a mapping $f^{-1} : \mathscr{B}(B) \to \mathscr{B}(A)$ via

$$f^{-1}(Y) := \{x \in A : f(x) \in Y\} \qquad (Y \in \mathscr{B}(B)).$$

The set $f^{-1}(Y)$ is called the *inverse image* or *pre-image* of Y under f, the map f^{-1} itself is referred to as the *pullback mapping* corresponding to f.

An *equivalence relation* on a set A is a correspondence Φ in A which is reflexive, symmetric, and transitive. Thus, we have

$$\Phi \supseteq \Delta_A, \ \Phi^{-1} = \Phi, \ \text{ and } \ \Phi \circ \Phi \subseteq \Phi.$$

We shall denote equivalence relations by lower case Gothic letters in this chapter. If $\mathfrak{q}$ is an equivalence relation on the set A, then for each $x \in A$ we define a subset $x^{\mathfrak{q}} \subseteq A$, the $\mathfrak{q}$-class of x, via

$$x^{\mathfrak{q}} := \{y \in A : (x, y) \in \mathfrak{q}\}.$$

Instead of '$(x, y) \in \mathfrak{q}$', we shall often write '$x \equiv y \pmod{\mathfrak{q}}$'. From the properties of $\mathfrak{q}$ it follows that $x \in x^{\mathfrak{q}}$, and that

$$x \equiv y \pmod{\mathfrak{q}} \iff x^{\mathfrak{q}} = y^{\mathfrak{q}}.$$

The $\mathfrak{q}$-classes form a *partition* $P_{\mathfrak{q}}$ of A, that is a decomposition of A into pairwise disjoint non-empty sets, the *classes* of the partition. Conversely, given a partition P of A, we obtain an equivalence relation $\mathfrak{q}_P$ by setting $x \equiv y \pmod{\mathfrak{q}_P}$ if, and only if, x and y are in the same class of P. The mapping given by $\mathfrak{q} \mapsto P_{\mathfrak{q}}$ is a bijection between the set $\mathscr{E}(A)$ consisting of all equivalence relations on A and the set $\Pi(A)$ of partitions of A, its inverse given by the map sending a partition P to $\mathfrak{q}_P$; cf. Exercise 7 below. For $\mathfrak{q} \in \mathscr{E}(A)$, the $\mathfrak{q}$-classes themselves are members of the Boolean $\mathscr{B}(A)$; the subset $P_{\mathfrak{q}}$ of $\mathscr{B}(A)$ consisting of all $\mathfrak{q}$-classes will be denoted by $A/\mathfrak{q}$, and is called the *quotient set* of A by $\mathfrak{q}$. Associating with each $x \in A$ its $\mathfrak{q}$-class $x^{\mathfrak{q}}$, we obtain a surjective mapping from A to $A/\mathfrak{q}$, the *natural projection* $\pi_{\mathfrak{q}}$ associated with $\mathfrak{q}$. We have the following basic decomposition result for mappings.

Proposition 2.3. (a) *Let $f: A \to B$ be a mapping, and set $\mathfrak{q} = f^{-1} \circ f$. Then $\mathfrak{q}$ is an equivalence relation on A, called the kernel of f, and denoted by* $\ker(f)$. *Furthermore, there is a canonical decomposition*

$$f = \iota \circ f' \circ \pi, \tag{2.7}$$

where $\pi : A \to A/\mathfrak{q}$ is surjective, $f' : A/\mathfrak{q} \to f(A)$ is a bijection, and $\iota : f(A) \to B$ is injective. In fact, we can take $\pi = \pi_{\mathfrak{q}}$ and $\iota = \iota_{f(A)}$.

(b) *Let $f : A \to B$ be a mapping, $\mathfrak{q}$ an equivalence relation on A, and suppose that $\mathfrak{q} \subseteq \ker(f)$. Then there exists a unique mapping $\bar{f} : A/\mathfrak{q} \to B$ such that the diagram*

$$\begin{array}{ccc} A & \xrightarrow{\pi_{\mathfrak{q}}} & A/\mathfrak{q} \\ {\scriptstyle f}\big\downarrow & & \big\downarrow{\scriptstyle \bar{f}} \\ B & \xrightarrow[1_B]{} & B \end{array} \tag{2.8}$$

commutes.

(c) *If $f = \iota_1 \circ f_1' \circ \pi_1$ is a second decomposition as in* (a), *with $\pi_1 : A \to A/\mathfrak{q}$ surjective, $f_1' : A/\mathfrak{q} \to f(A)$ bijective, and $\iota_1 : f(A) \to B$ injective, where $\mathfrak{q} = \ker(f)$, then there exist permutations α, β of $A/\mathfrak{q}$ respectively $f(A)$ such that $\pi_1 = \alpha \circ \pi_{\mathfrak{q}}$, $\iota_1 = \iota_{f(A)} \circ \beta$, and $f' = \beta \circ f_1' \circ \alpha$.*

Proof. (a) The situation may be illustrated by the commutative diagram

$$\begin{array}{ccc} A & \xrightarrow{f} & B \\ {\scriptstyle \pi_{\mathfrak{q}}}\big\downarrow & & \big\uparrow{\scriptstyle \iota_{f(A)}} \\ A/\mathfrak{q} & \xrightarrow[f']{} & f(A) \end{array} \tag{2.9}$$

Since f is a function,

$$\mathfrak{q} = f^{-1} \circ f \supseteq \Delta_A,$$

so $\mathfrak{q}$ is reflexive. Further,

$$\mathfrak{q}^{-1} = (f^{-1} \circ f)^{-1} = f^{-1} \circ (f^{-1})^{-1} = f^{-1} \circ f = \mathfrak{q},$$

thus $\mathfrak{q}$ is symmetric. Next, since $f \circ f^{-1} \subseteq \Delta_B$, we have

$$\mathfrak{q} \circ \mathfrak{q} = (f^{-1} \circ f) \circ (f^{-1} \circ f) = f^{-1} \circ (f \circ f^{-1}) \circ f \subseteq f^{-1} \circ \Delta_B \circ f = f^{-1} \circ f = \mathfrak{q},$$

so $\mathfrak{q}$ is transitive, and hence an equivalence relation on A.

In order to make the diagram (2.9) commute, we need to define $f' : A/\mathfrak{q} \to f(A)$ via $f'(x^{\mathfrak{q}}) := f(x)$. For $x, y \in A$ we have, by definition of $\mathfrak{q}$,

$$x \equiv y \pmod{\mathfrak{q}} \Longleftrightarrow (x,z) \in f \text{ and } (y,z) \in f \text{ for some } z \in B \Longleftrightarrow f(x) = f(y).$$

Hence, (i) f' is well defined, (ii) $f'(x^{\mathfrak{q}}) = f'(y^{\mathfrak{q}})$ implies $x \equiv y \pmod{\mathfrak{q}}$, and so $x^{\mathfrak{q}} = y^{\mathfrak{q}}$, thus f' is injective, and (iii) taking $z \in f(A)$ and choosing $x \in A$

such that $f(x) = z$, then $f'(x^{\mathfrak{q}}) = z$, so f' is surjective, hence a bijection as claimed.

(b) If a mapping $\bar{f}$ exists making the diagram (2.8) commute, then necessarily

$$\bar{f}(x^{\mathfrak{q}}) = f(x), \quad x \in A. \tag{2.10}$$

Hence, there can be at most one such mapping $\bar{f}$. On the other hand, if $x \equiv y \pmod{\mathfrak{q}}$, then $x \equiv y \pmod{\ker(f)}$, thus $f(x) = f(y)$, and $\bar{f}$ as given by (2.10) is indeed well defined.

(c) Suppose that $x \equiv y \pmod{\mathfrak{q}}$. Then $f(x) = f(y)$, which can be rewritten as

$$\iota_1 f_1'(\pi_1(x)) = \iota_1 f_1'(\pi_1(y)).$$

Since $\iota_1 f_1'$ is injective, it follows that $\pi_1(x) = \pi_1(y)$, that is,

$$x \equiv y \pmod{\ker(\pi_1)},$$

thus $\mathfrak{q} \subseteq \ker(\pi_1)$. Applying Part (b) with $f = \pi_1$, we obtain a map $\alpha : A/\mathfrak{q} \to A/\mathfrak{q}$ such that $\alpha\pi_{\mathfrak{q}} = \pi_1$. Moreover, α is surjective, since π_1 and $\pi_{\mathfrak{q}}$ are. Next, we have

$$\begin{aligned}\alpha(x^{\mathfrak{q}}) = \alpha(y^{\mathfrak{q}}) &\Longrightarrow \pi_1(x) = \pi_1(y) \Longrightarrow f(x) = f(y)\\ &\Longrightarrow x \equiv y \pmod{\mathfrak{q}} \Longrightarrow x^{\mathfrak{q}} = y^{\mathfrak{q}},\end{aligned}$$

so α is indeed a permutation of $A/\mathfrak{q}$. The equation

$$\iota_1 f_1' \alpha \pi_{\mathfrak{q}} = \iota_{f(A)} f' \pi_{\mathfrak{q}}$$

implies that

$$\iota_1 f_1' \alpha = \iota_{f(A)} f'$$

since $\pi_{\mathfrak{q}}$ is surjective, and setting $\beta := f'\alpha^{-1}(f_1')^{-1}$, we get a permutation of $f(A)$ such that $\iota_1 = \iota_{f(A)}\beta$ and $f' = \beta f_1' \alpha$. □

The quotient sets of a given set A are to some extent dual to the subsets of A, but this duality is not complete. For instance, while the relation: *B is a subset of A* is transitive, the relation: *B is a quotient set of A* is not. In fact, the set $\mathscr{B}(A)$ of all subsets of A is an ordered set under inclusion. The set $\Pi(A)$ of all partitions (quotient sets) of A may be ordered by setting

$$\begin{aligned}X \preceq Y :\Longleftrightarrow\ &\text{the kernel of the natural projection } A \to X \text{ is contained}\\ &\text{in the kernel of the natural projection } A \to Y.\end{aligned}$$

Thus, for two partitions X and Y of A, we have $X \preceq Y$ if, and only if, X is a refinement of Y, meaning that each X-class is completely contained in some Y-class. Hence, if $X \preceq Y$, there is a surjection $X \to Y$ which has some claim to be called natural, namely the one which sends an X-class to the unique Y-class it is contained in. There is a natural bijection between the set $\mathscr{E}(A)$ of equivalence relations on A and the set $\Pi(A)$, and the ordering $\preceq$ defined on $\Pi(A)$ corresponds under this bijection to the ordering by inclusion of $\mathscr{E}(A)$.

Corollary 2.4. *Let $\mathfrak{q}, \mathfrak{r}$ be equivalence relations on the set A, and suppose that $\mathfrak{q} \subseteq \mathfrak{r}$. Then there is a unique mapping $\Theta : A/\mathfrak{q} \to A/\mathfrak{r}$ such that $\Theta\pi_{\mathfrak{q}} = \pi_{\mathfrak{r}}$. If $\ker(\Theta)$ is denoted by $\mathfrak{r}/\mathfrak{q}$, then $\mathfrak{r}/\mathfrak{q}$ is an equivalence relation on $A/\mathfrak{q}$, and Θ induces a bijection $\Theta' : (A/\mathfrak{q})/(\mathfrak{r}/\mathfrak{q}) \to A/\mathfrak{r}$, such that the diagram*

$$\begin{array}{ccccc}
A & \xrightarrow{\pi_{\mathfrak{q}}} & A/\mathfrak{q} & \xrightarrow{\pi_{\mathfrak{r}/\mathfrak{q}}} & (A/\mathfrak{q})/(\mathfrak{r}/\mathfrak{q}) \\
{\scriptstyle \pi_{\mathfrak{r}}}\downarrow & & \downarrow{\scriptstyle \Theta} & & \downarrow{\scriptstyle \Theta'} \\
A/\mathfrak{r} & \xrightarrow[1_{A/\mathfrak{r}}]{} & A/\mathfrak{r} & \xrightarrow[1_{A/\mathfrak{r}}]{} & A/\mathfrak{r}
\end{array}$$

commutes.

Proof. To obtain Θ, apply Part (b) of Proposition 2.3 with $B = A/\mathfrak{r}$ and $f = \pi_{\mathfrak{r}}$. Then Θ is unique with the property that $\Theta\pi_{\mathfrak{q}} = \pi_{\mathfrak{r}}$, and surjective since $\pi_{\mathfrak{q}}$ and $\pi_{\mathfrak{r}}$ are. Now use Part (a) with $f = \Theta$ to obtain the bijection Θ' with $\Theta'\pi_{\mathfrak{r}/\mathfrak{q}} = \Theta$. □

Exercises for §2.

1. How many correspondences are there from the set $A = \{a, b\}$ to the set $B = \{x, y\}$?

2. Let Φ and Ψ be correspondences from the set A to the set B. Show that $(\Phi \cap \Psi)^{-1} = \Phi^{-1} \cap \Psi^{-1}$ and $(\Phi \cup \Psi)^{-1} = \Phi^{-1} \cup \Psi^{-1}$.

3. Prove the equalities (2.1)–(2.4).

4. Show that a correspondence $\Phi \subseteq A \times B$ from the set A to the set B is a function from A to B if, and only if, Φ satisfies Conditions (2.5) and (2.6).

5. How many n-ary relations are there on a set with m elements?

6. Let A be a set. Show that the map sending a subset $B \subseteq A$ to its characteristic function χ_B defines a bijection Φ from $\mathscr{B}(A)$ onto the set 2^A consisting of the maps $A \to 2$.

7. Let A be a set, and denote by $\mathscr{E}(A)$, respectively $\Pi(A)$, the set of equivalence classes on A, respectively the set of partitions of A. Moreover, let $\varphi : \mathscr{E}(A) \to \Pi(A)$ be the map given by $\mathfrak{q} \mapsto P_{\mathfrak{q}}$, and let $\psi : \Pi(A) \to \mathscr{E}(A)$ be the map sending a partition P of A to its corresponding equivalence relation $\mathfrak{q}_P$. Show that φ and ψ are bijective maps inverse to each other; that is, prove that

$$\mathfrak{q}_{P_{\mathfrak{q}}} = \mathfrak{q} \quad (\mathfrak{q} \in \mathscr{E}(A)) \quad \text{and} \quad P_{\mathfrak{q}_P} = P \quad (P \in \Pi(A)).$$

8. Let X be a set, and let $\{X_i\}_{i \in I}$ be a family of subsets such that $\bigcup_i X_i = X$ (that is, $\{X_i\}_{i \in I}$ is a covering of X). Suppose that, for each index $i \in I$, we are given a mapping $f_i : X_i \to Y$ into a fixed set Y, and that $f_i|_{X_i \cap X_j} = f_j|_{X_i \cap X_j}$ holds for all pairs of indices $(i, j) \in I^2$. Show that there exists precisely one mapping $f : X \to Y$, which is an extension of each f_i; that is, $f|_{X_i} = f_i$ for all $i \in I$.

9. List all equivalence relations on the set $3 = \{0, 1, 2\}$.

10. Let $\mathscr{E} \subseteq \mathscr{E}(A)$ be a non-empty set of equivalence relations on the set A. Show that $\bigcap \mathscr{E}$ is again an equivalence relation on A.

11. Let $\equiv_A$ be an equivalence relation on the set A, and let $\equiv_B$ be an equivalence relation on the set B. For $a_i \in A$ and $b_i \in B$, set

$$(a_1, b_1) \equiv (a_2, b_2) \quad :\Longleftrightarrow \quad a_1 \equiv_A a_2 \text{ and } b_1 \equiv_B b_2.$$

Show that $\equiv$ is an equivalence relation on the set $A \times B$.

§3. Ordered sets

Let A be an ordered set. We adopt the usual notation '$\leq$' for the ordering, so that the axioms read:

O1. *For every $x \in A$, we have $x \leq x$ (reflexivity).*

O2. *For any two elements $x, y \in A$, if $x \leq y$ and $y \leq x$, then $x = y$ (anti-symmetry).*

O3. *For any three elements $x, y, z \in A$, if $x \leq y$ and $y \leq z$, then $x \leq z$ (transitivity).*

As usual, we write '$x < y$' to mean '$x \leq y$ and $x \neq y$'; we also write '$x \geq y$' instead of '$y \leq x$', and '$x > y$' instead of '$y < x$'. The ordering $\leq$ is called *total* if, in addition, any two elements are *comparable*, that is, if we have

O4. *For all $x, y \in A$, either $x \leq y$ or $y \leq x$.*

We mention that our order relations are often called "partial orders" in the literature, in which case our total orders are usually referred to as (strict) "orders".

An abstract set may be regarded as a *totally unordered* set, in which $x \leq y$ holds only when $x = y$; thus, any two distinct elements are incomparable.

If A is an ordered set and $B \subseteq A$, then the ordering $\leq$ of A, when restricted to B, is an ordering $\leq_B$ of B, where $\leq_B = \leq \cap (B \times B)$. In this sense, every subset of an ordered set is itself an ordered set. If the ordering so defined on B is total, then B is called a *chain* in A.

Let B be any subset of an ordered (or, more generally, pre-ordered) set A. An element $a \in A$ with the property that $x \leq a$ for all $x \in B$ is called an *upper bound* for B in A. When such elements exist, we say that B is *bounded above* in A. *Lower bounds* of B are defined analogously. An ordered set in which every finite subset has an upper bound is said to be *directed upwards*. By induction on the number of elements, it suffices to require that any two elements in the set have an upper bound; see Exercise 1. A set *directed downwards* is defined similarly; when nothing to the contrary is said, 'directed' will always mean 'directed upwards'.

If the ordered set A itself has an upper bound a, then a is clearly the only upper bound; it is called the *greatest element* of A; cf. Exercise 2. If $a \in A$ is such that none of the upper bounds in A of the singleton $\{a\}$ exceed a, then a is said to be *maximal* in A. Thus, $a \in A$ is maximal in A, if $a \leq x$ with $x \in A$ implies $a = x$. Of course, A may have more than one maximal element, or none at all. If it has a greatest element, then this is also the unique maximal element. The converse does not hold in general, but it does hold in a directed set (Exercise 3): *a maximal element in a directed set is also the greatest element.* The *minimal* elements of A and the *least* element of A are defined correspondingly. If a subset B of the ordered set A has a least upper bound, this is called the *supremum* of B, denoted '$\sup B$'. Similarly, the greatest lower bound of B in A, if it exists, is called the *infimum* of B, written '$\inf B$'. All these definitions still make sense when the set A is only pre-ordered. We can now state our final axiom.

A11. (Zorn's Lemma) *An ordered set in which every chain has an upper bound, has a maximal element.*

An ordered set satisfying the hypothesis of Zorn's Lemma is termed *inductively ordered.* Instead of A11, one frequently postulates the

Axiom of Choice. *Given any set A, there exists a map $f : \mathscr{B}(A) \to A$ (called a choice function for A) such that $f(X) \in X$ for every $X \subseteq A$ with $X \neq \emptyset$.*

On the basis of the remaining axioms A1–A10, Zorn's Lemma and the Axiom of Choice are in fact *equivalent*; cf. §4.

An ordered set A is said to satisfy the *minimum condition*, if every non-empty subset of A has a minimal element. If this condition is satisfied and, moreover, the number of minimal elements of every non-empty subset of A is finite, then A is called *partly well-ordered*; and if every non-empty subset of A has a unique minimal element, then A is termed *well-ordered.* Clearly, a well-ordered set is totally ordered, so every non-empty subset of a well-ordered set has a least element; cf. Exercise 4. For sets with minimum condition, we have the following.

Proposition 3.1. (Noetherian induction) *Let $(A, \leq)$ be an ordered set satisfying the minimum condition, and let $B \subseteq A$ be a subset. Suppose that B contains any element $a \in A$ whenever it contains all elements $x \in A$ with $x < a$. Then $B = A$.*

Proof. This is left as an exercise for the interested reader; cf. Exercise 5. □

The special case of Proposition 3.1 when A is well-ordered is known as the *principle of transfinite induction.* Since the natural numbers $\mathbb{N}$ form a well-ordered set with respect to inclusion, transfinite induction generalises the principle of induction for natural numbers. We shall see in §4 that, on the basis of axioms A1–A11, every set can be well ordered.

An ordered set in which each pair of elements a, b has a supremum, denoted $a \vee b$, and an infimum $a \wedge b$ is called a *lattice*. It follows by induction that in a lattice every non-empty finite set has a supremum and an infimum.

If this is true for *every* subset, then the lattice is called *complete*. In particular, a complete lattice L always has a greatest element $1 := \sup L = \inf \emptyset$ and a least element $0 := \inf L = \sup \emptyset$.

Clearly, every lattice is directed. Also, every totally ordered set $(A, \leq)$ is a lattice (though not necessarily complete); here, $\inf\{a,b\} = \min\{a,b\}$ and $\sup\{a,b\} = \max\{a,b\}$ for $a, b \in A$. On the other hand, a finite lattice is always complete. An example of a complete lattice is $\mathscr{B}(A)$, for any set A; indeed, if $\mathscr{S} \subseteq \mathscr{B}(A)$, then $\bigcap \mathscr{S} = \inf \mathscr{S}$ and $\bigcup \mathscr{S} = \sup \mathscr{S}$.

More generally, let A be an ordered set. Then a subset X of A is called a *left segment* of A if, for every $x \in X$, $y \leq x$ implies $y \in X$. Similarly, if $y \in X$ whenever $y \geq x$ for some $x \in X$, then X is a *right segment* of A. The set $\mathscr{S}^-(A)$ of all left segments of A is a complete lattice under inclusion; indeed, for $\mathscr{S} \subseteq \mathscr{S}^-(A)$, again $\inf \mathscr{S} = \bigcap \mathscr{S}$ and $\sup \mathscr{S} = \bigcup \mathscr{S}$. Since any abstract set A may be regarded as a totally unordered set, and every subset of a totally unordered set is a left segment, we see that $\mathscr{S}^-(A)$ equals $\mathscr{B}(A)$ in this case.

The following result is useful in verifying that a given ordered set is a complete lattice.

Proposition 3.2. *Let A be an ordered set. Then the following assertions are equivalent:*

(i) *A is a complete lattice.*
(ii) *Every subset of A has an infimum.*
(iii) *Every subset of A has a supremum.*

Proof. Clearly, (i) implies (ii) and (iii). We show that (ii) implies (i); the proof of (iii) $\Rightarrow$ (i) being entirely analogous.

We have to show that every subset of A has a supremum. Given $X \subseteq A$, let Y be the set of all upper bounds of X in A, and set $y_0 := \inf Y$. Then every element of X is a lower bound for Y, so $x \leq y_0$ for all $x \in X$; that is, y_0 is an upper bound for X. Moreover, if z is any upper bound for X, then $z \in Y$, hence $z \geq y_0$; that is, $y_0 = \sup X$. □

A *sublattice* of a lattice L is a subset A of L closed under taking the supremum $a \vee b$ and infimum $a \wedge b$ of any two elements $a, b \in A$. Thus, a sublattice contains with any finite non-empty subset X also its supremum and infimum, taken in L. In a complete lattice, a sublattice is required to contain the supremum and infimum of any of its subsets.

Let A and B be ordered sets. A mapping $f : A \to B$ is termed *order-preserving* or an *order morphism*, if

$$x \leq y \Longrightarrow f(x) \leq f(y) \text{ for all } x, y \in A.$$

If $f: A \to B$ is order-preserving and bijective, and its inverse f^{-1} is also an order morphism (from B to A), then f is called an *order isomorphism* between A and B, and A and B are called *(order) isomorphic*, written '$A \cong B$' (see Exercise 7). If A is directed or totally ordered, then so is its image under any order morphism. However, if f is an order morphism between lattices A and B, then f need not preserve supremum or infimum; that is, it need not be true that

$$f(x \vee y) = f(x) \vee f(y) \text{ and } f(x \wedge y) = f(x) \wedge f(y) \text{ for all } x, y \in A; \tag{3.1}$$

cf. Exercise 8. If (3.1) is satisfied, f is called a *lattice homomorphism.* Correspondingly, a *lattice isomorphism* is a bijection f such that both f and f^{-1} are lattice homomorphisms. Every order isomorphism between lattices is in fact a lattice isomorphism; see Exercise 9.

In comparing ordered sets, we shall make use of the following remarkable property of complete lattices, first observed by Tarski; cf. [79].

Proposition 3.3. *Let A be a complete lattice, let $f: A \to A$ be an order morphism, and let $\mathscr{F}$ be the set of all fixed points of f. Then the set $\mathscr{F}$ is non-empty and, endowed with the induced ordering, forms a complete lattice. Moreover, we have*

$$\sup \mathscr{F} = \sup\{x \in A : x \leq f(x)\} \in \mathscr{F} \tag{3.2}$$

as well as

$$\inf \mathscr{F} = \inf\{x \in A : f(x) \leq x\} \in \mathscr{F}. \tag{3.3}$$

Proof. Set

$$\mathscr{I} := \{x \in A : x \leq f(x)\} \text{ and } a_0 := \sup \mathscr{I}.$$

For $x \in \mathscr{I}$, $x \leq f(x)$ and $x \leq a_0$; hence $x \leq f(a_0)$ for all $x \in \mathscr{I}$. Thus, $f(a_0)$ is an upper bound for $\mathscr{I}$, so

$$a_0 \leq f(a_0) \tag{3.4}$$

by definition of a_0, and thus $a_0 \in \mathscr{I}$. Applying f to (3.4), we now find that $f(a_0) \in \mathscr{I}$, so $f(a_0) \leq a_0$, which together with (3.4) shows that $f(a_0) = a_0$; hence, $\mathscr{F} \neq \emptyset$, and assertion (3.2) holds.

Next, consider the dual lattice $A' = (A, \geq)$. Like A, A' is complete, and f, considered as a map on A', is again order-preserving. Hence, applying to A' the result (3.2) established for A, we find that (3.3) holds true.

Finally, let X be any subset of $\mathscr{F}$. The interval $[\sup X, \sup A]$ of A is a complete lattice. For $x \in X$ we have $x \leq \sup X$, hence

$$x = f(x) \leq f(\sup X),$$

thus, $\sup X \leq f(\sup X)$. Consequently, $\sup X \leq z$ implies

$$\sup X \leq f(\sup X) \leq f(z);$$

hence we can restrict the domain of f to the interval $[\sup X, \sup A]$ to obtain an order-preserving map $f' : [\sup X, \sup A] \to [\sup X, \sup A]$. Applying (3.3) to the lattice $[\sup X, \sup A]$ and the map f', we conclude that the infimum u of all fixed points of f' is itself a fixed point of f'. Clearly, u is a fixed point of f, in fact the least fixed point of f which is an upper bound for X; i.e., u is the supremum of X in $\mathscr{F}$. Since X was an arbitrary subset of $\mathscr{F}$, we conclude from Proposition 3.2 that $(\mathscr{F}, \leq)$ is a complete lattice. □

The most important consequence of Proposition 3.3 is the existence of fixed points.

Corollary 3.4. *Every order morphism of a complete lattice into itself has a fixed point.*

Proposition 3.3 may be generalised by replacing f with a commuting set of order-preserving maps; see Exercise 10.

Remark 3.5. The property of complete lattices described in Corollary 3.4 can in fact be used to characterise them; cf. [27], as well as Section 7.2.

With the help of Corollary 3.4, we can now prove an important comparison result for ordered sets.

Proposition 3.6. *Let A and B be ordered sets such that A is order isomorphic to a left segment of B, and B is order isomorphic to a right segment of A. Then there exists a bijection $f : A \to B$ which weakly respects the ordering in the sense that*

$$x < y \Longrightarrow f(x) \not\geq f(y) \text{ for all } x, y \in A. \tag{3.5}$$

Proof. Let $g : A \to B_0$ and $h : B \to A_0$ be order isomorphisms, where B_0 is a left segment of B, and A_0 is a right segment of A. For any subset X of A, we denote by X' its complement in A, and likewise for subsets of B. Clearly, the complement of a left (right) segment is a right (left) segment, and the assignment $X \mapsto X'$ yields an order-reversing bijection of $\mathscr{S}^-(A)$ onto the set $\mathscr{S}^+(A)$ of right segments of A, whose inverse is also order-reversing. Define an order-preserving map $\Theta : \mathscr{S}^-(A) \to \mathscr{S}^-(A)$ via

$$\Theta(X) := (h((g(X))'))', \quad X \in \mathscr{S}^-(A).$$

The fact that Θ respects inclusion is obvious, so it remains to see that $\Theta(X) \in \mathscr{S}^-(A)$ for $X \in \mathscr{S}^-(A)$.

We claim that, if X is a left segment of A, then $g(X)$ is a left segment of B. To see this, let $y = g(x) \in g(X)$, and let $z \in B$ be such that $z \leq y$. We want to show that $z \in g(X)$. We have $y \in B_0$, so $z \in B_0$ since B_0 is a left segment of B, thus z can be written as $z = g(a)$ for some $a \in A$. Applying g^{-1} to the inequality $z \leq y$ now yields $a \leq x$, thus $a \in X$ since X is a left segment of A, and we see that $z = g(a) \in g(X)$ as required. We conclude that $(g(X))'$ is a right segment of B, so $h((g(X))')$ is a right segment of A by an argument similar to the one just given, hence $\Theta(X) \in \mathscr{S}^-(A)$.

Since $\mathscr{S}^-(A)$ is a complete lattice under inclusion, Θ has a fixed point by Corollary 3.4; that is, there exists a left segment A_1 of A such that

$$(h((g(A_1))'))' = A_1.$$

Thus, if $B_1 := g(A_1)$, then $h(B_1') = A_1'$. We now define a bijection $f : A \to B$ via

$$f(x) := \begin{cases} g(x), & x \in A_1 \\ h^{-1}(x), & x \in A_1'. \end{cases}$$

In verifying (3.5), it is enough to show that $x < y$ implies $f(x) \not> f(y)$. Suppose this is false. Then there exist elements $x_1, y_1 \in A$ such that

$$x_1 < y_1 \text{ and } f(x_1) > f(y_1). \tag{3.6}$$

If $y_1 \in A_1$, then $x_1 \in A_1$, and

$$f(x_1) = g(x_1) \leq g(y_1) = f(y_1),$$

which contradicts (3.6). Similarly, if $x_1 \in A_1'$, then $y_1 \in A_1'$, and

$$x_1 = h(f(x_1)) \geq h(f(y_1)) = y_1,$$

again contradicting (3.6). The only remaining possibility is that $x_1 \in A_1$ and $y_1 \in A_1'$. This implies that $f(x_1) \in B_1$ and $f(y_1) \in B_1'$; but B_1' is a right segment of B, thus, by (3.6), $f(x_1) \in B_1'$, which again is a contradiction. Hence, (3.6) cannot hold, and (3.5) is proven. □

If we apply Proposition 3.6 to totally unordered (i.e., abstract) sets, we obtain the following classical result.

Corollary 3.7. (Schröder-Bernstein theorem) *If A and B are sets, and if $g : A \to B$ and $h : B \to A$ are injections, then there exists a bijection between A and B.*

For totally ordered sets, we obtain the following.

Corollary 3.8. *Let A and B be ordered sets such that A is order isomorphic to a left segment of B, and B is order isomorphic to a right segment of A. If at least one of A, B is totally ordered, then A is order isomorphic to B.*

Proof. Without loss of generality, we may take B to be totally ordered; thus, in view of (3.5), the bijection $f : A \to B$ exhibited in Proposition 3.6 is an order morphism. Since in fact A is also totally ordered (being isomorphically embedded into B), f^{-1} is also order-preserving, so f is indeed an order isomorphism. To see that f^{-1} is order-preserving, suppose that $f(x) \leq f(y)$ and $y < x$ for some elements $x, y \in A$. Applying (3.5), we find that $f(y) \ngeq f(x)$, so $f(y) < f(x)$, contradicting our hypothesis that $f(x) \leq f(y)$. Hence, under this assumption, we have $y \nless x$, so $x \leq y$, as desired, since A is totally ordered. □

To every element a of an ordered set A we can associate the left segment $S_a := \{x \in A : x \leq a\}$; the mapping $A \to \mathscr{S}^-(A)$ given by $a \mapsto S_a$ is injective and order-preserving; thus, we find that A is always order isomorphic to a subset of $\mathscr{S}^-(A)$. Note that the embedding $A \to \mathscr{S}^-(A)$ just given is not surjective, as the empty segment is not in the image. The following result, due to Dilworth and Gleason [28], shows, in particular, that A can never be order isomorphic to $\mathscr{S}^-(A)$ itself.

Proposition 3.9. *Let A be an ordered set, let $A_0 \subseteq A$ be a subset, and let $f : A_0 \to \mathscr{S}^-(A)$ be an order morphism. Then f is not surjective.*

Proof. Suppose that f is surjective, set

$$B := \{x \in A_0 : x \notin f(x)\},$$

and let

$$\overline{B} = \{y \in A : y \leq x \text{ for some } x \in B\}$$

be the left segment of A generated by B. We note that, in particular, $B \subseteq \overline{B}$. By assumption, $\overline{B} = f(b)$ for some $b \in A_0$. If $b \notin \overline{B}$, then a fortiori $b \notin B$, while, by definition of the set B and the fact that $\overline{B} = f(b)$, we must have $b \in B$, a contradiction. Hence, we must have $b \in \overline{B}$, that is, $b \leq x$ for some $x \in B$, and so $x \in A_0$ and $\overline{B} = f(b) \subseteq f(x)$, since f is order-preserving. However, we have $x \in B \subseteq \overline{B} \subseteq f(x)$, so $x \in f(x)$, implying $x \notin B$, another contradiction. This final contradiction shows that f is not surjective, as claimed. □

The proof of Proposition 3.9 generalises Cantor's diagonal argument in [20] which shows that a set A is not equipotent to its Boolean $\mathscr{B}(A)$. In order to obtain the latter result, consider A as a totally unordered set; then $\mathscr{S}^-(A) = \mathscr{B}(A)$ and every map from A to $\mathscr{S}^-(A)$ is order-preserving. Thus, setting $A_0 = A$ in Proposition 3.9 yields the following.

Corollary 3.10. (G. Cantor, 1891) *If A is any set, there exists no surjective map from A onto $\mathscr{B}(A)$; in particular, A and $\mathscr{B}(A)$ are not equipotent.*

For ordered sets in general, we obtain the following.

Corollary 3.11. *If A is any ordered set, then A is not order isomorphic to $\mathscr{S}^-(A)$.*

We conclude this section with an interesting variation of Proposition 3.3, also due to Tarski. Let $(A,\leq)$ be a *continuously ordered* set, that is, $\leq$ is a total order on A, and A forms a complete lattice with respect to this order. Assume further that A is *densely ordered*, that is, for any two elements $x,y\in A$ with $x<y$ there exists $z\in A$ such that $x<z<y$. For instance, $\mathbb{R}\cup\{+\infty,-\infty\}$, or a compact interval $[a,b]$ of real numbers with the usual order is continuously and densely ordered. Proposition 3.3 applies to every continuously ordered set A. Under the additional assumption that A is densely ordered, we can improve Proposition 3.3 by introducing the concepts of lower and upper semi-continuous maps. Let A be a complete lattice and let B be an ordered set. A map $f:A\to B$ is called *upper semi-continuous* if, for each non-empty subset X of A and every $b\in B$,

$$(\forall x\in X: f(x)\geq b) \implies f(\sup X)\geq b \text{ and}$$
$$(\forall x\in X: f(x)\leq b) \implies f(\inf X)\leq b.$$

Similarly, $g:A\to B$ is called *lower semi-continuous* if it satisfies

$$(\forall x\in X: g(x)\leq b) \implies g(\sup X)\leq b \text{ and}$$
$$(\forall x\in X: g(x)\geq b) \implies g(\inf X)\geq b$$

for all X such that $\emptyset\neq X\subseteq A$ and all $b\in B$. A function $f:A\to B$ which is both upper and lower semi-continuous is called *continuous*. Clearly, every order morphism $A\to B$ is upper semi-continuous; also, the identity map 1_A on A and constant functions $f:A\to B$ are continuous.

Proposition 3.12. *Let $(A,\leq)$ be a continuously and densely ordered set, and let $(B,\leq)$ be totally ordered. Let $f:A\to B$ be upper semi-continuous, $g:A\to B$ lower semi-continuous, and suppose that $f(0)\geq g(0)$ and $f(1)\leq g(1)$, where 0 and 1 are, respectively, the least and greatest element of A. Then the equaliser*

$$\mathrm{Eq}(f,g):=\{x\in A: f(x)=g(x)\}$$

is non-empty and continuously ordered in the induced ordering. Moreover, we have

$$\sup\mathrm{Eq}(f,g)=\sup\{x\in A: f(x)\geq g(x)\}\in\mathrm{Eq}(f,g) \tag{3.7}$$

and

$$\inf\mathrm{Eq}(f,g)=\inf\{x\in A: f(x)\leq g(x)\}\in\mathrm{Eq}(f,g). \tag{3.8}$$

Proof. Let X be any subset of A such that

$$f(x)\geq g(x) \text{ for all } x\in X, \tag{3.9}$$

and suppose that

$$f(\sup X) < g(\sup X). \tag{3.10}$$

Since, by hypothesis, $f(0) \geq g(0)$, we see from (3.10) that $\sup X \neq 0$. We now distinguish two cases according to whether or not there exists some $b \in B$ with $f(\sup X) < b < g(\sup X)$.

(i) Suppose that such b exists, fix such an element, and let

$$Y := \{y \in A : y \leq \sup X \text{ and } g(y) \leq b\}.$$

Then $\sup Y \leq \sup X$ and $g(y) \leq b$ for all $y \in Y$. If $\sup Y = \sup X$, then $\sup Y \neq 0$; in particular $Y \neq \emptyset$. Since g is lower semi-continuous, we obtain

$$g(\sup X) = g(\sup Y) \leq b,$$

contradicting the case assumption. Hence, $\sup Y < \sup X$. Let

$$Z := \{z \in X : \sup Y < z\}.$$

If Z were empty, then we would have $x \leq \sup Y$ for all $x \in X$, thus $\sup X \leq \sup Y$, a contradiction. Hence, $Z \neq \emptyset$. Also, by definition of Z, $\sup Z = \sup X$. Since f is upper semi-continuous, it follows from

$$b > f(\sup X) = f(\sup Z)$$

that $b > f(z)$ for some $z \in Z$, for otherwise we would have $f(\sup Z) \geq b$. Thus, by (3.9) and the definition of Z, $\sup Y < z$ and $g(z) \leq b$ for some $z \in X$; in particular, $z \in Y$. However, the assertions $\sup Y < z$ and $z \in Y$ clearly contradict each other. This contradiction shows that (3.10) cannot hold in case (i).

(ii) Now suppose that $g(\sup X)$ covers $f(\sup X)$,[2] and let

$$Y := \{y \in A : y \leq \sup X \text{ and } g(y) < g(\sup X)\}.$$

Then $\sup Y \leq \sup X$ and $g(y) \leq f(\sup X)$ for all $y \in Y$, since B is totally ordered. We can now repeat the argument in case (i) with little change and again obtain an analogous contradiction.

From (i) and (ii) we conclude that

$$f(\sup X) \geq g(\sup X) \text{ for every subset } X \text{ of } \{x \in A : f(x) \geq g(x)\}. \tag{3.11}$$

Applying the result just obtained to the dual $A' = (A, \geq)$, and noting that $f : A \to B$ and $g : A \to B$ exchange their rôles (as do 0 and 1), we find that

$$f(\inf X) \leq g(\inf X) \text{ for every subset } X \text{ of } \{x \in A : f(x) \leq g(x)\}. \tag{3.12}$$

Now let U be any subset of $\mathrm{Eq}(f, g)$, and let

$$u := \sup\{x \in A : f(x) \geq g(x) \text{ and } x \leq \inf U\}.$$

[2] Let $(S, \leq)$ be an ordered set. If, for elements $a, b \in S$, we have $a < b$, and if there is no element $x \in S$ such that $a < x < b$, we say that b covers a.

By (3.11), (3.12), and the definition of u, we have

$$u \leq \inf U, \ f(u) \geq g(u), \ \text{and} \ f(\inf U) \leq g(\inf U). \tag{3.13}$$

Thus, if $u = \inf U$, we find at once that $f(u) = g(u)$, that is, $u \in \operatorname{Eq}(f,g)$. If, on the other hand, $u \neq \inf U$, then $u < \inf U$. Since A is densely ordered, we conclude that

$$u = \inf\{x \in A : u < x < \inf U\}. \tag{3.14}$$

We also see from the definition of u that $f(x) < g(x)$ for every element x of the set $\{x \in A : u < x < \inf U\}$. Hence, by (3.12) and (3.14), $f(u) \leq g(u)$, which, together with (3.13), again implies $u \in \operatorname{Eq}(f,g)$. Thus, we have shown that

for every subset U of $\operatorname{Eq}(f,g)$, *if* $u = \sup\{x \in A : f(x) \geq g(x) \text{ and } x \leq \inf U\}$, *then* $u \in \operatorname{Eq}(f,g)$. (3.15)

Clearly, u is the largest element of $\operatorname{Eq}(f,g)$ which is a lower bound for all elements of U; that is, u is the infimum of U in $(\operatorname{Eq}(f,g), \leq)$. Invoking Proposition 3.2 again, we see that $(\operatorname{Eq}(f,g), \leq)$ is a complete lattice; that is, $\operatorname{Eq}(f,g)$ is continuously ordered. Setting $U = \emptyset$ in (3.15), we get (3.7), and (3.8) is (3.7) when applied to the dual $A' = (A, \geq)$. □

Taking g as a constant function, we obtain the following.

Corollary 3.13. *Let $(A, \leq)$ be continuously and densely ordered, let $(B, \leq)$ be a totally ordered set, let $f : A \to B$ be upper semi-continuous, and let c be an element of B such that $f(0) \geq c \geq f(1)$. Then there exists some $x \in A$ such that $f(x) = c$.*

Proposition 3.12 and Corollary 3.13 can be applied to real functions defined on a compact interval of real numbers. Since, in the real domain, continuous functions in our terminology coincide with continuous functions in the usual sense, Corollary 3.13 contains, as a special case, Weierstraß' intermediate value theorem for continuous functions $f : [a,b] \to \mathbb{R}$.

Exercises for §3.

1. Let $(X, \leq)$ and $(Y, \leq')$ be ordered sets and, for $x_1, x_2 \in X$ and $y_1, y_2 \in Y$, set

$$(x_1, y_1) \preceq (x_2, y_2) \ :\Longleftrightarrow \ x_1 \leq x_2 \ \text{and} \ y_1 \leq' y_2.$$

Show that $\preceq$ is an order relation on the Cartesian product $X \times Y$.

2. Show that any non-empty intersection of order relations on a set is itself an order relation.

3. Let $(A, \leq)$ be an ordered set, such that every two-element subset has an upper bound. Show that $(A, \leq)$ is directed upwards.

4. Let $(A, \leq)$ be an ordered set, and let $a, b \in A$ be greatest elements of A. Show that $a = b$.

5. Prove that a maximal element of a directed set $(A, \leq)$ is the greatest element of A.

6. Show that a well-ordered set $(A, \leq)$ is totally ordered.

7. Prove Proposition 3.1.

8. Show that, in a lattice, every non-empty finite subset has an infimum and a supremum.

9. Exhibit an example of a bijective order morphism $f : A \to B$ between two ordered sets A and B, such that the inverse map $f^{-1} : B \to A$ is not order-preserving.

10. Exhibit an example of an order morphism $f : A \to B$ between lattices A and B, which is not a lattice homomorphism.

11. Show that an order isomorphism $f : A \to B$ between lattices A and B is in fact a lattice isomorphism.

12. This exercise provides a generalisation of Proposition 3.3.

    Let $(A, \leq)$ be a complete lattice, and let F be a set of order-preserving maps from A into itself, which commute in pairs; that is, we have $fg = gf$ for all $f, g \in F$. Moreover, let

    $$\mathscr{F} := \{a \in A : f(a) = a \text{ for all } f \in F\}$$

    be the set of common fixed points of F. Show that $\mathscr{F} \neq \emptyset$, that the analogues of Equations (3.2) and (3.3) hold, and that $\mathscr{F}$, endowed with the induced ordering, is a complete lattice.

13. Let A be an ordered set. Show that the set $\mathscr{S}^-(A)$ of left segments of A is a complete lattice under inclusion.

14. Provide the missing arguments in Case (ii) of the proof of Proposition 3.12.

§4. Around the axiom of choice

Our principal goal here is to establish the following fundamental result.

Theorem 4.1. *On the basis of axioms* (A1)–(A10), *the following assertions are equivalent.*

(a) *The axiom of choice.*

(b) *Zorn's Lemma (Axiom* (A11)*).*

(c) *If A is a set of sets ordered by inclusion, and if A contains as a member the union of every chain in A, then A has a maximal element.*[3]

(d) *If $\mathfrak{P}$ is a property of finite type for subsets of a set A, then A has a subset which has property $\mathfrak{P}$, and is maximal among such subsets (Teichmüller, Tukey).*

(e) *Every ordered set has a maximal chain (Hausdorff, Birkhoff).*

(f) *Every set can be well ordered (Zermelo).*

Here, a property $\mathfrak{P}$ for subsets of a set A, that is, a subset $\mathfrak{P} \subseteq \mathscr{B}(A)$, is called of *finite type*, if

$$\forall S \in \mathscr{B}(A) : S \in \mathfrak{P} \Longleftrightarrow T \in \mathfrak{P} \text{ for all finite subsets } T \text{ of } S.$$

We shall prove Theorem 4.1 in a sequence of lemmas. For further studies concerning the axiom of choice and its role within contemporary mathematics the reader is referred to [36] and [46].

Lemma 4.2. *The axiom of choice implies Zorn's Lemma.*

Proof. Let $(A, \leq)$ be an inductively ordered set (we recall that this means that each chain in A has an upper bound). Since the empty chain has an upper bound, A is not empty. Let $f : \mathscr{B}(A) \to A$ be a choice function for A (that is, $f(X) \in X$ for $X \subseteq A$ and $X \neq \emptyset$). Given a chain $C \subseteq A$, denote by B_C the set of *proper* upper bounds of C in A (i.e., upper bounds not contained in C). Set

$$C' := \begin{cases} C \cup \{f(B_C)\}, & B_C \neq \emptyset \\ C, & B_C = \emptyset. \end{cases}$$

Then C' is again a chain in A, and $C \subseteq C'$. We shall show that there exists a chain C_0 in A such that $C_0' = C_0$. An upper bound for C_0 (which is guaranteed to exist by our hypothesis) is then the unique greatest element of C_0, and a maximal element of A.

Call a set $\mathscr{C}$ of chains in A *admissible*, if

[3] A result closely resembling Part (c), but with an extra hypothesis concerning existence of choice functions, was obtained by Kuratowski [52]; see also [53, Chap. VII, § 8, Theorem 6].

(i) $\emptyset \in \mathscr{C}$,
(ii) $C \in \mathscr{C} \Longrightarrow C' \in \mathscr{C}$,
(iii) the union of every chain in $\mathscr{C}$ is a member of $\mathscr{C}$.

Such admissible sets do exist; for instance, the collection of all chains in A has these properties. Denote by $\mathscr{I}$ the intersection of all admissible sets of chains in A. Then $\mathscr{I}$ is the unique smallest admissible set of chains in A. It will be enough to show that $\mathscr{I}$ is *totally ordered by inclusion*; for then the union U taken over all members of $\mathscr{I}$ is a chain in A, which is contained in $\mathscr{I}$ by Property (iii), hence U is the greatest element of $\mathscr{I}$, and since $U' \in \mathscr{I}$ by (ii) we must have $U' \subseteq U$, and so $U = U'$ by definition of the map $C \mapsto C'$ as required.

Call an element $C \in \mathscr{I}$ *regular*, if C is compatible with every member of $\mathscr{I}$, our task being now to show that every element of $\mathscr{I}$ is regular. For $C \in \mathscr{I}$ regular, set

$$\mathscr{F}_C := \{X \in \mathscr{I} : X \subseteq C \text{ or } C' \subseteq X\}.$$

Note that $\emptyset, C, C' \in \mathscr{F}_C$; in particular, $\mathscr{F}_C$ satisfies Property (i) in the definition of an admissible set of chains in A. Also, if $X \in \mathscr{F}_C$, then $X \subseteq C$ or $C' \subseteq X$; in the second case a fortiori $C' \subseteq X'$, so $X' \in \mathscr{F}_C$. Now consider the first case, that is, $X \subseteq C$. By regularity of C, $X' \subseteq C$ or $C \subseteq X'$. If $X' \subseteq C$, then again $X' \in \mathscr{F}_C$; and if $C \subseteq X'$, then $X \subseteq C \subseteq X'$, so $X = C$ or $X' = C$, since $C' \setminus C$ has at most one element. In each case, $X' \in \mathscr{F}_C$, so $\mathscr{F}_C$ satisfies Property (ii). Moreover, if $\{X_i : i \in I\}$ is a chain in $\mathscr{F}_C$, then, in particular, it is a chain in $\mathscr{I}$, thus $X_0 := \bigcup_i X_i \in \mathscr{I}$, since $\mathscr{I}$ meets Condition (iii). Also, for each i, either $X_i \subseteq C$ or $C' \subseteq X_i$. If $X_i \subseteq C$ holds for all $i \in I$, then $X_0 \subseteq C$, implying $X_0 \in \mathscr{F}_C$. Otherwise, we have $C' \subseteq X_i$ for at least one $i \in I$, thus $C' \subseteq X_0$, which again implies $X_0 \in \mathscr{F}_C$. Hence, $\mathscr{F}_C$ also satisfies Property (iii), thus is an admissible set of chains in A. By definition of $\mathscr{F}_C$, we have $\mathscr{F}_C \subseteq \mathscr{I}$, while $\mathscr{I} \subseteq \mathscr{F}_C$ holds by minimality of $\mathscr{I}$. Thus, we have demonstrated that

$$\mathscr{F}_C = \mathscr{I} \text{ for every regular element } C \text{ of } \mathscr{I}. \tag{4.1}$$

We now finish the proof by showing that *the collection $\mathscr{R}$ of regular elements of $\mathscr{I}$ is itself an admissible set of chains in* A. First, the empty chain $\emptyset$ is regular. Second, if $C \in \mathscr{I}$ is regular, then so is C', since by (4.1), $X \in \mathscr{I}$ satisfies $X \subseteq C$, so a fortiori $X \subseteq C'$, or we have $C' \subseteq X$. Finally, let $\{X_i : i \in I\} \subseteq \mathscr{R}$ be a chain of regular elements. Since $\mathscr{R} \subseteq \mathscr{I}$, we have $X_0 := \bigcup_i X_i \in \mathscr{I}$. Given $X \in \mathscr{I}$, for each index i, we have $X_i \subseteq X$ or $X \subseteq X_i$, since X_i is regular. If $X_i \subseteq X$ holds for all $i \in I$, then $X_0 \subseteq X$; otherwise, $X \subseteq X_0$, hence $X_0 \in \mathscr{R}$. Consequently, $\mathscr{R}$ is indeed an admissible set of chains in A, so $\mathscr{R} = \mathscr{I}$ by minimality of $\mathscr{I}$ plus the fact that $\mathscr{R} \subseteq \mathscr{I}$ by definition. It follows that $\mathscr{I}$ is totally ordered by inclusion, as required, and the proof is complete. □

Lemma 4.3. $(c) \Longrightarrow (d)$.

Proof. This is left as an exercise for the reader; cf. Exercise 1 below. □

Let (e') be the assertion that *every chain in an ordered set* $(A, \leq)$ *is contained in a maximal chain of* A.

Lemma 4.4. $(d) \Longrightarrow (e')$.

Proof. Let $(A, \leq)$ be an ordered set, and let C be any chain in A. Consider the set S_C of all elements in A which are compatible with every element of C. Clearly, a subset of an ordered set is a chain if, and only if, each of its two element subsets is a chain; hence, the property of a subset of A to be a chain is of finite type. Applying (d) to $(S_C, \leq)$, we obtain existence of a maximal chain C^* in S_C. Since $C \subseteq S_C$ by definition of S_C, $C \subseteq C^*$ by maximality of C^*, and C^* is a maximal chain in A. □

Applying (e') to the empty chain, we obtain the following.

Corollary 4.5. $(d) \Longrightarrow (e)$.

Lemma 4.6. $(e) \Longrightarrow (f)$.

Proof. Let A be an abstract set, and consider the set $\mathscr{W}$ of all pairs $(S, \leq)$, where $S \subseteq A$ and $\leq$ is a well-ordering on S. We define a binary relation $\preceq$ on $\mathscr{W}$ as follows:

$$(S_1, \leq_1) \preceq (S_2, \leq_2)$$

if, and only if,

(i) $S_1 \subseteq S_2$,

(ii) $\forall x, y \in S_1 : x \leq_1 y \Longleftrightarrow x \leq_2 y$,

(iii) $\forall x, y \in S_2 : x \in S_1$ and $y \in S_2 \setminus S_1 \Longrightarrow x \leq_2 y$.

Then $\preceq$ is an order relation on the set $\mathscr{W}$; cf. Exercise 2. Consequently, our hypothesis (e) guarantees existence of a maximal chain $\mathscr{C} = \{(S_i, \leq_i)\}_{i \in I}$ in $\mathscr{W}$. Set $S^* := \bigcup_i S_i$, and define a binary relation $\leq^*$ on S^* as follows: given elements $x, y \in S^*$, find an index $i \in I$ such that $x, y \in S_i$, and define

$$x \leq^* y :\Longleftrightarrow x \leq_i y.$$

This definition is independent of the S_i used by (ii), and is clearly a total order on S^*. Moreover, each S_i is a left segment in $(S^*, \leq^*)$. Indeed, let $x, y \in S^*$, $y \in S_i$, and $x \leq^* y$, and suppose that $x \notin S_i$. Then $x \leq_j y$ in S_j for some $j \in I$, and $(S_i, \leq_i) \preceq (S_j, \leq_j)$ by (i) together with the chain property of $\mathscr{C}$. It follows by (iii) that $y \leq_j x$, so $x = y \in S_i$, a contradiction. Thus, if $T \subseteq S^*$ is non-empty, then there is an index $i \in I$ with $T \cap S_i \neq \emptyset$, and the least element x_0 in $T \cap S_i$ is least in T. It follows that $\leq^*$ is a well-ordering on S^*, so $(S^*, \leq^*) \in \mathscr{W}$, and we have $(S_i, \leq_i) \preceq (S^*, \leq^*)$ for all i by construction.

If we had $S^* \subset A$, then we could adjoin to S^* an element from $A \setminus S^*$ as greatest element, obtaining a well-ordered set $(S^{**}, \leq^{**}) \in \mathscr{W}$ such that

$$(S^*, \leq^*) \prec (S^{**}, \leq^{**}).$$

We could then adjoin $(S^{**}, \leq^{**})$ to $\mathscr{C}$, in this way obtaining a chain in $\mathscr{W}$ strictly containing $\mathscr{C}$, contradicting the maximality of $\mathscr{C}$. Thus, $S^* = A$, so that A can be well ordered. □

To finish the proof of Theorem 4.1, we note that (c) is an immediate consequence of Zorn's Lemma (b) (since the union of a chain $\mathscr{C}$ in A is an upper bound for $\mathscr{C}$), and that $(f) \Longrightarrow (a)$ is easy: if A is any set, introduce a well-ordering in A via (f), and define a choice function φ on A by sending a non-empty subset S of A to its unique minimal element $\varphi(S)$.

Remark 4.7. The axiom of choice also implies the following formal strengthening of Zorn's Lemma: *If $(S, \leq)$ is an inductively ordered set, and if $a \in S$ is a given element, then there exists a maximal element s_0 in S with $a \leq s_0$.*

The axiom of choice is a powerful tool in all parts of mathematics not entirely confined to the study of finite or countable structures. In an algebraic context, Zorn's Lemma, Zermelo's well- ordering theorem, and the Teichmüller-Tukey axiom (d) appear to be the most natural and useful versions. To conclude this section, we describe some classical applications. Further results of an algebraic nature involving transfinite constructions form the topic of §8–§14. It has to be noted though that the axiom of choice is also the source of some rather unpleasant results, for instance, the existence of bounded non-measurable sets of real numbers; cf. §7.1.

A theorem of Krull

We shall deal here in detail with the case of right ideals, but the reader will have no problem adapting the result or the proof to the case of left or full (left and right) ideals.

An element e of a ring Λ is called a *left identity element*, if $ea = a$ for all $a \in \Lambda$. A *right ideal* of Λ is a subring $\mathfrak{I}$ of Λ such that $ab \in \mathfrak{I}$ whenever $a \in \mathfrak{I}$ and $b \in \Lambda$. An *ideal* in Λ is a subring $\mathfrak{I}$ such that $ab \in \mathfrak{I}$ whenever at least one of the factors is in $\mathfrak{I}$. A right ideal (ideal) $\mathfrak{I}$ is called *proper* if $\mathfrak{I} \neq \Lambda$. Finally, a proper right ideal (ideal) $\mathfrak{I}$ of Λ is *maximal*, if $\mathfrak{I} \leq \mathfrak{J} \leq \Lambda$ with any right ideal (ideal) $\mathfrak{J}$ implies $\mathfrak{J} = \mathfrak{I}$ or $\mathfrak{J} = \Lambda$.

Proposition 4.8. *Let Λ be a ring having a left identity element e. Then every proper right ideal of Λ is contained in a maximal right ideal of Λ.*

Proof. Given a proper right ideal $\mathfrak{I}$ of Λ, consider the set $\mathfrak{S}$ of all right ideals $\mathfrak{J}$ in Λ such that $\mathfrak{I} \subseteq \mathfrak{J}$ and $e \notin \mathfrak{J}$. $\mathfrak{S}$ is not empty, for instance $\mathfrak{I} \in \mathfrak{S}$, and is ordered by inclusion. If $\mathscr{C}$ is a non-empty chain in $(\mathfrak{S}, \subseteq)$, then $\mathfrak{J}_0 := \bigcup_{\mathfrak{J} \in \mathscr{C}} \mathfrak{J}$ is again in $\mathfrak{S}$; that is, $(\mathfrak{S}, \subseteq)$ is inductively ordered. By Zorn's Lemma, $(\mathfrak{S}, \subseteq)$ has a maximal element $\mathfrak{M}$. By definition of $\mathfrak{S}$, $\mathfrak{M}$ is a proper right ideal of Λ containing $\mathfrak{I}$, and if $\mathfrak{M} \subset \mathfrak{J}$ for some right ideal $\mathfrak{J}$ of Λ, then $\mathfrak{J} \notin \mathfrak{S}$, thus $e \in \mathfrak{J}$; that is, $\mathfrak{J} = \Lambda$, so $\mathfrak{M}$ is a maximal right ideal of Λ. □

Remark 4.9. Somewhat surprisingly, it turns out that Krull's ideal theorem is actually *equivalent* to Zorn's Lemma. In fact, Wilfrid Hodges [42] has shown that, on the basis of Zermelo–Fraenkel set theory, a special case of Krull's theorem already implies an equivalent of the axiom of choice; cf. §7.3 for the precise result and its proof.

Existence of algebraic closures

A field K is called *algebraically closed*, if every polynomial $f(x) \in K[x]$ of degree at least 1 has a root in K, or, equivalently, every non-constant polynomial with coefficients in K decomposes over K as a product of linear factors; cf. Exercise 3. The complex numbers $\mathbb{C}$ provide a first example of an algebraically closed field.[4] If $K \leq L$ is a field extension, then an element $a \in L$ is called *algebraic over K*, if a satisfies a polynomial equation

$$a^n + c_1 a^{n-1} + \cdots + c_{n-1} a + c_n = 0$$

with coefficients $c_1, \ldots, c_n \in K$. The extension $K \leq L$ is called *algebraic*, if every element of L is algebraic over K. If $K \leq \overline{K}$ is a field extension such that $\overline{K}$ is algebraically closed and algebraic over K, then $\overline{K}$ is called an *algebraic closure* of K. One can show that such a field $\overline{K}$ is determined up to an isomorphism inducing the identity on K; cf., for instance, Theorem 2 in [57, Chap. 7, §2]. The following result, asserting the *existence* of algebraic closures, is of fundamental importance in Galois Theory.

Proposition 4.10. *Every field admits an algebraic closure.*

The main part of the argument, which goes back to Steinitz [75, §17–21] deals with the existence of an algebraically closed extension field; it is here that transcendental methods necessarily enter (in our case via Krull's maximal ideal theorem, which in turn is based on Zorn's Lemma, as we have seen). Once such a field extension has been constructed, existence of an algebraic closure (i.e., of an *algebraic* extension which is algebraically closed) follows by standard arguments of a purely algebraic nature. We

[4]This famous result is called the *Fundamental Theorem of Algebra*. See §7.4 for an elementary proof.

shall only sketch this last part. For the proof we need the following simple but important observation, which goes back to Kronecker.

Lemma 4.11. *Let K be a field, and let $f \in K[x]$ be a polynomial of degree at least one. Then there exists an extension field L over K, such that f has a root in L.*

Proof. Choose an irreducible factor $f_1 = f_1(x) = \sum_{\nu=0}^{n} c_\nu x^\nu$ of f over K. Then $\langle f_1 \rangle$ is a maximal ideal in $K[x]$, hence $L := K[x]/\langle f_1 \rangle$ is a field; cf. Exercises 4 and 5. The composite map

$$K \hookrightarrow K[x] \overset{\pi}{\twoheadrightarrow} K[x]/\langle f_1 \rangle = L$$

is an embedding (as homomorphism between fields), hence, identifying K with its image in L, L becomes an extension field over K. Setting $y := \pi(x)$, we find that

$$f_1(y) = \sum_{\nu=0}^{n} c_\nu y^\nu = \sum_{\nu=0}^{n} c_\nu \pi(x)^\nu = \pi\Big(\sum_{\nu=0}^{n} c_\nu x^\nu\Big) = \pi(f_1) = 0,$$

so y is a root of f_1, and hence of f. $\square$

Proof of Proposition 4.10. Let K be a field, consider the system of variables $\mathfrak{X} = \{x_f\}_{f \in I}$ indexed by the set

$$I = \big\{f \in K[x] : \deg(f) \geq 1\big\},$$

and form the polynomial ring $K[\mathfrak{X}]$. Let

$$\mathfrak{I} = \big\langle f(x_f) : f \in I \big\rangle$$

be the ideal generated by all polynomials $f(x_f)$, where the variable x of f has been replaced by the corresponding variable x_f. Suppose that $\mathfrak{I} = K[\mathfrak{X}]$. Then there exist finitely many polynomials $f_1, \ldots, f_n \in I$ and elements $g_1, \ldots, g_n \in K[\mathfrak{X}]$ such that

$$g_1 f_1(x_{f_1}) + g_2 f_2(x_{f_2}) + \cdots + g_n f_n(x_{f_n}) = 1. \tag{4.2}$$

Applying Lemma 4.11 successively to the polynomials $f_1, \ldots, f_n$, we can find an extension field K' of K such that each f_ν has a root α_ν in K'. Setting $x_{f_\nu} = \alpha_\nu$ for $\nu = 1, \ldots, n$ in (4.2) then leads to the contradiction that $0 = 1$. Hence, $\mathfrak{I} < K[\mathfrak{X}]$. By Krull's Theorem (Proposition 4.8), there exists a maximal ideal $\mathfrak{M}$ in $K[\mathfrak{X}]$ containing $\mathfrak{I}$. Then $L_1 := K[\mathfrak{X}]/\mathfrak{M}$ is a field, which can be viewed as an extension of K via the canonical maps

$$K \hookrightarrow K[\mathfrak{X}] \overset{\pi}{\twoheadrightarrow} K[\mathfrak{X}]/\mathfrak{M} = L_1.$$

As in the proof of Lemma 4.11, we see that $f(y_f) = 0$, where $y_f = \pi(x_f)$ is the residue class in L_1 corresponding to the variable x_f:

$$f(y_f) = f(\pi(x_f)) = \pi(f(x_f)) = 0.$$

Hence, every non-constant polynomial over K has a root in L_1. Iterating this construction, we obtain an ascending chain of fields

$$K = L_0 \leq L_1 \leq L_2 \leq \cdots,$$

such that every polynomial $f \in L_n[x]$ of degree $\deg(f) \geq 1$ has a root in L_{n+1}. Set

$$L := \bigcup_{n \geq 0} L_n.$$

Then L is a field extension of K; moreover, we claim that L is algebraically closed. Indeed, let $f \in L[x]$ be a non-constant polynomial. Then there exists some $n \in \mathbb{N}$ such that $f \in L_n[x]$, so f has a root in L_{n+1}, and thus in L.

To finish the proof, set

$$\overline{K} := \{a \in L : a \text{ is algebraic over } K\}.$$

Then $\overline{K}$ is a field, and hence an algebraic extension of K, since for $a, b \in \bar{K}$, we have $K(a,b) \subseteq \overline{K}$; furthermore, if $f \in \overline{K}[X]$ is a polynomial of degree $\deg(f) \geq 1$, then f has a root α in L, α is algebraic over $\overline{K}$, thus algebraic over K, so $\alpha \in \overline{K}$; that is, $\overline{K}$ is algebraically closed. □

Existence of bases in vector spaces

Let V be a vector space over a skew field K (an algebraic structure satisfying all the axioms of a field, except possibly the commutative law of multiplication). A subset $S \subseteq V$ is called *linearly independent*, if every relation of the form

$$\alpha_1 s_1 + \alpha_2 s_2 + \cdots + \alpha_n s_n = 0$$

with $\alpha_\nu \in K$ and $s_\nu \in S$ implies $\alpha_1 = \cdots = \alpha_n = 0$. Linear independence is thus a property of finite type for the subsets of V. By Part (d) of Theorem 4.1, V has at least one maximal linearly independent set B.

A subset S of V is called a *generating system* for V, if $\langle S \rangle = V$, that is, if S is not contained in a proper subspace of V. A *basis* of V is a linearly independent generating system of V.

Let B be a maximal linearly independent set, and let $v \in V$ be an arbitrary vector. If $v \notin B$, then, by maximality of B, there exist finitely many elements $b_1, \ldots, b_n \in B$ and scalars $\alpha_0, \alpha_1, \ldots, \alpha_n \in K$, not all zero, such that

$$\alpha_0 v + \alpha_1 b_1 + \cdots + \alpha_n b_n = 0. \tag{4.3}$$

Since B is linearly independent, $\alpha_0 \neq 0$, so we can rewrite (4.3) in the form

$$v = -\alpha_0^{-1}\alpha_1 b_1 - \alpha_0^{-1}\alpha_2 b_2 - \cdots - \alpha_0^{-1}\alpha_n b_n.$$

Hence, $v \in \langle B \rangle$, and since v was arbitrary, $\langle B \rangle = V$; that is, B is a basis of V.

Summarising, we have shown the following.

Proposition 4.12. *Every vector space over a skew field has a basis.*

The (algebraic) dimension $\dim_K(V)$ of a vector space V over a skew field K is defined as the cardinality of a basis of V. This, of course, raises the problem of showing that any two bases of V have the same cardinality. If V is finitely generated, the result follows from the Steinitz exchange lemma, which is a well-known and extremely valuable tool for various aspects of linear algebra; cf., for instance, Lemma 4 and Theorem 5 in [23, Sec. 4.3]. If, however, V is not finitely generated, the result is still true, but the argument is necessarily quite different, making use of cardinal arithmetic; see Exercise 5.10 and Corollary 11.5.

Existence of spanning trees in connected graphs

By an S-graph[5] X we shall mean a quintuple $X = (V, E, \partial_0, \partial_1, ^-)$, where $V = V(X)$ and $E = E(X)$ are sets (the set of vertices and edges of X, respectively), and $\partial_0, \partial_1, ^-$ are maps such that $\partial_i : E \to V$ and $^- : E \to E$. We require that $\overline{\overline{e}} = e$, $\overline{e} \neq e$, and that $\partial_0(e) = \partial_1(\overline{e})$ for all $e \in E$. The vertices $\partial_0(e)$ and $\partial_1(e)$ are usually referred to as *origin* and *terminus* of the edge e, respectively. For $e \in E$, the pair $\{e, \overline{e}\}$ is called a *geometric edge* of X, and an *orientation* $\mathscr{O}_X$ of X is a subset $\mathscr{O}_X \subseteq E$, such that $E = \mathscr{O}_X \sqcup \overline{\mathscr{O}_X}$; that is, $\mathscr{O}(X)$ consists of a choice of exactly one member from each geometric edge.

By a path p in an S-graph X is meant a sequence of edges $p = e_1, e_2, \ldots, e_r$, such that $\partial_1(e_i) = \partial_0(e_{i+1})$ for $1 \leq i < r$. Setting $v_0 := \partial_0(e_1)$ and $v_j = \partial_1(e_j)$ for $1 \leq j \leq r$, we call $(v_0, v_1, \ldots, v_r)$ the vertex sequence of the path p, and we say that p connects the vertex v_0 to the vertex v_r, or runs from v_0 to v_r. The positive integer r is called the *length* of the path p. For each vertex v of X, we also allow the *empty path* $p = 1_v$ which has length 0, and begins and ends in v. A pair of edges of the form $(e_i, e_{i+1}) = (e_i, \overline{e_i})$ in a path p is termed a *backtracking*. By deleting a backtracking occurring in a path p of length r running from v_0 to v_r, we obtain a path of length $r-2$ from v_0 to v_r. Deleting, one after the other (say from left to right), all backtrackings in p, we obtain a *reduced path*, that is, a path without backtracking, still connecting v_0 to v_r. Should this process lead to a path with 0 edges, then we have $v_0 = v_r$, and the result is the empty path 1_{v_0}. An S-graph X is called *connected*, if to any pair $(v, v') \in V(X) \times V(X)$, there exists a path p in X connecting v to v'.

Given an S-graph $X = (V, E, \partial_0, \partial_1, ^-)$, a quintuple $X' = (V', E', \partial_0', \partial_1', ^{-'})$ with $\partial_i' : E' \to V'$ and $^{-'} : E' \to E'$ is called a *subgraph of* X, denoted $X' \leq X$, if $V' \subseteq V$, $E' \subseteq E$, and ∂_0', ∂_1', $^{-'}$ are the restrictions of the maps ∂_0, ∂_1, and

[5] The concept of graph used here follows Serre's book [73]; see, in particular, [73, Chap. I, §2].

$^{-}$, respectively. The maximal connected subgraphs of X (under the relation $\leq$) are termed the *connected components* of the S-graph X.

Let $r \geq 1$ be an integer. A *circuit of length* r in an S-graph X is a reduced path $p = e_1 e_2 \ldots e_r$ in X, such that the vertices $v_j = \partial_1(e_j)$ with $1 \leq j \leq r$ are pairwise distinct, and such that $v_r = \partial_0(e_1)$. An *S-tree* is a connected S-graph without circuits.

Let $X = (V, E, \partial_0, \partial_1, {}^{-})$ be an S-graph, let $v \in V$ be a vertex, and let

$$S_v := \{e \in E : \partial_1(e) = v\}.$$

The (cardinal) number $s := |S_v|$ is called the *index* of v. If $s = 0$, we say that v is *isolated* (if X is connected, this is only possible for $V = \{v\}$ and $E = \emptyset$). If $s \leq 1$, we say that v is a *terminal* vertex of X. We denote by $X - v$ the subgraph of X with vertex set $X - \{v\}$ and edge set $E - (S_v \cup \overline{S_v})$.

Lemma 4.13. *Let v be a non-isolated terminal vertex of an S-graph X. Then we have the following.*

(i) *X is connected if, and only if, $X - v$ is connected.*
(ii) *Every circuit of X is contained in $X - v$.*
(iii) *X is an S-tree if, and only if, $X - v$ is an S-tree.*

Proof. The hypothesis says that v is the end-vertex (terminus) of a unique edge e. Assertion (i) is immediate. Also, every vertex belonging to a circuit has index at least 2, so does not involve v, whence (ii). Assertion (iii) follows from (i) and (ii). □

A *subtree* of a non-empty S-graph X is a subgraph $T \leq X$, which is an S-tree in its own right. A subtree $T \leq X$ of a connected non-empty S-graph X is called *spanning*, if $V(T) = V(X)$; that is, if T contains all vertices of the graph X.

Proposition 4.14. *Every non-empty connected S-graph contains a spanning subtree.*

Proof. Let $X = (V, E, \partial_0, \partial_1, {}^{-})$ be a non-empty connected S-graph. The collection of all subtrees of X becomes an ordered set under the relation $\leq$ (is a subgraph of). Let $\mathfrak{T} = (T_i)_{i \in I}$ be a chain of subtrees in X, and let $V_i = V(T_i)$, $E_i = E(T_i)$. Set

$$V' := \bigcup_i V_i \text{ and } E' := \bigcup E_i.$$

The pair (V', E') is clearly closed under the maps ∂_i and $^{-}$, so forms a subgraph X' of X such that $T_i \leq X'$ for all $i \in I$. We claim that X' is a subtree of X. Indeed, if $v_1', v_2' \in V'$ are two vertices of X', say, $v_1' \in V_i$

and $v_2' \in V_j$ for $i, j \in I$, and (to fix ideas) $T_i \leq T_j$, then $v_1', v_2' \in V_j$, so there exists a path in T_j connecting v_1' to v_2', since T_j is connected. Hence, X' is connected. Next, suppose that $c = e_1 e_2 \ldots e_r$ is a circuit in X', where $e_j \in E_{i_j}$. Using the fact that $\mathfrak{T}$ is a chain under $\leq$, an immediate induction shows that $e_1, e_2, \ldots, e_r \in E_i$ for some i, so that c lies in T_i, a contradiction since T_i is an S-tree, thus has no circuits. Hence, X' has no circuits, so is a subtree of X, implying that the set of all subtrees of X is inductively ordered under $\leq$. By Zorn's Lemma, there exists a subtree T of X maximal among all subtrees, and we claim that T is a spanning subtree. If $V(T) \subset V$ then, since X is connected, there exists an edge $e \in E$ with $\partial_0(e) \in V(T)$ and $\partial_1(e) \in V \setminus V(T)$. According to Lemma 4.13, the subgraph T' obtained from T by adjoining the vertex $\partial_1(e)$ and the edges $e, \bar{e}$ is again a subtree of X, and we have $T < T'$, contradicting maximality of T. Hence, T is a spanning subtree of X, as claimed. □

Cauchy's functional equation

In 1821, Cauchy published an investigation into the functional equation

$$f(x+y) = f(x) + f(y). \tag{4.4}$$

His memoir is of fundamental importance, and may be viewed as a starting point for the theory of functional equations. Assuming continuity everywhere on the real line, Cauchy showed that the solutions $f(x)$ of (4.4) are all of the form

$$f(x) = cx, \quad -\infty < x < \infty \tag{4.5}$$

for some constant $c \in \mathbb{R}$. Subsequently, in 1880, Darboux was able to improve Cauchy's result. He showed that if $f(x)$ is (a) continuous at a point, or (b) bounded on an arbitrarily small interval towards one side, then $f(x)$ is of the form (4.5); cf. [24]–[25], as well as [1, p. 8]. On the basis of these results, it was conjectured by Darboux that *every real function* $f(x)$ *satisfying the functional equation* (4.4) *on the whole real line is in fact of the form* (4.5). This assertion was finally disproved in 1905 by Hamel [35] using transfinite methods; his result and argument are given below.

We are seeking all real functions $f: \mathbb{R} \to \mathbb{R}$ satisfying the functional equation (4.4) everywhere. An immediate induction using (4.4) shows that

$$f(x_1 + \cdots + x_n) = f(x_1) + \cdots + f(x_n), \quad n \in \mathbb{N};$$

in particular, setting $x_1 = \cdots = x_n = x$,

$$f(nx) = nf(x), \quad n \in \mathbb{N}. \tag{4.6}$$

Next, we observe that Equation (4.6) also holds for negative integral values of n. Indeed, let $n \in \mathbb{Z} \setminus \mathbb{N}$. Then $-n \in \mathbb{N}$, and, by (4.4) and (4.6),

$$0 = f(0) = f(nx + (-nx)) = f(nx) + f(-nx) = f(nx) - nf(x),$$

so that

$$f(nx) = nf(x)$$

as required. Hence, we have

$$f(mx) = mf(x), \quad m \in \mathbb{Z}. \tag{4.7}$$

Now let $x = m/n$ be a rational number, $m, n \in \mathbb{Z}$, and $n \neq 0$. Then $nx = m$, hence by (4.7),

$$f(nx) = nf(x) = f(m) = mf(1),$$

thus,

$$f(x) = xf(1).$$

Setting $f(1) = c$, we thus have

$$f(x) = cx, \quad x \in \mathbb{Q}.$$

Assuming that $f(x)$ is continuous everywhere, the fact that $\mathbb{Q}$ is dense in $\mathbb{R}$ now gives $f(x) = cx$ for all $x \in \mathbb{R}$; that is, Cauchy's result. In the general case, there are, however, much more complicated solutions. Consider the field $\mathbb{R}$ as a vector space $\mathfrak{V}$ over the field $\mathbb{Q}$ of rational numbers. The decisive observation is now that a real function $f(x)$ satisfying (4.4) everywhere on the real line is in fact a *linear endomorphism* of $\mathfrak{V}$. For, if $y = \frac{m}{n}x$ with $m, n \in \mathbb{Z}$ and $x \in \mathbb{R}$, then, by (4.7),

$$ny = mx, \; nf(y) = mf(x), \; f(y) = \frac{m}{n}f(x),$$

that is,

$$f\left(\frac{m}{n}x\right) = \frac{m}{n}f(x),$$

which in conjunction with (4.4) establishes our claim. Conversely, every linear endomorphism of $\mathfrak{V}$ is a solution of (4.4). Making use of the fact, contained in Proposition 4.12, that $\mathfrak{V}$ has a basis, we now obtain the following.

Proposition 4.15. (G. K. W. Hamel, 1905) *The most general solution f which satisfies Cauchy's functional equation* (4.4) *for all real x, y may be constructed from a basis B of $\mathfrak{V}$ by choosing arbitrary values of f for the points $b \in B$, and defining $f(x)$ for any real number*

$$x = r_1 b_1 + r_2 b_2 + \cdots + r_n b_n \quad (b_\nu \in B, r_\nu \in \mathbb{Q})$$

via

$$f(x) = r_1 f(b_1) + r_2 f(b_2) + \cdots + r_n f(b_n).$$

Such a solution is continuous if, and only if, there exists a constant c such that

$$f(b) = cb, \quad b \in B,$$

in which case we have $f(x) = cx$ identically.

Hausdorff's Co-finality Theorem

To conclude this section, we describe an application of Zermelo's well-order theorem to arbitrary ordered sets. Let A be an ordered set, B a subset of A. Then A is called *co-initial* with B, if every lower bound of B in A is a member of B (that is, there is no strict lower bound for B in A). Dually, A is called *co-final* with B, if every upper bound of B in A is a member of B, so that there does not exist a strict upper bound for B in A. Note that A is both co-initial and co-final with each of its maximal chains. We have the following useful transitivity property.

Proposition 4.16. *Let A be an ordered set, let B be a chain in A, and let C be a subset of B.*

(i) *If A is co-initial with B, and B is co-initial with C, then A is co-initial with C.*

(ii) *If A is co-final with B, and B is co-final with C, then A is co-final with C.*

Proof. (i) Since B is totally ordered, and B is co-initial with C, for every $b \in B$, there exists an element $c \in C$ such that $c \leq b$. Hence, if $a < c$ for some $a \in A$ and all $c \in C$, we would have $a < b$ for all $b \in B$, contradicting our assumption that A is co-initial with B.

(ii) Dual to (i). □

Proposition 4.17. *Every totally ordered set is co-final with some well-ordered subset.*

Proof. Let $(A, \leq)$ be the totally ordered set, introduce a well-ordering $\preceq$ on A in accordance with Theorem 4.1, and let

$$B := \{b \in A : a \leq b \text{ for all } a \in A \text{ such that } a \preceq b\}.$$

Clearly, the totally ordered set $(B, \leq)$ and the well-ordered set $(B, \preceq)$ are order-isomorphic; that is, B is a well-ordered subset of $(A, \leq)$. Suppose that A contains a strict upper bound for B, and let a_0 be the smallest of these bounds with respect to the well-ordering $\preceq$. Since $a_0 \notin B$, a_0 is not the least element of $(A, \preceq)$, so there exists at least one element $a_1 \in A$ with $a_1 \prec a_0$. Moreover, since $a_0 \notin B$ and A is totally ordered under $\leq$, there exists at least one such element a_1 with $a_1 \prec a_0$, such that $a_1 > a_0$. However, since a_0 is a strict upper bound for B in A, so is a_1, contradicting the minimality in the definition of a_0. Hence, A does not contain a strict upper bound for B, thus is co-final with its well-ordered subset B. □

Exercises for §4.

1. Prove Lemma 4.3.

2. Show that the binary relation $\preceq$ introduced in the proof of Lemma 4.6 is an order relation.

3. Let $f(x) \in K[x]$ be a polynomial over the field K, and suppose that $\xi \in K$ is a root of f; that is, $f(\xi) = 0$. Show that there exists a polynomial $g(x) \in K[x]$ such that $f(x) = (x - \xi)g(x)$.

4. Let K be a field, and let $K[x]$ be the ring (principal ideal domain) of polynomials over K in the variable x. If $f \in K[x]$ is irreducible, show that $\langle f \rangle$, the principal ideal generated by f, is maximal in $K[x]$.

5. Let Λ be a commutative ring with identity element 1, and let $\mathfrak{J}$ be a maximal ideal in Λ. Show that the quotient ring $\bar{\Lambda} = \Lambda/\mathfrak{J}$ is a field.

6. Let ρ be an anti-symmetric relation on the set S. Show by means of an example that ρ may not extend to an order relation on S; that is, there is no order relation ρ' on S with $\rho \subseteq \rho'$.

7. Let ρ be an order relation on the set S. Show that there exists a total order $\bar{\rho}$ on S with $\rho \subseteq \bar{\rho}$.

8. Prove the following refinement of Proposition 4.12.

Proposition 4.18. *Let V be a vector space over a skew field, and let $A \subseteq V$ be a given set of linearly independent vectors. Then V has a basis B such that $B \supseteq A$.*

9. Let X be a finite non-empty connected S-graph with n vertices and $2m$ edges. Show that $m \geq n - 1$, with equality occurring in this inequality if, and only if, X is a tree.

10. Let $X = (V, E, \partial_0, \partial_1, ^-)$ and $X' = (V', E', \partial_0', \partial_1', ^{-'})$ be S-graphs. By definition, a *graph morphism* $\Phi : X \to X'$ is a pair of maps $\Phi = (\varphi, \psi)$, where $\varphi : V \to V'$ and $\psi : E \to E'$, such that $\varphi(\partial_0(e)) = \partial_0'(\psi(e))$ and $\psi(\bar{e}) = \overline{\psi(e)}'$ holds for all $e \in E$. A graph morphism $\Phi = (\varphi, \psi) : X \to X'$ as above is a *graph isomorphism*, if there exists a graph morphism $\Phi' = (\varphi', \psi') : X' \to X$, such that $\Phi' \circ \Phi := (\varphi' \circ \varphi, \psi' \circ \psi) = (1_V, 1_E)$ and $\Phi \circ \Phi' := (\varphi \circ \varphi', \psi \circ \psi') = (1_{V'}, 1_{E'})$. Let $\Phi = (\varphi, \psi) : X \to X'$ be a graph morphism as above. Show the following.

(i) Φ satisfies $\varphi(\partial_1(e)) = \partial_1'(\psi(e))$ for all $e \in E$.

(ii) Φ is a graph isomorphism if, and only if, both maps $\varphi : V \to V'$ and $\psi : E \to E'$ are bijections.

§5. Cardinals and ordinals

Unless otherwise stated, all sets are taken to lie in a fixed (but arbitrary) universe U. With every ordered set $A \in U$ there is associated an object called its *order type*, denoted by $o(A)$, such that

$$o(A) = o(B) \iff A \text{ is order-isomorphic to } B.$$

In detail, this means that we partition the set of all ordered sets (in the universe U) into equivalence classes of pairwise order-isomorphic ones, and with each such class we associate a member of the given universe U (the equivalence class itself will in general not be a member of U, so cannot be used). Two cases are of particular importance:

(i) A is an abstract set, regarded as totally unordered. In this case, we write '$|A|$' instead of $o(A)$, and call it the *cardinal (number)* of A.

(ii) A is a well-ordered set. Then $o(A)$ is called the *ordinal (number)* of A.

In general, by a cardinal number, or an ordinal number, we mean the order type of a totally unordered set, or of a well-ordered set, respectively.

We define a pre-ordering $\leq$ on the class of order types occurring in U via

$$o(A) \leq o(B) :\iff A \text{ is order-isomorphic to a left segment of } B; \qquad (5.1)$$

cf. Exercise 1. In general, $\leq$ is not an ordering; we do, however, obtain an ordering, indeed a total order, if we limit ourselves to cardinals, or to ordinals. In preparation for this argument, we need two auxiliary results.

Lemma 5.1. *If A is an ordered set with minimum condition, and $f : A \to A$ is injective and order-preserving, then $f(x) \not< x$ for all $x \in A$.*

Proof. Suppose there exists some $x \in A$ such that $f(x) < x$, and let $a \in A$ be a minimal element with this property. Then $f(a) < a$, so $f^2(a) < f(a)$, since f is order-preserving and injective, which shows that $f(a)$ also has this property, contradicting the minimality of a. □

We have already noted earlier that the set $\mathscr{S}^-(A)$ of left segments of an ordered set A forms a complete lattice under inclusion; see Exercise 13 in §3. For well-ordered sets A, this assertion has the following interesting counterpart.

Lemma 5.2. *The set $\mathscr{S}^-(A)$ of left segments of a well-ordered set A is well-ordered under inclusion.*

Proof. Mapping $a \in A$ to $J_a := \{x \in A : x < a\}$ gives an injective order-preserving map $\varphi : A \to \mathscr{S}^-(A)$; cf. Exercise 2. Note also that $J_a \neq A$, since $a \notin J_a$. Next, let $J \in \mathscr{S}^-(A)$, suppose that $J \neq A$, and let a be the least element of $A \setminus J$. Then

$$x \in A \setminus J \Longleftrightarrow x \geq a,$$

since J is a left segment of A by hypothesis, implying that $J = J_a$, as A is totally ordered. Hence, we have

$$\text{image}(\varphi) = \mathscr{S}^-(A) \setminus \{A\},$$

and the inverse map φ^{-1} mapping $J = J_a \in \mathscr{S}^-(A) \setminus \{A\}$ to $a \in A$ is clearly also order-preserving, again since A is totally ordered. It follows that $\mathscr{S}^-(A) \setminus \{A\}$ is order-isomorphic to A, thus well-ordered under inclusion, and since $\mathscr{S}^-(A)$ results from image(φ) by adjoining a greatest element A, $\mathscr{S}^-(A)$ itself is also well-ordered. □

Proposition 5.3. (i) *If α and β are two cardinal numbers, or two ordinal numbers, such that $\alpha \leq \beta$ and $\beta \leq \alpha$, then $\alpha = \beta$.*

(ii) *The ordinal numbers are well-ordered with respect to the relation $\leq$.*

(iii) *The cardinal numbers are well-ordered with respect to the relation $\leq$.*

Proof. (i) For cardinals this follows from Corollary 3.7. If α, β are ordinals, say $\alpha = o(A), \beta = o(B)$ with well-ordered sets A, B, then A is order-isomorphic to a left segment of B, and B is order-isomorphic to a left segment of A. Combining these isomorphisms, we obtain an order isomorphism $f : A \to A_0$, where A_0 is a left segment of A. The desired conclusion (that $\alpha = \beta$) will follow if we can show that $A_0 = A$. Suppose that $A_0 \neq A$, and let $a \in A \setminus A_0$. If we had $f(a) \geq a$, then $a \in A_0$, since A_0 is a left segment of A and $f(a) \in A_0$, a contradiction. Thus, $f(a) < a$ (as A is totally ordered), which contradicts Lemma 5.1. Hence, we have $A_0 = A$, so that B is indeed order-isomorphic to A, and therefore $\alpha = \beta$.

(ii) We first show that the ordering of ordinal numbers is total. Let A and B be well-ordered sets, and let F be the set of all functions defining an order isomorphism between a left segment of A and a left segment of B. Then F is non-empty (since $\emptyset \in F$) and ordered by inclusion. Moreover, since any union of left segments of A is again a left segment of A (and likewise for B), F is inductively ordered; see Exercise 3. By Zorn's Lemma, F has a maximal element f_0, a function defining an order isomorphism between a left segment A_0 of A and a left segment B_0 of B, say. If both $A_0 \neq A$ and $B_0 \neq B$, let a be the least element of $A \setminus A_0$ and let b be the least element of $B \setminus B_0$. We can then replace f_0 by $f_0 \cup \{(a,b)\}$ to obtain an order isomorphism from $A_0 \cup \{a\}$ onto $B_0 \cup \{b\}$, properly extending f_0, contradicting maximality of f_0; see Exercise 4. Hence, either $A_0 = A$ or $B_0 = B$ (or both), and accordingly $o(A) \leq o(B)$ or $o(B) \leq o(A)$.

To see that the ordinal numbers are well-ordered with respect to $\leq$, consider a non-empty set $\mathscr{O}$ of ordinals, and let $\alpha = o(A) \in \mathscr{O}$, where A is well-ordered. Suppose without loss of generality that α is not minimal in $\mathscr{O}$ (otherwise α is the sought least element of $\mathscr{O}$ since $\leq$ is total). Every $\beta \in \mathscr{O}$ which precedes α is represented by a well-ordered set order-isomorphic to some left segment J of A; let $\beta_0 \in \mathscr{O}$ correspond to the smallest of these left segments J (which exists by Lemma 5.2). Then β_0 is the least element of $\mathscr{O}$.

(iii) If $\mathscr{C}$ is any non-empty set of cardinals, their corresponding (totally unordered) sets can be well-ordered by Theorem 4.1, giving a non-empty set $\mathscr{O}$ of ordinals; applying part (ii), there exists $\beta_0 \in \mathscr{O}$ such that $\beta_0 \leq \beta$ for all $\beta \in \mathscr{O}$. Hence, the abstract set A corresponding to β_0 satisfies $|A| \leq \mathfrak{c}$ for every cardinal $\mathfrak{c} \in \mathscr{C}$. □

With each ordinal α we can unambiguously associate a cardinal number $|\alpha|$, namely the cardinal number of a well-ordered set of type α, and, by Theorem 4.1, every cardinal occurs in this way. Often one identifies a cardinal number with the least ordinal to which it belongs; however, the cardinal number of $\mathbb{N}$, denoted by $\aleph_0$ (read: aleph-nought), is usually associated with the set of natural numbers $\mathbb{N}$ itself. A set of cardinal $\aleph_0$, that is, a set which is in $1-1$ correspondence with the natural numbers, is said to be *countably infinite*. A set is called *countable*, if it is finite or countably infinite. $\aleph_0$ is the smallest transfinite cardinal; that is, every infinite set contains a countably infinite subset. Indeed, let X be an infinite set, and let φ be a choice function for X, that is, a map $\varphi : \mathscr{B}(X) \to X$ such that $\varphi(Y) \in Y$ for every non-empty subset Y of X. Then define an injective function $f : \mathbb{N} \to X$ via

$$\begin{aligned} f(0) &= \varphi(X) \\ f(n) &= \varphi(X \setminus \{f(0), f(1), \ldots, f(n-1)\}), \quad n \geq 1, \end{aligned}$$

(note that, at each stage, $X \setminus \{f(0), f(1), \ldots, f(n-1)\} \neq \emptyset$ since X is infinite).

The existence of uncountable sets follows from our next result.

Proposition 5.4. *For any ordered set A, we have $o(A) \neq o(\mathscr{S}^-(A))$. Moreover, if A is well-ordered, then $o(A) < o(\mathscr{S}^-(A))$, and if A is totally unordered, then $|A| < |\mathscr{B}(A)|$. In particular, among the ordinal numbers of a given non-empty universe (and likewise among the cardinal numbers) there exists no greatest one.*

Proof. The first claim follows from Corollary 3.11. It remains to show that $o(A) \leq o(\mathscr{S}^-(A))$ if A is well-ordered, and $|A| \leq |\mathscr{B}(A)|$ if A is totally unordered.

Thus, for a well-ordered set A, we have to define an order isomorphism from A onto a left segment of $\mathscr{S}^-(A)$; however, the map φ defined in the proof of Lemma 5.2 is an order isomorphism identifying A with

$$\mathscr{S}^-(A) \setminus \{A\} = \{X \in \mathscr{S}^-(A) : X \subset A\},$$

a left segment of $\mathscr{S}^-(A)$.

If A is totally unordered, we set $f(a) = \{a\}$ to obtain an order isomorphism between A and the collection of all singletons, which is a left segment in $\mathscr{B}(A) \setminus \{\emptyset\}$. Hence, $o(A) \leq o(\mathscr{B}(A) \setminus \{\emptyset\})$, and therefore

$$|A| \leq |\mathscr{B}(A) \setminus \{\emptyset\}| \leq |\mathscr{B}(A)|,$$

as required. The particular statement follows now immediately from the definition of a universe. □

Cardinal arithmetic

If A and B are two sets of cardinality α and β, respectively, then $|A \times B|$ depends only on α, β, and not on the sets A, B themselves; cf. Exercise 5. We write $|A \times B| = \alpha\beta$, and call $\alpha\beta$ the *product* of α and β. Similarly, the *sum* $\alpha + \beta$ is defined as $|A \cup B|$, where A and B are disjoint sets such that $|A| = \alpha$ and $|B| = \beta$; cf. Exercise 6.[6] Straight from these definitions, we have

$$\begin{aligned} \alpha + \beta &= \beta + \alpha \\ \alpha\beta &= \beta\alpha \\ \alpha + (\beta + \gamma) &= (\alpha + \beta) + \gamma \\ \alpha(\beta\gamma) &= (\alpha\beta)\gamma, \end{aligned}$$

that is, sum and product of cardinals are commutative as well as associative. Also, if $\alpha \leq \alpha'$ and $\beta \leq \beta'$, then $\alpha + \beta \leq \alpha' + \beta'$ and $\alpha\beta \leq \alpha'\beta'$ (monotonicity). We note that $\aleph_0^2 = \aleph_0$, since, for instance, the map sending a pair (m, n) of natural numbers to the natural number $2^m \cdot 3^n$ embeds $\mathbb{N} \times \mathbb{N}$ into the set $\mathbb{N}$.

If α, β are cardinals, then we define β^α as $|B^A|$, the cardinal of the set of all maps $A \to B$, where $|A| = \alpha, |B| = \beta$. For finite cardinals, addition, multiplication, and exponentiation reduce to the usual arithmetic of the natural numbers. In the general situation, the usual rules for exponentiation still

[6] Note that, if the given sets A and B representing the cardinals α and β, respectively, are not disjoint, then we can replace them by equipotent sets A', B' satisfying $A' \cap B' = \emptyset$ by setting

$$A' := \{(a,1) : a \in A\} \quad \text{and} \quad B' := \{(b,2) : b \in B\}.$$

apply; that is, we have

$$\begin{aligned} \alpha^{\beta+\gamma} &= \alpha^{\beta} \cdot \alpha^{\gamma} && (5.2)\\ (\alpha\beta)^{\gamma} &= \alpha^{\gamma} \cdot \beta^{\gamma} && (5.3)\\ \alpha^{\beta\gamma} &= (\alpha^{\beta})^{\gamma}; && (5.4) \end{aligned}$$

cf. Exercise 7.

Part (b) of Proposition 5.5 below summarises the main results of cardinal arithmetic; Part (a) isolates the core of the argument.

Proposition 5.5. (a) *If $\alpha \geq \aleph_0$ is a transfinite cardinal, then $\alpha^2 = \alpha$.*

(b) *Let α, β be non-zero cardinal numbers, at least one of which is transfinite. Then*

(i) $\alpha + \beta = \alpha \cdot \beta = \max\{\alpha, \beta\}$.

(ii) *If $\alpha < \beta$, then $\beta - \alpha = \beta$; that is, removal of a set having a smaller cardinal number does not reduce the cardinal number of the given set.*

(iii) *If $|2| \leq \alpha \leq \beta$, then $\alpha^{\beta} = |2|^{\beta}$.*

(iv) *We have $\alpha^{|n|} = \alpha$ for each $n \in \mathbb{N} \setminus \{0\}$ and every transfinite cardinal α.*

Proof. (a) We first show that, if $\alpha^2 = \alpha$ for some transfinite cardinal α, then $|2| \cdot \alpha = |3| \cdot \alpha = \alpha$. Indeed,

$$\alpha = \alpha + |0| \leq \alpha + \alpha = |2| \cdot \alpha \leq |3| \cdot \alpha \leq \alpha^2 = \alpha.$$

In order to show that $\alpha^2 = \alpha$ for some transfinite cardinal α, it suffices to exhibit, for some set M with $|M| = \alpha$, a bijection $M \times M \to M$. Let X be any set with $|X| = \alpha$, and let $\mathscr{A}$ be the set of all pairs (A, φ_A), where $A \subseteq X$ and $\varphi_A : A \times A \to A$ is a bijection. Clearly, $\mathscr{A} \neq \emptyset$; for instance, X contains a subset A_0 with $|A_0| = \aleph_0$, and we have seen that $\aleph_0^2 = \aleph_0$, so that there exists a bijection $\varphi_{A_0} : A_0 \times A_0 \to A_0$; cf. p. 49, first paragraph. Define an order relation $\preceq$ on $\mathscr{A}$ via

$$(A, \varphi_A) \preceq (B, \varphi_B) :\Longleftrightarrow A \subseteq B \text{ and } \varphi_B|_A = \varphi_A; \qquad (5.5)$$

cf. Exercise 8. Then each non-empty chain $\mathscr{C} = \{(A_i, \varphi_{A_i})\}_{i \in I}$ in $\mathscr{A}$ has an upper bound $(\tilde{A}, \tilde{\varphi}_{\tilde{A}}) \in \mathscr{A}$, where $\tilde{A} = \bigcup_i A_i$ and $\tilde{\varphi}_{\tilde{A}}$ is defined in the natural way (see Exercise 9); that is, $(\mathscr{A}, \preceq)$ is inductively ordered. It follows that the subset

$$\mathscr{A}_0 := \{(A, \varphi_A) \in \mathscr{A} : (A_0, \varphi_{A_0}) \preceq (A, \varphi_A)\}$$

with (A_0, φ_{A_0}) as above is also non-empty and inductively ordered with respect to $\preceq$, consequently, has a maximal element (M, φ_M) by Zorn's Lemma, and $|M| \geq \aleph_0$. Since $\varphi_M : M \times M \to M$ is bijective, we only need to show that $|M| = \alpha$. Since $M \subseteq X$, we have $|M| \leq \alpha$. Suppose that $|M| < |X|$. Then we must have

$$|M| < |X \setminus M|. \tag{5.6}$$

For, if $|X \setminus M| \leq |M|$, then

$$|X| = |M| + |X \setminus M| \leq |M| + |M| = |2| \cdot |M| = |M|$$

(the last equation following since $|M|^2 = |M| \geq \aleph_0$), contradicting our assumption that $|M| < |X|$. In view of (5.6), there exists a bijective map from M onto a subset Y of X with $Y \cap M = \emptyset$. Since $|Y|^2 = |M|^2 = |M| = |Y| \geq \aleph_0$, the set

$$(M \times Y) \cup (Y \times M) \cup (Y \times Y)$$

has cardinality $|3| \cdot |Y| = |Y|$ (by our initial remark); consequently, there exists a bijection

$$(M \times Y) \cup (Y \times M) \cup (Y \times Y) \longrightarrow Y,$$

which combines with φ_M to yield a bijection

$$\varphi : (M \cup Y) \times (M \cup Y) \longrightarrow (M \cup Y).$$

Since $M \neq \emptyset$, we have $Y \neq \emptyset$, and so

$$(A_0, \varphi_{A_0}) \preceq (M, \varphi_M) \prec (M \cup Y, \varphi),$$

contradicting the maximality of (M, φ_M). Hence, $|M| < |X|$ is impossible, and we conclude that $|M| = |X| = \alpha$ and, consequently, $\alpha^2 = \alpha$. Since α was an arbitrary transfinite cardinal, Part (a) is established.

(b) Let α, β be cardinal numbers with $\max\{\alpha, \beta\} \geq \aleph_0$.

(i) Assume without loss that $\alpha \leq \beta$. Then $\beta = \max\{\alpha, \beta\} \geq \aleph_0$, and so, by Part (a),

$$\beta \leq \alpha \cdot \beta \leq \beta^2 = \beta,$$

i.e., $\alpha \cdot \beta = \beta$. Similarly, for $\alpha \leq \beta$,

$$\beta \leq \beta + \alpha \leq \beta + \beta = |2| \cdot \beta \leq \beta^2 = \beta,$$

that is, $\alpha + \beta = \beta$.

(ii) Let $|A| = \alpha$, $|B| = \beta$, and $A \subseteq B$. Since $B = A \cup (B \setminus A)$ is a disjoint union,

$$|B| = |A| + |B \setminus A| = \max\{|A|, |B \setminus A|\}$$

by Part (b)(i); and since $\alpha < \beta$, we must have

$$\beta = |B| = |B \setminus A| = \beta - \alpha.$$

(iii) By Proposition 5.4, we always have $\alpha < |2|^\alpha$; hence,

$$|2|^\beta \leq \alpha^\beta \leq (|2|^\alpha)^\beta = |2|^{\alpha \cdot \beta} = |2|^\beta$$

by Part (i); that is, $|2|^\beta = \alpha^\beta$, as required.

(iv) This follows by induction on n: clearly, $\alpha^{|1|} = \alpha$, and if $\alpha^{|n|} = \alpha$, then

$$\alpha^{|n+1|} = \alpha^{|n|+|1|} = \alpha^{|n|} \cdot \alpha^{|1|} = \alpha^2 = \alpha,$$

the last step using Part (a). □

Ordinal Numbers Again

Recall that ordinal numbers were introduced as order types of well-ordered sets (in a given universe U). An alternative, more explicit, way of defining them is to distinguish, for every equivalence class of order-isomorphic well-ordered sets, one element in this class, and to regard this set as representing the class as well as the corresponding ordinal. As it turns out, all the different isomorphism types of well-ordered sets can be obtained by starting from the empty set, adjoining one element (as new greatest element) at a time, and taking unions of ascending chains. In particular, the finite ordinal numbers are just the natural numbers, and the first transfinite ordinal is the well-ordered set $\omega = (\mathbb{N}, \leq)$ of all natural numbers. For this construction it holds true that *each ordinal α coincides with the initial segment consisting of all ordinals strictly less than α*. We shall not enter into the technicalities of this construction here; but we shall prove the following result, Part (c) of which makes this construction at least plausible.

Proposition 5.6. (a) *Let A, B be well-ordered sets, and let $\varphi : A \to B$ be an order isomorphism onto a left segment of B. Then each injective order-preserving map $f : A \to B$ satisfies $\varphi(a) \leq f(a)$ for all $a \in A$.*

(b) *An initial segment $J_a = \{x \in A : x < a\}$ of a well-ordered set A cannot be order-isomorphic to A.*

(c) *A well-ordered set A of order type α is order-isomorphic to the set of all ordinal numbers strictly less than α, ordered according to their magnitude.*

Proof. (a) Let $f : A \to B$ be injective and order-preserving. We will show that the assumption

$$\{a \in A : f(a) < \varphi(a)\} \neq \emptyset$$

implies that $\varphi(A)$ is not a left segment of B, contradicting our hypothesis. Since A is totally ordered, this will prove our claim.

Indeed, since A is well-ordered, there would then have to be a smallest element $a_0 \in A$ with $f(a_0) < \varphi(a_0)$. Thus, if $a \in A$ is such that $a < a_0$, then $\varphi(a) \leq f(a) < f(a_0)$; and if $a_0 \leq a$, then $f(a_0) < \varphi(a_0) \leq \varphi(a)$. It follows that φ does not attain the value $f(a_0)$, so that $\varphi(A)$ cannot be a left segment of B.

(b) Let J_a be an initial segment of A, and suppose that there exists an order isomorphism $\varphi : J_a \to A$ onto A. Comparing φ with the natural inclusion map $\iota : J_a \to A$, we see that, by Part (a), $\varphi(x) \leq x$ must hold for all $x \in J_a$; in particular, $\varphi(x) < a$ for all $x \in J_a$, so that φ is not surjective. This contradiction shows that J_a and A cannot be order-isomorphic.

(c) We have already seen (in the proof of Lemma 5.2) that the map $a \mapsto J_a$ defines an order isomorphism φ of A onto the set $\mathscr{S}^-(A) \setminus \{A\}$ of proper left segments of A, ordered by inclusion. Each of these segments J_a is itself a well-ordered set via the induced ordering, thus has an ordinal ξ_a as its order type. We have $\xi_a < \alpha$ by the definition of $\leq$ and Part (b), hence mapping J_a to $\xi_a = o(J_a)$ gives an order-preserving map ψ from $\mathscr{S}^-(A) \setminus \{A\}$ to the set

$$\{\xi : \xi \text{ ordinal}, \xi < \alpha\}. \tag{5.7}$$

Let $a, b \in A$ be elements such that $o(J_a) = o(J_b)$. Assuming without loss of generality that $a \leq b$, we see that $J_a \subseteq J_b$; and if we had $a < b$, then J_a would be an initial segment of J_b, which cannot be the case by Part (b). Hence $a = b$, so that $J_a = J_b$, and we conclude that ψ is injective. Next, consider any ordinal number ξ with $\xi < \alpha$, and let B be a well-ordered set with $o(B) = \xi$. Then B is order-isomorphic to a proper left segment J_a of A, so $o(J_a) = o(B) = \xi$; hence, ψ is surjective as well. Finally, we claim that the map ψ^{-1} is also order preserving. For, if ξ, η are ordinals such that $\xi \leq \eta < \alpha$, and if J_a, J_b are initial segments of A such that $o(J_a) = \xi$ and $o(J_b) = \eta$, then we must have $J_a \subseteq J_b$. Indeed, if we had $a > b$, then J_b would be an initial segment of J_a, implying

$$\eta = o(J_b) < o(J_a) = \xi$$

again by Part (b), which is a contradiction. Consequently, $a \leq b$, so $J_a \subseteq J_b$ as claimed. Hence, ψ is an order isomorphism of $\mathscr{S}^-(A) \setminus \{A\}$ onto the set (5.7), and combining the maps φ and ψ we obtain the desired order isomorphism. □

Exercises for §5.

1. Show that the relation $\leq$ as defined in (5.1) is a pre-ordering on the class of order types occurring in the universe U; that is, $\leq$ is reflexive and transitive.

2. Let $(A, \leq)$ be a totally ordered set. Show that the map $\varphi : A \to \mathscr{S}^-(A)$ sending $a \in A$ to $J_a = \{x \in a : x < a\}$ is an injective order-preserving map.

3. Let A and B be ordered sets, and let F be the set of all functions defining an order isomorphism between a left segment of A and a left segment of B. Show that F, ordered by inclusion, is inductively ordered.

4. This exercise concerns the map $f_0' := f_0 \cup \{(a,b)\} : A_0' \to B_0'$ constructed in the proof of Part (ii) of Proposition 5.3, where $A_0' := A_0 \cup \{a\}$ and $B_0' := B_0 \cup \{b\}$. Show that A_0', B_0' are left segments of A and B, respectively, and that the map $f_0' : A_0' \to B_0'$ is order-preserving.

5. Show that the product of two cardinal numbers is well-defined. In detail, prove that if A, A', B, and B' are (abstract) sets such that $|A| = |A'|$ and $|B| = |B'|$, then $|A \times B| = |A' \times B'|$.

6. Show that the sum $\alpha + \beta$ of two cardinal numbers α, β is well-defined. In detail, if A, A', B, B' are (abstract) sets such that $|A| = |A'|$, $|B| = |B'|$, and $A \cap B = \emptyset = A' \cap B'$, prove that $|A \cup B| = |A' \cup B'|$.

7. (i) Show that exponentiation of cardinals is well-defined.

   (ii) For cardinals α, β, and γ, prove Equations (5.2)–(5.4); that is, show that $\alpha^{\beta+\gamma} = \alpha^\beta \cdot \alpha^\gamma$, $(\alpha\beta)^\gamma = \alpha^\gamma \cdot \beta^\gamma$, and $\alpha^{\beta\gamma} = (\alpha^\beta)^\gamma$.

8. Show that the binary relation $\preceq$ on the set $\mathscr{A}$ defined in (5.5) is an order relation.

9. Describe in detail the definition of the map $\tilde{\varphi}_{\tilde{A}} : \tilde{A} \times \tilde{A} \to \tilde{A}$ in the proof of Part (a) of Proposition 5.5. In particular, show that $\tilde{\varphi}_{\tilde{A}}$ is well-defined and a bijection.

10. We assume that the reader is acquainted with the standard argument based on Steinitz's exchange lemma, proving invariance of dimension for finitely generated vector spaces V; see also Part (iv) of Exercise 11.3. Here, we consider the case where V is not finitely generated. Let B_1, B_2 be two bases of V. Show that $|B_1| = |B_2|$.

§6. First order logic and the axioms of set theory revisited

At the end of our rather informal and cursory introduction to axiomatic set theory, it is perhaps fitting to take a brief look back, and give a more precise meaning to some of the concepts used and statements made in a somewhat leisurely fashion in Sections §1–§5.

As was already remarked in §1, some care needs to be taken as to what should constitute an *admissible sentence* (property) to be used in the formation of sets and classes via Axiom **A2**. The framework usually employed in the formulation of admissible sentences is a suitable form of *first order predicate logic*. More precisely, the expressions of set theory are built up from *atomic formulae*, that is, formulae of the form

$$X \in Y \quad \text{or} \quad X = Y$$

by means of *connectives*

$$\varphi \wedge \psi, \quad \varphi \vee \psi, \quad \neg\varphi, \quad \varphi \rightarrow \psi, \quad \varphi \leftrightarrow \psi$$

(that is, *conjunction*, *disjunction*, *negation*, *implication*, and *equivalence*)

and *quantifiers*

$$\forall X\, \varphi \quad \text{and} \quad \exists X\, \varphi$$

(that is, the *all quantifier*, and the *existence quantifier*). For instance,

$$\exists X\, \forall Y\, \neg(Y \in X)$$

is a syntactically correct formula in this first order language, defining the empty class (set). In practice, other symbols are used as well in formulae, for example defined predicates, defined operators, and constants; however, it is tacitly understood that each such formula can be rewritten in a form only requiring $\in$ and $=$ as non-logical symbols.

A formula $\varphi(X_1,\ldots,X_n)$ may contain *bound variables* (variables acted upon by a quantifier), as well as *free variables* (variables not referred to by a quantifier). A *sentence* is a formula without free variables. For instance, Axiom **A2** now takes the form:

If $\varphi(X,P_1,\ldots,P_n)$ *is a first order formula, then* $\mathbf{C} = \{X : \varphi(X,P_1,\ldots,P_n)\}$ *is a class.*

The elements of the class **C** are precisely those sets X which satisfy the predicate $\varphi(X,P_1,\ldots,P_n)$. We say that **C** is *definiable* from $P_1,\ldots,P_n$. Moreover, the statement that two classes Y and Z are equal if they have the same elements (the axiom of extensionality) is expressed in the form:

$$\forall X\, (X \in Y \leftrightarrow X \in Z) \rightarrow Y = Z.$$

The converse, to the effect that, if $Y = Z$, then

$$X \in Y \leftrightarrow X \in Z$$

is an axiom of predicate calculus. Thus, taken together, we have

$$Y = Z \leftrightarrow \forall X(X \in Y \leftrightarrow X \in Z).$$

This last formula expresses the basic idea of a class: A class is determined by its members (elements).

The axiom of pairing (implicitly introduced on page 4): *For any A and B, there exists a set $\{A, B\}$ precisely containing A and B* becomes

$$\forall A \, \forall B \, \exists C \, \forall X \, (X \in C \leftrightarrow X = A \vee X = B).$$

By extensionality, the set C is uniquely determined, and we may define the pair $\{A, B\}$ as the unique C such that

$$\forall X \, (X \in C \leftrightarrow X = A \vee X = B).$$

The singleton $\{A\}$ may now be defined as the set

$$\{A\} = \{A, A\},$$

etc. At this point, the reader will most likely be willing to concede that the axioms of set theory can all be formulated in precise terms by making use of the first order predicate language sketched above; the interested reader is referred to [76] for a full and detailed treatment of this aspect of set theory.

§7. Some excursions

The sections below provide a potpourri of little gems illustrating, rounding off, applying, or complementing topics discussed in §1–§5. In §7.1, we describe the construction, by means of the axiom of choice, of a bounded not Lebesgue-measurable set of real numbers; this provides an illustration of the fact that the axiom of choice, apart from being an extremely valuable tool, also has some unwanted and rather unpleasant consequences.[7] In Section §7.2 we present an argument due to Anne Davis [27], a student of Alfred Tarski, establishing the converse of Tarski's fixed point theorem for complete lattices; cf. Proposition 3.3 and Corollary 3.4. Section §7.3 discusses a rather surprising result of Wilfrid Hodges [42] to the effect that, in Zermelo-Fraenkel set theory, (a special case of) Krull's maximal ideal theorem implies the axiom of choice. The last section of §7 aims to explain the, apparently little known, construction by Novotný [63] of the collection of all maps $\varphi : S \to T$ respecting given self-maps $\sigma : S \to S$ and $\tau : T \to T$ on the sets S and T, respectively; that is, maps φ satisfying the identity

$$\varphi(\sigma(s)) = \tau(\varphi(s)), \quad (s \in S).$$

Sections §7.2 and §7.5 require most of the set theory contained in §1–§5. The only topic whose appearance in this context might beg some justification is Argand's proof of the fundamental theorem of algebra. Admittedly, this result has little to do with set theory; however, Steinitz's existence theorem for algebraic closures is one of the applications of the axiom of choice (in the form of Krull's maximal ideal theorem) discussed in §4, and the field of complex numbers is the unavoidable first example of an algebraically closed field, so I decided to include a proof of this fact in §7 and, once I became aware of Argand's approach, which is quite different from other known arguments, I could not resist the temptation to present his proof, which only uses elementary real analysis.[8]

§7.1. A bounded non-measurable set of real numbers

So far, we have already seen a number of highly desirable results (and we shall see many more of them), whose proof requires the axiom of choice (or some equivalent assumption). However, the axiom of choice is not only an extremely valuable tool in large parts of mathematics, it is also the source of some rather unpleasant results. We confine ourselves here to one particularly fundamental phenomenon: the existence of bounded non-measurable sets.

---

[7]The existence of non-measurable sets of real numbers was apparently first established by Giuseppe Vitali in 1905; cf. [80].

[8]Apart from Argand's original paper [7], this particular proof is, for instance, discussed in Remmert's chapter on the fundamental theorem of algebra in [30].

Let $\mu(X)$ denote the Lebesgue measure of a set X of real numbers. As is well known, μ is countably additive, translation invariant, and satisfies

$$\mu([a,b]) = b - a$$

for each closed interval $[a,b]$; cf., for instance, [34] for definitions and background information on measure theory. In an ideal world, every bounded set of real numbers would have a well-defined finite Lebesgue measure but, due to the axiom of choice, this is not so, as the following example demonstrates.

Define an equivalence relation $\sim$ on the unit interval $[0,1]$ via

$$x \sim y :\Longleftrightarrow x - y \in \mathbb{Q}, \quad x, y \in [0,1].$$

By the axiom of choice, we can choose one element from each equivalence class $[x]$ for $x \in [0,1]$, thus obtaining a system R of representatives for $\sim$ on the set $[0,1]$. Consequently, there is a set R of real numbers $R \subseteq [0,1]$, with the property that, for each real number x, there exists a unique element $y \in R$ and a unique rational number ρ, such that $x = y + \rho$. Setting

$$R_\rho := \{y + \rho : y \in R\} = R + \rho, \quad \rho \in \mathbb{Q},$$

we obtain countably many pairwise disjoint translates of the set R. Indeed, if $x \in R_{\rho_1} \cap R_{\rho_2}$ for some $\rho_1, \rho_2 \in \mathbb{Q}$, then

$$x = y_1 + \rho_1 = y_2 + \rho_2$$

for some $y_1, y_2 \in R$, implying

$$y_1 - y_2 = \rho_2 - \rho_1 \in \mathbb{Q}.$$

It follows that $y_1 \sim y_2$, so that $y_1 = y_2$, forcing $\rho_1 = \rho_2$. Hence, the sets R_ρ are pairwise disjoint, as claimed.

Now consider the sets R_ρ for $\rho \in [-1,1]$, each of which is contained in the interval $[-1,2]$. We obtain the inclusions

$$[0,1] \subseteq R^* := \bigcup_{\substack{\rho \in \mathbb{Q} \\ -1 \le \rho \le 1}} R_\rho \subseteq [-1,2].$$

Suppose that the set R is measurable, say $\mu(R) = c$. Then, on the one hand, the assumption that $c = 0$ implies that $\mu(R^*) = 0$, contradicting the fact that $[0,1] \subseteq R^*$, with $[0,1]$ of measure 1. On the other hand, if $c > 0$, then

$$\mu(R^*) = \sum_{\substack{\rho \in \mathbb{Q} \\ -1 \le \rho \le 1}} c = \infty,$$

contradicting the fact that $R^* \subseteq [-1,2]$, with $[-1,2]$ having finite measure. We conclude that R is a bounded non-measurable set of real numbers.

§7.2. A characterisation of complete lattices

We have seen in §3 that every order-preserving map of a complete lattice into itself has a fixed point; cf. Corollary 3.4. Here, we describe Anne Davis' argument proving the converse of this assertion, thus establishing the following beautiful result.

Proposition 7.1. *A lattice $(A, \leq)$ is complete if, and only if, every order-preserving map $f : A \to A$ has a fixed point.*

If α is any (finite or transfinite) ordinal, we denote by $(a_\xi)_{\xi<\alpha}$ the (generalised) sequence $a_0, a_1, \ldots, a_\xi, \ldots$ with $\xi < \alpha$, and by $\{a_\xi : \xi < \alpha\}$ its set of terms. The sequence $(a_\xi)_{\xi<\alpha}$ is *increasing* (respectively *strictly increasing*), if $a_\xi \leq a_\eta$ (respectively $a_\xi < a_\eta$) for every pair (ξ, η) of ordinals such that $\xi < \eta < \alpha$. Decreasing and strictly decreasing sequences are defined analogously. The following remarkable property of incomplete lattices is the key to proving the converse of Corollary 3.4.

Lemma 7.2. *Let $(A, \leq)$ be an incomplete lattice. Then there exist sequences $(b_\xi)_{\xi<\beta}$ and $(c_\eta)_{\eta<\gamma}$ of elements in A such that*

(i) *$b_\xi \leq c_\eta$ for each $\xi < \beta$ and every $\eta < \gamma$,*

(ii) *the sequence $(b_\xi)_{\xi<\beta}$ is increasing, while $(c_\eta)_{\eta<\gamma}$ is decreasing,*

(iii) *the sequences $(b_\xi)_{\xi<\beta}$ and $(c_\eta)_{\eta<\gamma}$ cannot be separated in A; that is, there is no element $a \in A$ such that $b_\xi \leq a \leq c_\eta$ holds for all $\xi < \beta$ and all $\eta < \gamma$ simultaneously.*

Let us first see how Lemma 7.2 leads to the construction of a fixed-point-free order morphism $f : A \to A$ in any incomplete lattice A.

Let $(b_\xi)_{\xi<\beta}$ and $(c_\eta)_{\eta<\gamma}$ be two sequences in A satisfying the conclusions (i)–(iii) of Lemma 7.2. In order to define f for an element $x \in A$, we distinguish two cases, depending on whether or not x is a lower bound of $\{c_\eta : \eta < \gamma\}$. If it is, then, by (iii), x is not an upper bound for $\{b_\xi : \xi < \beta\}$, hence the set of ordinals

$$\Phi(x) := \{\xi : \xi < \beta \text{ and } b_\xi \not\leq x\}$$

is non-empty, and we put

$$\phi(x) := \min \Phi(x) \text{ and } f(x) := b_{\phi(x)}.$$

In the second case, the set of ordinals

$$\Psi(x) := \{\eta : \eta < \gamma \text{ and } x \not\leq c_\eta\}$$

is non-empty, and we put

$$\psi(x) := \min \Psi(x) \text{ and } f(x) := c_{\psi(x)}.$$

We have thus defined a map $f : A \to A$ such that $f(x) \not\leq x$ or $x \not\leq f(x)$ for every $x \in A$, so f has no fixed point. Let $x, y \in A$ be elements such that $x \leq y$. If x is a lower bound of $\{c_\eta : \eta < \gamma\}$ but y is not, then, by definition of f and conclusion (i) of Lemma 7.2, $f(x) \leq f(y)$. If both x and y are lower bounds of $\{c_\eta : \eta < \gamma\}$, then $\Phi(y) \subseteq \Phi(x)$; hence, $\phi(x) \leq \phi(y)$, and thus, by conclusion (ii) of Lemma 7.2,

$$f(x) = b_{\phi(x)} \leq b_{\phi(y)} = f(y).$$

Finally, if x is not a lower bound of $\{c_\eta : \eta < \gamma\}$, then y is not either, we have $\Psi(x) \subseteq \Psi(y)$, thus $\psi(y) \leq \psi(x)$, and hence

$$f(x) = c_{\psi(x)} \leq c_{\psi(y)} = f(y),$$

since $(c_\eta)_{\eta<\gamma}$ is decreasing by conclusion (ii) of Lemma 7.2. We conclude that f is order-preserving and fixed-point-free, and Proposition 7.1 follows from this together with Corollary 3.4.

Proof of Lemma 7.2. By Proposition 3.2, since $(A, \leq)$ is not complete, there exists at least one subset of A without supremum. Hence, since the cardinal numbers are well-ordered, there exists a subset B of A such that

$$\sup B \text{ does not exist;} \tag{7.8}$$

and

$$\text{if } X \subseteq A \text{ and } |X| < |B|, \text{ then } \sup X \text{ exists.} \tag{7.9}$$

Let β be the initial ordinal of the same cardinality as B (that is, the smallest ordinal such that the set of all preceding ordinals has cardinality $|B|$). We may have $\beta = 0$; otherwise β is transfinite and, since it is initial, it has no predecessor, i.e., $\xi < \beta$ implies $\xi + 1 < \beta$. The set B can be (bijectively) parametrised in the form

$$B = \{b'_\xi : \xi < \beta\}.$$

For every ordinal $\xi < \beta$, the set $\{b'_\eta : \eta < \xi + 1\}$ is of cardinality $< |B|$, hence its least upper bound

$$b_\xi := \sup \{b'_\eta : \eta < \xi + 1\}$$

exists by (7.9). The sequence $(b_\xi)_{\xi<\beta}$ is clearly increasing (though not necessarily strictly increasing). By construction,

$$\text{for every } b \in B \text{ there exists an ordinal } \xi < \beta \text{ such that } b \leq b_\xi; \tag{7.10}$$

moreover,

$$\text{for every } \xi < \beta, \text{ there exists a subset } X_\xi \text{ of } B \text{ such that } b_\xi = \sup X_\xi. \tag{7.11}$$

By (7.10) and (7.11), if the least upper bound of $\{b_\xi : \xi < \beta\}$ were to exist, it would have to coincide with $\sup B$; hence, by (7.8),

$$\sup\{b_\xi : \xi < \beta\} \text{ does not exist.} \tag{7.12}$$

Let

$$C := \{c \in A : c \geq b_\xi \text{ for all } \xi < \beta\}$$

be the set of all upper bounds for $\{b_\xi : \xi < \beta\}$ in A, and let $C' = (C, \geq)$ be the dual of C. Fix a maximal chain M' in C' according to Theorem 4.1, and let W' be a well-ordered subset of M' such that M' is co-final with W' (such a set W' exists by Proposition 4.17). Let γ be the order type of W', fix an order isomorphism $\eta \mapsto c_\eta$ of the set of ordinals $\{\eta : \eta < \gamma\}$ onto W', and write W' in the form

$$W' = \{c_\eta : \eta < \gamma\}.$$

Then $(c_\eta)_{\eta<\gamma}$ is a decreasing sequence in C; moreover, by Proposition 4.16, C is co-initial with $\{c_\eta : \eta < \gamma\}$. If $\inf\{c_\eta : \eta < \gamma\}$ were to exist, it would be an upper bound for $\{b_\xi : \xi < \beta\}$; but since $\sup\{b_\xi : \xi < \beta\}$ does not exist, there would be an element $c \in C$ such that

$$\inf\{c_\eta : \eta < \gamma\} \not\leq c.$$

This would imply

$$\inf\{c, \inf\{c_\eta : \eta < \gamma\}\} \in C$$

and

$$\inf\{c, \inf\{c_\eta : \eta < \gamma\}\} < \inf\{c_\eta : \eta < \gamma\},$$

contradicting the fact that C is co-initial with $\{c_\eta : \eta < \gamma\}$. Consequently,

$$\inf\{c_\eta : \eta < \gamma\} \text{ does not exist.} \tag{7.13}$$

The sequences $(b_\xi)_{\xi<\beta}$ and $(c_\eta)_{\eta<\gamma}$ clearly satisfy conditions (i) and (ii) of the lemma. In order to see that condition (iii) also holds, suppose that an element $a \in A$ is both an upper bound for $\{b_\xi : \xi < \beta\}$ and a lower bound for $\{c_\eta : \eta < \gamma\}$. By definition, $a \in C$. Since C is co-initial with $\{c_\eta : \eta < \gamma\}$, we must have $a \in \{c_\eta : \eta < \gamma\}$, and therefore

$$a = \inf\{c_\eta : \eta < \gamma\},$$

contradicting (7.13). □

§7.3. Krull's maximal ideal theorem implies Zorn's Lemma

As we have seen in §4, by the axiom of choice (or rather its equivalent, Zorn's Lemma), every commutative ring with $1 \neq 0$ has a maximal ideal; cf. Proposition 4.8 for a slightly stronger result. It is a natural problem, apparently first raised by Dana Scott in [72], whether the converse holds as well: If every commutative ring with $1 \neq 0$ has a maximal ideal, then the axiom of choice holds true. This turns out to be correct. In fact, as we shall see presently, the following stronger result holds.

Proposition 7.3. (W. Hodges [42]) *In Zermelo-Fraenkel set theory, the statement "Every unique factorisation domain has a maximal ideal" implies the axiom of choice.*

We begin the proof by suitably paraphrasing the axiom of choice. By a *generalised tree* (gtree for short), we mean an ordered set $(T, \leq)$ such that, for each element $t \in T$, the segment

$$\hat{t} := \{x \in T : x \leq t\}$$

is totally ordered. By a *branch* in the gtree T we shall mean a maximal totally ordered subset of T.

Let **GTree** be the statement: *Every gtree has a branch.*

Lemma 7.4. GTree *is equivalent to the axiom of choice.*

Proof. The fact that *AC* (in the form of Zorn's Lemma) implies **GTree** is straightforward. For the converse, we shall show that, assuming **GTree**, every set can be well-ordered.

Indeed, let S be any set, and let T be the set of injective maps $f : \alpha \to S$, with α an ordinal. For two such functions f and g, we set $f \leq g$ if, and only if, g extends f. Then $(T, \leq)$ is a gtree, which must have a branch B by our hypothesis. The union $\bigcup B$ is an injective map $h : \beta \to S$ for some ordinal β. However, h is also surjective, since otherwise we could extend h within T, contradicting the maximality of B. Transferring the order structure of β via the bijection h now gives a well-ordering on S, as required. □

Now let $(T, \leq)$ be a gtree. We construct an associated ring $\mathfrak{R}_T$ as follows. Let $\mathbb{Q}[T]$ be the polynomial ring over the rationals with the elements of T as indeterminates. It is well known that such a ring is a unique factorisation domain (the axiom of choice is not needed for this). If $G \subseteq T$, then the complex product $G\mathbb{Q}[T]$ is a prime ideal of $\mathbb{Q}[T]$.[9] Let $\mathscr{L}$ be the set of totally ordered subsets of T, and set

$$\mathfrak{S} := \mathbb{Q}[T] - \bigcup_{G \in \mathscr{L}} G\mathbb{Q}[T].$$

Then $\mathfrak{S}$ is the complement of a union of prime ideals, thus is multiplicatively closed. Inverting $\mathfrak{S}$, we obtain the ring

$$\mathfrak{R}_T = \mathfrak{S}^{-1}\mathbb{Q}[T].$$

Every element $c \in \mathfrak{R}_T$ is of the form $c = a/s$ with $a \in \mathbb{Q}[T]$ and $s \in S$, and if common factors are cancelled, then x and s are unique up to factors in $\mathbb{Q}$. The element $c \in \mathfrak{R}_T$ is invertible in $\mathfrak{R}_T$ if, and only if, $a \notin \hat{t}\mathbb{Q}[T]$ for all $t \in T$. Moreover, the ring $\mathfrak{R}_T$ is a unique factorisation domain (again, *AC* is not needed for this).

---

[9]See, for instance, Section 6.1 in [69] for the definition and properties of prime ideals in a commutative ring.

Suppose that $\mathfrak{R}_T$ has a maximal ideal $\mathfrak{M}$. Let $c \in \mathfrak{M}$ be arbitrary, express c as $c = a/s$ so that a and s have no common non-scalar factors. Then a can be expressed in the form $a = r_1\mathfrak{m}_1 + \cdots + r_n\mathfrak{m}_n$ with $r_1, \ldots, r_n \in \mathbb{Q}$ and distinct monomials $\mathfrak{m}_1, \ldots, \mathfrak{m}_n$ over T. Since c is not invertible, there exists a finite totally ordered set $A \subseteq T$, such that (i) each monomial $\mathfrak{m}_i$ has a factor in A, and (ii) each element of A occurs as a factor of some $\mathfrak{m}_i$. The set A is not necessarily unique, but there are at most finitely many choices for it, say, $A_1, A_2, \ldots, A_k$. Set

$$E_c := \{\max A_i : 1 \leq i \leq k\}.$$

If $t \in E_c$, then $c \in \hat{t}\mathfrak{R}_T$. If $c = 0$, then necessarily $A = \emptyset$; if, on the other hand, $c \neq 0$, and if $d \in \mathfrak{M}$ is any other element involving the monomials $\mathfrak{m}_1, \ldots, \mathfrak{m}_n$ (and possibly others as well), then there exists, for each $t' \in E_d$, some $t \in E_c$ with $t \leq t'$. Define $D \subseteq T$ to be the set of those $t \in T$ such that, for every $c \in \mathfrak{M} \setminus \{0\}$, there is some $t' \in E_c$ which is comparable with t.

Lemma 7.5. *We have $D \subseteq \mathfrak{M}$.*

Proof. Let $t \in D$, and suppose for a contradiction that $t \notin \mathfrak{M}$. Then, since $\mathfrak{M}$ is maximal, there exist elements $a \in \mathfrak{R}_T$ and $c \in \mathfrak{M}$, such that $at + c = 1$. As $\{t\} \in \mathscr{L}$, t is not invertible in $\mathfrak{R}_T$, thus $c \neq 0$. By definition of the set D, there exists some $t' \in E_c$ which is comparable with t; thus $c \in \hat{t'}\mathfrak{R}_T$. If $t' \leq t$ then $c, t \in \hat{t}\mathfrak{R}_T$, and so $1 \in \hat{t}\mathfrak{R}_T$, a contradiction. If, on the other hand, $t \leq t'$, then $c, t \in \hat{t'}\mathfrak{R}_T$, another contradiction to the definition of the ring $\mathfrak{R}_T$. Hence, $t \in \mathfrak{M}$, as claimed. □

Lemma 7.6. *The set D is a totally ordered left segment of T.*

Proof. If $t, w \in D$ then, by Lemma 7.5, $t + w \in \mathfrak{M}$, so that $t + w$ is not invertible. It follows from the construction of the ring $\mathfrak{R}_T$ that t and w are comparable, thus D is a totally ordered subset of T. Suppose that $v \leq t$ in T; we want to show that $v \in D$. Let c be a non-zero element of $\mathfrak{M}$. As $t \in D$, there exists some $t' \in E_c$ which is comparable with t. If $t \leq t'$, then $v \leq t'$; in particular, v and $t' \in E_c$ are comparable. If, on the other hand, $t' \leq t$, then t' and v are comparable, since they are both elements of the totally ordered set $\hat{t}$. Hence, since $c \in \mathfrak{M} \setminus \{0\}$ was arbitrary, we conclude that $v \in D$, showing that D is a left segment of T. □

Lemma 7.7. *We have $\mathfrak{M} \subseteq D\mathfrak{R}_T$.*

Proof. Suppose that $\mathfrak{M} - D\mathfrak{R}_T \neq \emptyset$, and let c be an arbitrary element of this set. Since $0 \in D\mathfrak{R}_T$, we have $c \neq 0$, and we may suppose without loss of generality that $c \in \mathbb{Q}[T]$. We claim that $D \cap E_c = \emptyset$. Indeed, if $t \in D \cap E_c$, then $c \in \hat{t}\mathfrak{R}_T \subseteq D\mathfrak{R}_T$ by Lemma 7.6, a contradiction. Now let

$$E_c = \{t_1, t_2, \ldots, t_k\},$$

where the t_j are pairwise distinct. Since none of the t_j are in D, there is, for each $j \in [k]$, a non-zero element $c_j \in \mathfrak{M}$, such that t_j is incomparable with all the elements of E_{c_j}. We may suppose without loss of generality that $c_j \in \mathbb{Q}[T]$ for each $j \in [k]$, since the definition of the set E_c does not depend on a possible denominator of c. Since $\mathbb{Q}$ is an infinite field, we may choose scalars $q_1, \ldots, q_k \in \mathbb{Q}$ in such a way that in the element

$$x = c + q_1 c_1 + \cdots q_k c_k \in \mathbb{Q}[T]$$

no monomial which occurs with non-zero coefficient in c or in one of the c_j vanishes in x. Suppose that $t' \in E_x$. Then, by the choice of the numbers q_j, there exists some $j \in [k]$ such that $t_j \leq t'$. However, for the same reason, there also exists some $t'' \in E_{c_j}$ with $t'' \leq t'$. Hence, t_j and t'' are comparable, since both lie in the totally ordered subset $\hat{t'}$ of T. This contradicts the choice of the element c_j. □

Proof of Proposition 7.3. We assume that every unique factorisation domain has a maximal ideal, and want to prove assertion **GTree** which, as we have seen in Lemma 7.4, is equivalent to the axiom of choice.

Let $(T, \leq)$ be a gtree, and let $\mathfrak{R}_T$ be the associated ring defined as above. By assumption, $\mathfrak{R}_T$ has a maximal ideal $\mathfrak{M}$. By Lemmas 7.5 and 7.7, plus maximality of $\mathfrak{M}$, we have $\mathfrak{M} = D\mathfrak{R}_T$, where D is as above. Suppose that D is not a branch of T. Then, by Lemma 7.6, there exists some $t' \in T$, such that $t' > t$ for all $t \in D$. It follows that

$$\mathfrak{M} = D\mathfrak{R}_T \subset \hat{t'}\mathfrak{R}_T \subset \mathfrak{R}_T,$$

contradicting the maximality of $\mathfrak{M}$. Hence, $(T, \leq)$ has the branch D. Since T was an arbitrary gtree, this proves assertion **GTree**, as desired. □

§7.4. An elementary proof of the fundamental theorem of algebra

We assume that the reader has seen one-dimensional Weierstraß-theory, in particular the boundedness and intermediate value properties for continuous real-valued functions on closed bounded intervals. Here, we shall need a version of the boundedness theorem for continuous functions of two real variables. We begin by defining the terms: *bounded*, *closed*, and *compact* for subsets of $\mathbb{R}^2$.

Definition 7.8. (a) A subset $D \subseteq \mathbb{R}^2$ is called *bounded*, if D is contained in some finite disc around the origin; that is, if there exists some real number $r > 0$ such that

$$D \subseteq \{\mathbf{x} \in \mathbb{R}^2 : x_1^2 + x_2^2 \leq r^2\},$$

where $\mathbf{x} = (x_1, x_2) \in \mathbb{R}^2$.

(b) A set $D \subseteq \mathbb{R}^2$ is called *closed*, if every convergent sequence $(d_n) \subseteq D$ converges to a point in D.

(c) The set $D \subseteq \mathbb{R}^2$ is called *compact*, if it is closed and bounded.

With these definitions, we can now state the 2-dimensional boundedness theorem for continuous functions.

Theorem 7.9. *Let $D \subseteq \mathbb{R}^2$ be a compact set, and let $f : D \to \mathbb{R}$ be a continuous function. Then f assumes both a maximum and a minimum value on D.*

For a proof, the reader may consult, for instance, Theorems 2 and 3 in [60, 2nd volume, §119] or [6, Theorem 4-20]. Apart from Theorem 7.9, we shall need two elementary facts: (i) that every complex polynomial defines a continuous function on $\mathbb{C}$, and (ii) the existence of roots for complex numbers.

The idea of the proof is as follows: we first show, by means of a growth argument, that the modulus $|f(z)|$ of a complex polynomial $f(z)$ always assumes a minimum on $\mathbb{C}$ (minimum theorem of Cauchy). Second, we prove an inequality due to Argand, which, third, leads to the d'Alembert-Gauß theorem, stating that, for a non-constant polynomial $f(z)$, the minimum in Cauchy's minimum theorem is actually zero, whence the fundamental theorem.

Step 1. *Cauchy's minimum theorem.* Given a polynomial

$$f(z) = a_0 + a_1 z + \cdots + a_n z^n$$

with complex coefficients $a_0, a_1, \ldots, a_n$, there exists some $c \in \mathbb{C}$ such that $|f(c)| = \inf_{z \in \mathbb{C}} |f(z)|$.

Proof. We may suppose that $a_n \neq 0$ with $n \geq 1$ (otherwise, any $c \in \mathbb{C}$ would do). Now we claim the following:

$(*)$ *There exists some real number r, such that $|f(z)| > |f(0)|$ for all $z \in \mathbb{C}$ with $|z| > r$.*

Proof of $(*)$. For $z \neq 0$, we have

$$|f(z)| = |z|^n \cdot |a_n + h(z^{-1})|,$$

where $h(w) := a_{n-1} w + \cdots + a_0 w^n$. Since h is continuous at 0, there exists some $\delta > 0$, such that

$$|h(w) - h(0)| = |h(w)| \leq \frac{1}{2}|a_n| \text{ for } |w| < \delta.$$

Hence,

$$|f(z)| \geq |z|^n \cdot (|a_n| - |h(z^{-1})|) \geq \frac{1}{2}|a_n| \cdot |z|^n, \quad |z| > \frac{1}{\delta}.$$

Thus, we only need to choose $r > 1/\delta$ such that $|a_n| \cdot r^n > 2|a_0|$, to ensure that $(*)$ follows. □

We can now finish the proof of Cauchy's minimum theorem. Since $f(z)$ is continuous on $\mathbb{C}$, the function $|f(z)| : \mathbb{R}^2 \to \mathbb{R}$ is continuous on $\mathbb{R}^2$. Also, by $(*)$,

$$\inf_{z \in \mathbb{C}} |f(z)| = \inf_{|z| \le r} |f(z)|.$$

Since the disc $D = \{z \in \mathbb{C} : |z| \le r\}$ is compact, the 2-dimensional boundedness theorem ensures that, for some point $c \in D$,

$$|f(c)| = \inf_{z \in D} |f(z)|,$$

whence the result. □

Step 2. *Argand's inequality.* This inequality is the key idea in the present elementary proof of the fundamental theorem of algebra. In order to prove it, we shall need the following.

Lemma 7.10. *Let $k \ge 1$ be a natural number, and let*

$$h(z) := 1 + bz^k + z^k g(z),$$

where $b \in \mathbb{C} \setminus \{0\}$ and $g(z)$ is a complex polynomial such that $g(0) = 0$ (i.e., g is without constant term). Then there exists some $u \in \mathbb{C}$ with $|h(u)| < 1$.

Proof. Let $\zeta \in \mathbb{C}$ be a k-th root of $-1/b$, so that $b\zeta^k = -1$. For all real numbers t with $0 < t \le 1$, we then have

$$\begin{aligned}
|h(\zeta t)| &= |1 + b(\zeta t)^k + (\zeta t)^k g(\zeta t)| \\
&= |1 - t^k + \zeta^k t^k g(\zeta t)| \\
&\le |1 - t^k| + |\zeta^k t^k g(\zeta t)| \\
&= 1 - t^k + t^k |\zeta^k g(\zeta t)|.
\end{aligned}$$

Since g is continuous at 0 and $g(0) = 0$, there exists some δ with $0 < \delta < 1$ and

$$|\zeta^k g(\zeta t)| < \frac{1}{2} \text{ for } 0 < t < \delta.$$

For such t, we thus have

$$|h(\zeta t)| \le 1 - t^k + t^k |\zeta^k g(\zeta t)| < 1 - t^k + \frac{1}{2} t^k = 1 - \frac{1}{2} t^k < 1,$$

whence the result. □

We can now formulate and prove Argand's inequality.

Argand's inequality. Let $f(z)$ be a complex polynomial of degree at least 1. Then, for each point $c \in \mathbb{C}$ with $f(c) \neq 0$, there exists some $c' \in \mathbb{C}$ such that $|f(c')| < |f(c)|$.

Proof of Argand's inequality. Since $f(z)$ is not constant, neither is the polynomial

$$h(z) := f(c+z)/f(c).$$

We have

$$h(z) \;=\; 1 + b_k z^k + b_{k+1} z^{k+1} + \cdots + b_n z^n$$

with $1 \le k \le n$ and $b_k \neq 0$. Let

$$g(z) := b_{k+1} z + \cdots + b_n z^{n-k},$$

so that

$$h(z) = 1 + b_k z^k + z^k g(z),$$

where $g(0) = 0$. By Lemma 7.10, there exists some $u \in \mathbb{C}$ such that $|h(u)| < 1$. For $c' := c + u$, we thus have

$$|f(c')| \;=\; |h(u)| \cdot |f(c)| \;<\; |f(c)|,$$

whence Argand's inequality. □

Step 3. *Proof of the Fundamental Theorem of Algebra.* Let

$$f(z) = a_0 + a_1 z + \cdots + a_n z^n$$

be a non-constant polynomial of degree n with complex coefficients. By Cauchy's minimum theorem, there exists some $c \in \mathbb{C}$, such that $|f(c)| \le |f(z)|$ for all $z \in \mathbb{C}$. If $f(c) \neq 0$ then, by Argand's inequality, there exists some $c' \in \mathbb{C}$ such that $|f(c')| < |f(c)|$, which is absurd. Hence, $f(c) = 0$, so that $f(z)$ has a complex root.

§7.5. On the functional equation $\varphi(\sigma(s)) = \tau(\varphi(s))$

Let S and T be sets, and let $\sigma : S \to S$ and $\tau : T \to T$ be given maps. Following [63], we are going to construct the set of all maps $\varphi : S \to T$ such that the equation of the title holds true for each $s \in S$.

Define the iterates σ^μ of σ recursively via

$$\sigma^\mu = \begin{Bmatrix} 1_S, & \mu = 0 \\ \sigma \circ \sigma^{\mu-1}, & \mu \ge 1 \end{Bmatrix} \quad (\mu \in \mathbb{N}),$$

and introduce a binary relation ρ_σ on the set S via

$$s \rho_\sigma s' :\Longleftrightarrow \sigma^m(s) = \sigma^n(s') \text{ for some integers } m, n \ge 0, \quad (s, s' \in S). \tag{7.14}$$

It is easy to see that ρ_σ is an equivalence relation on S, thus defines a partition π_σ of S; cf. Exercise 1 below. Clearly, we have $s \rho_\sigma s'$ if, and only if, the sequences $(\sigma^\mu(s))_{\mu \ge 0}$ and $\sigma^\mu(s'))_{\mu \ge 0}$ contain the same elements, with the possible exception of finitely many elements.

Lemma 7.11. *If $P \in \pi_\sigma$, then $\sigma(P) \subseteq P$ and $\sigma^{-1}(P) \subseteq P$.*

The straightforward proof of Lemma 7.11 is left to the reader; cf. Exercise 2 below.

Let $s \in S$, and suppose that there exist integers m and n with $0 \leq m < n$ and $\sigma^m(s) = \sigma^n(s)$. Then the set $C_\sigma(s)$ of elements occurring infinitely often in the sequence $(\sigma^\mu(s))_{\mu \geq 0}$ is finite and non-empty. We call $C_\sigma(s)$ the *cycle* of the element s, and $c_\sigma(s) = |C_\sigma(s)|$ the *rank* of s. If the terms of the sequence $(\sigma^\mu(s))_{\mu \geq 0}$ are pairwise distinct, then we set $c_\sigma(s) = 0$. To be more precise, if $\sigma^m(s) = \sigma^n(s)$ with $0 \leq m < n$, if $\sigma^m(s)$ is the first member of the sequence $(\sigma^\mu(s))_{\mu \geq 0}$ to be repeated later on, and if $\sigma^n(s)$ is the first repetition of $\sigma^m(s)$, then the sequence $(\sigma^\mu(s))_{\mu \geq 0}$ is ultimately periodic with pre-period $\sigma^0(s), \sigma^1(s), \ldots, \sigma^{m-1}(s)$ of length m, and minimal period $\omega = n - m$, and we have $C_\sigma(s) = \{\sigma^m(s), \sigma^{m+1}(s), \ldots, \sigma^{n-1}(s)\}$. We note that, if $s, s' \in P \in \pi_\sigma$, then $c_\sigma(s) = c_\sigma(s')$ and $C_\sigma(s) = C_\sigma(s')$ for $c_\sigma(s) > 0$. We define

$$c_\sigma(P) := c_\sigma(s) \text{ for } s \in P,$$

and call this number the *rank* of the class P. If $c_\sigma(P) > 0$, then the cycle $C_\sigma(s)$ common to all $s \in P$ is termed the *cycle* of P, denoted $C_\sigma(P)$.

Definition 7.12. We say that an element $s \in S$ has property $\mathfrak{p}$, if there exists a sequence $(s_n)_{n \geq 0}$ such that $s_0 = s$ and $\sigma(s_{n+1}) = s_n$ for all $n \geq 0$. The set consisting of those elements $s \in S$ having property $\mathfrak{p}$ is denoted by $S_\mathfrak{p}$, and we set $S'_\mathfrak{p} := S - S_\mathfrak{p}$.

We define subsets $S'_\mathfrak{p}(\beta)$ of the set $S'_\mathfrak{p}$ for each ordinal β via

$$S'_\mathfrak{p}(0) := \{s \in S'_\mathfrak{p} : \sigma^{-1}(s) = \emptyset\}, \tag{7.15}$$

$$S'_\mathfrak{p}(\alpha) := \{s \in S'_\mathfrak{p} - \bigcup_{\beta < \alpha} S'_\mathfrak{p}(\beta) : \sigma^{-1}(s) \subseteq \bigcup_{\beta < \alpha} S'_\mathfrak{p}(\beta)\}, \tag{7.16}$$

assuming in (7.16) that the sets $S'_\mathfrak{p}(\beta)$ are already defined for all ordinals β with $\beta < \alpha$. We note that, as a consequence of this definition, the sets $S'_\mathfrak{p}(\beta)$ are pairwise disjoint.

Lemma 7.13. *There exists an ordinal number $\gamma < \infty$, where ∞ denotes the least ordinal satisfying $|\infty| > \max\{|S|, |T|\}$, such that $S'_\mathfrak{p} = \bigsqcup_{\beta < \gamma} S'_\mathfrak{p}(\beta)$.*

Proof. Let γ be the least ordinal such that $S'_\mathfrak{p}(\gamma) = \emptyset$ (such γ exists by definition of the set $S'_\mathfrak{p}$). Then $|\gamma| \leq |S'_\mathfrak{p}|$ and $\bigcup_{\beta < \gamma} S'_\mathfrak{p}(\beta) \subseteq S'_\mathfrak{p}$. Disjointness being clear, it remains to see that $S'_\mathfrak{p} = \bigcup_{\beta < \gamma} S'_\mathfrak{p}(\beta)$. Suppose for a contradiction that $S'_\mathfrak{p} - \bigcup_{\beta < \gamma} S'_\mathfrak{p}(\beta) \neq \emptyset$, and let $s \in S'_\mathfrak{p} - \bigcup_{\beta < \gamma} S'_\mathfrak{p}(\beta)$. By definition of $S'_\mathfrak{p}(\gamma)$, the set $\sigma^{-1}(s)$ contains at least one element $s_1 \in S'_\mathfrak{p} - \bigcup_{\beta < \gamma} S'_\mathfrak{p}(\beta)$. Replacing s by s_1, we now find in the same way that $\sigma^{-1}(s_1)$ contains at least one element $s_2 \in S'_{\mathfrak{p}} - \bigcup_{\beta<\gamma} S'_{\mathfrak{p}}(\beta)$, etc. It follows that s has property $\mathfrak{p}$, thus $s \in S_{\mathfrak{p}}$, the desired contradiction. □

The partition of the set $S'_{\mathfrak{p}}$ described in Lemma 7.13 gives rise to a concept of order for the elements of the set S.

Definition 7.14. The *order* $O_\sigma(s)$ of an element $s \in S$ is defined by

$$O_\sigma(s) = \begin{cases} \infty, & s \in S_{\mathfrak{p}}, \\ \beta, & s \in S'_{\mathfrak{p}}(\beta). \end{cases}$$

We note the following relation between the orders of an element $s \in S$ and that of its image $\sigma(s)$.

Lemma 7.15. *If $O_\sigma(s) \neq \infty$, then $O_\sigma(\sigma(s)) > O_\sigma(s)$. If $O_\sigma(s) = \infty$, then also $O_\sigma(\sigma(s)) = \infty$.*

Proof. If $O_\sigma(s) = \infty$ then, since $\gamma < \infty$, we have $s \in S_{\mathfrak{p}}$, so that s has property $\mathfrak{p}$. Then $\sigma(s)$ also has property $\mathfrak{p}$, thus $\sigma(s) \in S_{\mathfrak{p}}$ and, consequently, $O_\sigma(\sigma(s)) = \infty$. Next, if $O_\sigma(s) \neq \infty$ and $O_\sigma(\sigma(s)) = \infty$, our claim holds again. Finally, suppose that $O_\sigma(s), O_\sigma(\sigma(s)) \neq \infty$, and let $\sigma(s) \in S'_{\mathfrak{p}}(\alpha)$, with $S'_{\mathfrak{p}}(\alpha)$ as given in (7.16). Then $\sigma^{-1}(\sigma(s)) \subseteq \bigcup_{\beta<\alpha} S'_{\mathfrak{p}}(\beta)$, thus $s \in S'_{\mathfrak{p}}(\beta)$ for some ordinal $\beta < \alpha$. Hence,

$$O_\sigma(s) = \beta < \alpha = O_\sigma(\sigma(s)),$$

as claimed. □

Given a part $P \in \pi_\sigma$ and an element $s_0 \in P$, we define a sequence $(P_n(s_0))_{n\geq 0}$ of subsets $P_n(s_0)$ of S via

$$P_0(s_0) := \{\sigma^m(s_0) : m = 0, 1, 2, \dots\}$$

$$P_{n+1}(s_0) := \sigma^{-1}(P_n(s_0)) - \bigcup_{\nu=0}^{n} P_\nu(s_0), \quad n \geq 0.$$

According to this definition, the sets $P_n(s_0)$ are pairwise disjoint, using Lemma 7.11, we see that $P_n(s_0) \subseteq P$, and an immediate induction on n shows that

$$\bigcup_{\nu=0}^{n} P_\nu(s_0) = \{s \in P : \sigma^\nu(s) \in P_0(s_0) \text{ for some } \nu \text{ with } 0 \leq \nu \leq n\}, \quad n \geq 0.$$

Since the elements of P are all ρ_σ-equivalent to s_0, we obtain, for each $P \in \pi_\sigma$ and every element $s_0 \in P$, the decomposition

$$P = \bigsqcup_{n \geq 0} P_n(s_0). \tag{7.17}$$

As before, let S, T be sets, and let σ, τ be maps defined on S and T, respectively. As we have seen, the map σ induces an equivalence relation ρ_σ with corresponding partition π_σ, a rank function c_σ, and an order function O_σ on the set S, and we denote by ρ_τ, π_τ, c_τ, O_τ the corresponding entities induced by τ on the set T.

Definition 7.16. Given a class $P \in \pi_\sigma$, we term the class $Q \in \pi_\tau$ *P-admissible*, if either

(a) $c_\tau(Q) \neq 0$ and $c_\tau(Q) \mid c_\sigma(P)$, or

(b) $c_\sigma(P) = 0 = c_\tau(Q)$, and there exist elements $s_0 \in P$, $s'_0 \in Q$, such that $O_\sigma(\sigma^m(s_0)) \leq O_\tau(\tau^m(s'_0))$ for all integers $m \geq 0$.

If $\varphi : S \to T$ is such that

$$\varphi(\sigma(s)) = \tau(\varphi(s)), \quad s \in S, \tag{7.18}$$

we shall say that φ is (σ, τ)-*equivariant*, or *equivariant* for short. We note that, by iteration, the functional equation (7.18) implies

$$\varphi(\sigma^m(s)) = \tau^m(\varphi(s)), \quad (s \in S, m \geq 0).$$

In Lemmas 7.17–7.20 and Corollary 7.21 below, we shall always assume, without explicitly mentioning it, that the map φ is (σ, τ)-equivariant.

Lemma 7.17. *For each class $P \in \pi_\sigma$ there exists a class $Q \in \pi_\tau$ such that $\varphi(P) \subseteq Q$.*

Proof. If $s'_1, s'_2 \in \varphi(P)$, then there exist elements $s_1, s_2 \in P$ with $\varphi(s_i) = s'_i$ for $i = 1, 2$. Also, since $s_1 \rho_\sigma s_2$, there exist integers $m, n \geq 0$ such that $\sigma^m(s_1) = \sigma^n(s_2)$. It follows that

$$\tau^m(s'_1) = \tau^m(\varphi(s_1)) = \varphi(\sigma^m(s_1)) = \varphi(\sigma^n(s_2)) = \tau^n(\varphi(s_2)) = \tau^n(s'_2),$$

implying $s'_1 \rho_\tau s'_2$. Consequently, fixing s'_1 and letting s'_2 vary over $\varphi(P)$, we see that there exists a class $Q \in \pi_\tau$ with $\varphi(P) \subseteq Q$, namely the ρ_τ-class of the element s'_1. □

Lemma 7.18. *Let $P \in \pi_\sigma$, and let $Q \in \pi_\tau$ be the class for which $\varphi(P) \subseteq Q$. Then either*

(a) $c_\tau(Q) \neq 0$ *and* $c_\tau(Q) \mid c_\sigma(P)$, *or*

(b) $c_\sigma(P) = 0 = c_\tau(Q)$.

Proof. Suppose first that $c_\tau(Q) > 0$. If $c_\sigma(P) = 0$, our claim holds; thus, we may assume that $c_\sigma(P) > 0$. Then the cycle of P is mapped under φ onto the cycle of Q, which implies $c_\tau(Q)\big|c_\sigma(P)$. Now let $c_\tau(Q) = 0$. If we had $c_\sigma(P) > 0$, then the image of the cycle of P would be a cycle of Q, showing that $c_\tau(Q) > 0$, a contradiction. Hence, $c_\sigma(P) = 0$ in this second case. □

Lemma 7.19. *For each $s \in S$, we have $O_\sigma(s) \leq O_\tau(\varphi(s))$.*

Proof. We use transfinite induction on $O_\sigma(s)$. If $O_\sigma(s) = 0$, then our claim holds true . Suppose that our claim holds for all $s \in S$ such that $O_\sigma(s) < \alpha$, where $0 < \alpha \leq \infty$, and ∞ is defined as in Lemma 7.13. We also suppose (for a contradiction) that there exists an element $s_0 \in S$ such that $O_\sigma(s_0) = \alpha$ and $O_\sigma(s_0) > O_\tau(\varphi(s_0))$. There are two cases to consider.

(i) $\alpha < \infty$. Pick $s_1 \in \sigma^{-1}(s_0)$ arbitrarily. By Lemma 7.15, we have $O_\sigma(s_1) < O_\sigma(s_0) = \alpha$. Hence,

$$O_\sigma(s_1) \leq O_\tau(\varphi(s_1)) \leq O_\tau(\tau(\varphi(s_1))) = O_\tau(\varphi(\sigma(s_1))) = O_\tau(\varphi(s_0))$$
$$< O_\sigma(s_0) = \alpha < \infty,$$

where we have used the induction hypothesis for the first inequality, Lemma 7.15 for the second estimate, the functional equation for φ in the next step, the fact that $s_1 \in \sigma^{-1}(s_0)$ in Step 4, and our assumptions on s_0 and the ordinal α after that. Applying again Lemma 7.15 with σ replaced by τ and s replaced by $\varphi(s_1)$, we find that $O_\tau(\varphi(s_1)) < O_\tau(\tau(\varphi(s_1)))$. From these relations we conclude that

$$O_\sigma(s_1) \leq O_\tau(\varphi(s_1)) < O_\tau(\tau(\varphi(s_1))) = O_\tau(\varphi(s_0))$$

for each $s_1 \in \sigma^{-1}(s_0)$. Since $O_\sigma(s_0) = \alpha$, it follows that $O_\tau(\varphi(s_0)) \geq \alpha = O_\sigma(s_0)$, a contradiction.

(ii) $\alpha = \infty$. Then s_0 has property $\mathfrak{p}$, thus $\varphi(s_0)$ also has property $\mathfrak{p}$. Consequently, $O_\tau(\varphi(s_0)) = \infty = O_\sigma(s_0)$, contradicting our hypotheses on s_0. Hence, our claim holds for all elements $s \in S$ with $O_\sigma(s) = \alpha$, and the proof is complete. □

Corollary 7.20. *Let $P \in \pi_\sigma$, and let $Q \in \pi_\tau$ be the class for which $\varphi(P) \subseteq Q$. Then Q is P-admissible.*

Proof. By Lemma 7.18, we either have $c_\tau(Q) \neq 0$ and $c_\tau(Q)\big|c_\sigma(P)$, or $c_\sigma(P) = c_\tau(Q) = 0$. In the latter case, choose $s_0 \in P$ arbitrarily, and set $s_0' := \varphi(s_0) \in Q$. Then, for each integer $m \geq 0$, we have

$$O_\sigma(\sigma^m(s_0)) \leq O_\tau(\varphi(\sigma^m(s_0))) = O_\tau(\tau^m(s_0'))$$

by Lemma 7.19. Hence, the class Q is P-admissible in the sense of Definition 7.16, as claimed. □

In Lemmas 7.21–7.22 and Definition 7.23 below we assume that S,T are sets, that $\sigma : S \to S$ is a map defined on S, and that $\tau : T \to T$ is a map defined on T.

Lemma 7.21. *Let $P \in \pi_\sigma$, and let $Q \in \pi_\tau$ be P-admissible in the sense of Definition 7.16. Then there exist elements $s_0 \in P$ and $s_0' \in Q$, such that*

$$O_\sigma(\sigma^m(s_0)) \le O_\tau(\tau^m(s_0')), \; m \ge 0.$$

Proof. If $c_\tau(Q) = 0$, then our claim follows from Definition 7.16. If $c_\tau(Q) > 0$, then we choose $s_0' \in C_\tau(Q)$ and $s_0 \in P$ arbitrarily, noting that $O_\tau(\tau^m(s_0')) = \infty$ for all $m \ge 0$, since the elements of $C_\tau(Q)$ have property $\mathfrak{p}$. □

Lemma 7.22. *Let $P \in \pi_\sigma$, and let $Q \in \pi_\tau$ be P-admissible. Moreover, let $s \in P$ and $s' \in Q$ be elements such that $O_\sigma(s) \le O_\tau(s')$, and let $s_1 \in \sigma^{-1}(s)$ be chosen arbitrarily. Then there exists an element $s_1' \in \tau^{-1}(s')$ such that $O_\sigma(s_1) \le O_\tau(s_1')$.*

Proof. If $O_\tau(s') = \infty$, then s' has property $\mathfrak{p}$, implying $\tau^{-1}(s') \ne \emptyset$ and $O_\tau(s_1') = \infty$ for some element $s_1' \in \tau^{-1}(s')$; thus our claim holds in this case, and we may suppose that $O_\sigma(s) \le O_\tau(s') < \infty$, in particular $O_\sigma(s_1) < \infty$. If we had $O_\tau(s_1') < O_\sigma(s_1)$ for all $s_1' \in \tau^{-1}(s')$, then Lemma 7.15 would yield

$$O_\tau(s_1') < O_\sigma(s_1) < O_\sigma(\sigma(s_1)) = O_\sigma(s) \le O_\tau(s') < \infty$$

for each element $s_1' \in \tau^{-1}(s')$. Consequently, it would follow that

$$O_\tau(s') \le O_\sigma(s_1) < O_\tau(s'),$$

which is impossible. Hence, there must exist some element $s_1' \in \tau^{-1}(s')$ such that $O_\sigma(s_1) \le O_\tau(s_1')$, as required. □

Definition 7.23. (Construction (C)) Let $P \in \pi_\sigma$ be an arbitrary class, and let $Q \in \pi_\tau$ be P-admissible. Let $s_0 \in P$ and $s_0' \in Q$ be chosen such that $O_\sigma(\sigma^m(s_0)) \le O_\tau(\tau^m(s_0'))$ for all integers $m \ge 0$ (such elements exist by Lemma 7.21). We set $\varphi(\sigma^m(s_0)) := \tau^m(s_0')$. Then $\varphi(P_0(s_0)) = Q_0(s_0')$, and we have $O_\sigma(s) \le O_\tau(\varphi(s))$ for each $s \in P_0(s_0)$. Suppose that φ has already been defined on $\bigcup_{\nu=0}^n P_\nu(s_0)$ in such a way that $\varphi(\bigcup_{\nu=0}^n P_\nu(s_0)) \subseteq \bigcup_{\nu=0}^n Q_\nu(s_0')$, and such that $O_\sigma(s) \le O_\tau(\varphi(s))$ for all $s \in \bigcup_{\nu=0}^n P_\nu(s_0)$. If $P_{n+1}(s_0)$ is non-empty, then we define φ on this set as follows: if $s \in P_{n+1}(s_0)$, then $\sigma(s) \in P_n(s_0)$, thus $s' := \varphi(\sigma(s)) \in \bigcup_{\nu=0}^n Q_\nu(s_0')$ is defined, we have $O_\sigma(\sigma(s)) \le O_\tau(s')$, and we set $\varphi(s) := s_1'$, where $s_1' \in \tau^{-1}(s') \subseteq \bigcup_{\nu=0}^{n+1} Q_\nu(s_0')$ is chosen such that $O_\sigma(s) \le O_\tau(s_1')$ (such s_1' exists by Lemma 7.22). Then $\varphi(\bigcup_{\nu=0}^{n+1} P_\nu(s_0)) \subseteq \bigcup_{\nu=0}^{n+1} Q_\nu(s_0')$ and $O_\sigma(s) \le O_\tau(\varphi(s))$ for all $s \in \bigcup_{\nu=0}^{n+1} P_\nu(s_0)$. Proceeding in this way, we obtain a map $\varphi_P : P \to Q$ such that $O_\sigma(s) \le O_\tau(\varphi_P(s))$ for each $s \in P$.

Lemma 7.24. *Any map $\varphi_P : P \to Q$ defined via Construction (C) satisfies $\varphi_P(\sigma_P(s)) = \tau_Q(\varphi_P(s))$ for all $s \in P$, where $\sigma_P : P \to P$ and $\tau_Q : Q \to Q$ are the restrictions of σ and τ to P and Q, respectively.*

Proof. If $s \in P_0(s_0)$ then $s = \sigma^m(s_0)$ for some integer $m \geq 0$. Thus, by definition of φ_P,

$$\varphi_P(\sigma_P(s)) = \varphi_P(\sigma_P^{m+1}(s_0)) = \tau_Q^{m+1}(s_0') = \tau_Q(\tau_Q^m(s_0')) = \tau_Q(\varphi_P(s)),$$

so that our claim holds for $s \in P_0(s_0)$. Suppose that $\varphi_P(\sigma_P(s)) = \tau_Q(\varphi_P(s))$ holds for all $s \in \bigcup_{\nu=0}^n P_\nu(s_0)$, and let $s \in P_{n+1}(s_0)$. Then, again by definition of φ_P, we have $\varphi_P(s) = s_1' \in \tau_Q^{-1}(\varphi_P(\sigma_P(s))$, hence $\tau_Q(\varphi_P(s)) = \varphi_P(\sigma_P(s))$, as desired, and our claim follows by induction on n. □

Lemma 7.25. *Let σ be a map defined on the set S, and let τ be a map defined on the set T. Moreover, let $P \in \pi_\sigma$, $Q \in \pi_\tau$, and let $\varphi : P \to Q$ be a map which is (σ_P, τ_Q)-equivariant. Then φ_P results from Construction (C).*

Proof. By Corollary 7.20, the class Q is admissible for P. Let $s_0 \in P$ be an arbitrary element, and set $s_0' := \varphi_P(s_0) \in Q$. Then $O_\sigma(\sigma^m(s_0)) \leq O_\tau(\tau^m(s_0'))$ for each integer $m \geq 0$ by Lemma 7.19. Moreover, we have

$$\varphi_P(\sigma^m(s_0)) = \tau^m(\varphi_P(s_0) = \tau^m(s_0'), \quad m \geq 0,$$

in accordance with Construction (C). Let $s \in P_{n+1}(s_0)$, set $s' := \varphi_P(\sigma_P(s))$, and let $\varphi_P(s) = s_1'$. Then

$$\tau(s_1') = \tau(\varphi_P(s)) = \varphi_P(\sigma_p(s)) = s',$$

so that $s_1' \in \tau^{-1}(s')$, and we have $O_\sigma(s) \leq O_\tau(s_1')$ by Lemma 7.19. Hence, φ_P does indeed arise from Construction (C), as claimed. □

Definition 7.26. Let S and T be sets, and let π be a partition of S. Suppose that, for each part $P \in \pi$, there is given a map $\varphi_P : P \to T$. Then we denote by $\varphi = \bigoplus_{P \in \pi} \varphi_P$ the map $\varphi : S \to T$ satisfying $\varphi(s) = \varphi_P(s)$ for each $P \in \pi$ and every $s \in P$.

We now obtain the main result of Novotný's paper.

Theorem 7.27. (Hauptsatz 2.14 in [63]) *Let S, T be non-empty sets, and let $\sigma : S \to S$ and $\tau : T \to T$ be maps defined on S and T, respectively.*

(A) *Let $\Phi : \pi_\sigma \to \pi_\tau$ be a map such that $\Phi(P)$ is admissible for P, for each class $P \in \pi_\sigma$. For each $P \in \pi_\sigma$, let $\varphi_P : P \to \Phi(P)$ be a map defined via Construction* (C)*. Then the map $\varphi := \bigoplus_{P \in \pi_\sigma} \varphi_P : S \to T$ is (σ, τ)-equivariant.*

(B) *Every (σ, τ)-equivariant map $\varphi : S \to T$ arises in the way described in* (A).

Proof. This follows from Lemmas 7.24 and 7.17, Corollary 7.20, and Lemma 7.25. □

Remark 7.28. In the context of Universal Algebra, a set S equipped with a unary operation, that is, a map $\sigma : S \to S$, is termed a *mono-unary algebra*. If (S,σ) and (T,τ) are mono-unary algebras, then a map $\varphi : S \to T$ is called an *algebra morphism*, if it satisfies the equation $\varphi(\sigma(s)) = \tau(\varphi(s))$ for all $s \in S$; that is, if it is (σ,τ)-equivariant; cf. [59]. Theorem 7.27 thus describes the set of all algebra morphisms $\varphi : (S,\sigma) \to (T,\tau)$.

If $\sigma : S \to S$ is bijective, i.e., a permutation on S, then σ^m is a permutation for each integer $m \geq 0$. Setting $\sigma^{-m} := (\sigma^m)^{-1}$, the inverse of σ^m, for $m \geq 0$, we have defined the map σ^m for each integer m. Suppose now that S is a finite set, and let σ be a permutation on S. Then each class $P \in \pi_\sigma$ is a cycle, thus $O_\sigma(s) = \infty$ and $c_\sigma(s) > 0$ for each $s \in S$. Also, we have $P = P_0(s)$ for each class $P \in \pi_\sigma$ and every element $s \in P$. Theorem 7.27 thus yields the following.

Corollary 7.29. ([82]) *Let S be a non-empty finite set, and let σ be a permutation of S. Let $\Phi : \pi_\sigma \to \pi_\sigma$ be such that $c_\sigma(\Phi(P))$ divides $c_\sigma(P)$ for each $P \in \pi_\sigma$. For each $P \in \pi_\sigma$, let $s_P \in P$ and $s'_P \in \Phi(P)$ be arbitrary elements, and define a map $\varphi : S \to S$ via $\varphi(\sigma^m(s_P)) := \sigma^m(s'_P)$ for each integer $m \geq 0$ and every part $P \in \pi_\sigma$. Then φ satisfies the equation*

$$\varphi(\sigma(s)) = \sigma(\varphi(s)), \quad s \in S. \tag{7.19}$$

Moreover, each map φ on S satisfying Equation (7.19) *is obtained in this way.*

Clearly, a map φ as constructed in Corollary 7.29 is a permutation on S if, and only if, (i) the map Φ is injective, and (ii) we have $c_\sigma(P) = c_\sigma(\Phi(P))$ for each cycle $P \in \pi_\sigma$. Consequently, Corollary 7.29 yields the following set-theoretic description of the centraliser $C_{\mathrm{Sym}(S)}(\sigma)$ for each finite set S and every permutation $\sigma \in \mathrm{Sym}(S)$.

Corollary 7.30. *Let S be a non-empty finite set, and let σ be a permutation on S. Let Φ be a permutation of the set π_σ such that $c_\sigma(P) = c_\sigma(\Phi(P))$ for each cycle P of σ. Choosing arbitrary elements $s_P \in P$ and $s'_P \in \Phi(P)$, we set $\varphi(\sigma^m(s_P)) = \sigma^m(s'_P)$ for $m \geq 0$. Then φ is a permutation of S commuting with σ, and each permutation $\sigma \in \mathrm{Sym}(S)$ commuting with the given permutation σ can be obtained in this way.*

Exercises for §7.5.

1. Show that the binary relation ρ_σ defined as in (7.14) on the set S is an equivalence relation.
2. Prove Lemma 7.11.

3. Let S be a non-empty finite set of order $|S| = n$, let $G = \mathrm{Sym}(S)$ be the symmetric group on S, and let $\sigma \in G$ be a permutation on S having λ_ν cycles of length ν in its disjoint cycle decomposition for $1 \leq \nu \leq n$. We define the *centraliser* $C_G(\sigma)$ of σ in the group G as

$$C_G(\sigma) := \{\varphi \in G : \varphi\sigma = \sigma\varphi\}.$$

Compute the order of $C_G(\sigma)$.

Part II: Topics in Transfinite Algebra

The second part of this book may be viewed as one continuous advertisement for the applicability of the axiom of choice (mostly, but not always, in the form of Zorn's Lemma) in the context of algebraic theories. The content ranges from questions of orderability of groups, rings, and fields, over semi-simplicity and injectivity of modules, to characterisations of the Jacobson radical of a ring and matroid theory in arbitrary cardinality. The last section takes up from the characterisation of orderable fields due to Artin and Schreier, which is discussed, among other things, in §9, and describes Artin's solution of Hilbert's 17th problem concerning the decomposition of definite rational functions over $\mathbb{Q}$ as a sum of squares of rational functions. As far as the author is aware, ours is the first treatment of this beautiful and subtle proof in English.

§8. Group and ring structures on non-empty sets

In this section, we will consider the question whether group or ring structures may be defined on an arbitrary non-empty set. We shall see that the answer to this question is in the affirmative, and we will obtain more detailed information on the precise structure of such groups and rings.

§8.1. Group structures on non-empty sets

Does every non-empty set G carry a group structure? Certainly, if G is finite, $|G| = n$ say, we can find a bijection $\varphi : G \to C_n$, where C_n is the cyclic group of order n, inducing the structure of a cyclic group on G itself via

$$g_1 + g_2 := \varphi^{-1}(\varphi(g_1) + \varphi(g_2)), \quad g_1, g_2 \in G.$$

Let G be a set with $|G| = \mathfrak{m} \geq \aleph_0$. Recall that an additive abelian group is an *elementary abelian 2-group* if $2x = 0$ for every element x of the group. Such a group is a direct sum of cyclic groups of order 2. Let $\mathscr{H}$ be the set of all pairs $(H, +_H)$, where H is a non-empty subset of G, and $+_H$ is a binary operation on H giving H the structure of an elementary abelian 2-group. The set $\mathscr{H}$ is non-empty, for instance choose elements $a, b \in G$ with $a \neq b$, and define a binary operation $+_S$ on $S = \{a, b\}$ via $a +_S a = a$, $a +_S b = b +_S a = b$, and $b +_S b = a$. Then $(S, +_S) \cong C_2$, so $(S, +_S) \in \mathscr{H}$.

Define an order relation $\preceq$ on $\mathscr{H}$ via

$$(H, +_H) \preceq (K, +_K) :\Longleftrightarrow H \subseteq K \text{ and } +_K \text{ extends } +_H; \tag{8.20}$$

cf. Exercise 1. More explicitly, the second condition in (8.20) means that

$$h_1 +_K h_2 = h_1 +_H h_2, \quad (h_1, h_2 \in H).$$

Let $\mathscr{C} = \{(H_\nu, +_{H_\nu}) : \nu \in N\}$ be a non-empty chain in $(\mathscr{H}, \preceq)$, and set $\tilde{H} := \bigcup_{\nu \in N} H_\nu$. Then $\tilde{H}$ is a non-empty subset of G, and the operations $+_{H_\nu}$ induce a well-defined binary operation $+_{\tilde{H}}$ on $\tilde{H}$ via

$$\tilde{h}_1 +_{\tilde{H}} \tilde{h}_2 := \tilde{h}_1 +_{H_\nu} \tilde{h}_2, \quad (\tilde{h}_1, \tilde{h}_2 \in H_\nu), \tag{8.21}$$

giving $\tilde{H}$ the structure of an elementary abelian 2-group; cf. Exercise 2. Hence, $(\tilde{H}, +_{\tilde{H}})$ is an upper bound for $\mathscr{C}$ in $\mathscr{H}$, so that $\mathscr{H}$ is inductively ordered. By Zorn's lemma, $\mathscr{H}$ contains a maximal element, say $(H^*, +_{H^*})$.

If $|H^*| = |G| \, (= \mathfrak{m})$, then there exists a bijection $\varphi : H^* \to G$, inducing the structure of an elementary abelian 2-group on G in the obvious way. Suppose for a contradiction that $\mathfrak{n} := |H^*| < \mathfrak{m}$, and let $\mathfrak{n}'$ be the cardinality of the difference set $G - H^*$. Since (see Part (b)(i) of Proposition 5.5)

$$\mathfrak{m} = \mathfrak{n} + \mathfrak{n}' = \max\{\mathfrak{n}, \mathfrak{n}'\},$$

we must have $\mathfrak{n} \leq \mathfrak{n}'$. Hence, there exists an injective map $\psi : H^* \to G - H^*$. Set

$$G^* := H^* \sqcup \psi(H^*) \subseteq G,$$

and define a binary operation $+_{G^*}$ on G^* via

$$x +_{G^*} y := \begin{cases} x +_{H^*} y, & x, y \in H^* \\ \psi^{-1}(x) +_{H^*} \psi^{-1}(y), & x, y \in \psi(H^*) \\ \psi(x +_{H^*} \psi^{-1}(y)), & x \in H^*, y \in \psi(H^*) \\ \psi(\psi^{-1}(x) +_{H^*} y), & x \in \psi(H^*), y \in H^*. \end{cases}$$

Let $\langle \zeta \rangle$ be a cyclic group of order 2 generated by a symbol ζ, and consider the direct sum

$$\Sigma := (H^*, +_{H^*}) \oplus \langle \zeta \rangle,$$

an elementary abelian 2-group $(\Sigma, +_\Sigma)$. The elements of Σ are of the form (x, ξ) with $x \in H^*$ and $\xi \in \{0, \zeta\}$, and the addition in Σ is given explicitly by

$$(x_1, \xi_1) +_\Sigma (x_2, \xi_2) = (x_1 +_{H^*} x_2, \xi_1 +_{\langle \zeta \rangle} \xi_2), \quad (x_1, x_2 \in H^*; \xi_1, \xi_2 \in \langle \zeta \rangle).$$

Define a map $\chi : \Sigma \to G^*$ via

$$\chi(x, \xi) := \left\{ \begin{matrix} x, & \xi = 0 \\ \psi(x), & \xi = \zeta \end{matrix} \right\} \quad ((x, \xi) \in \Sigma).$$

We claim that χ is an isomorphism of elementary abelian 2-groups, so that $(G^*, +_{G^*}) \in \mathscr{H}$ and $(H^*, +_{H^*}) \prec (G^*, +_{G^*})$, contradicting maximality of the element $(H^*, +_{H^*}) \in \mathscr{H}$. This contradiction would then serve to show that $\mathfrak{n} = \mathfrak{m}$, as desired. Indeed, we have

$$\chi((x_1, \xi_1) +_\Sigma (x_2, \xi_2)) = \begin{cases} x_1 +_{H^*} x_2, & \xi_1 +_{\langle \zeta \rangle} \xi_2 = 0 \\ \psi(x_1 +_{H^*} x_2), & \xi_1 +_{\langle \zeta \rangle} \xi_2 = \zeta, \end{cases}$$

while

$$\chi(x_1, \xi_1) +_{G^*} \chi(x_2, \xi_2) = \begin{cases} x_1 +_{H^*} x_2, & \xi_1 = \xi_2 = 0 \\ x_1 +_{G^*} \psi(x_2) = \psi(x_1 +_{H^*} x_2), & \xi_1 = 0, \xi_2 = \zeta \\ \psi(x_1) +_{G^*} x_2 = \psi(x_1 +_{H^*} x_2), & \xi_1 = \zeta, \xi_2 = 0 \\ \psi(x_1) +_{G^*} \psi(x_2) = x_1 +_{H^*} x_2, & \xi_1 = \xi_2 = \zeta, \end{cases}$$

hence

$$\chi((x_1, \xi_1) +_\Sigma (x_2, \xi_2)) = \chi(x_1, \xi_1) +_{G^*} \chi(x_2, \xi_2),$$

that is, χ is a homomorphism. Moreover, χ is clearly surjective, and

$$\chi(x_1, \xi_1) = \left\{ \begin{matrix} x_1, & \xi_1 = 0 \\ \psi(x_1), & \xi_1 = \zeta \end{matrix} \right\} = \left\{ \begin{matrix} x_2, & \xi_2 = 0 \\ \psi(x_2), & \xi_2 = \zeta \end{matrix} \right\} = \chi(x_2, \xi_2)$$

yields $\xi_1 = \xi_2$ since $H^* \cap \psi(H^*) = \emptyset$, and therefore either $x_1 = x_2$ or $\psi(x_1) = \psi(x_2)$, the latter implying again $x_1 = x_2$ by injectivity of ψ. Hence, χ is injective, thus an isomorphism as claimed.

We have shown the following.

Proposition 8.1. *Every non-empty set G can be given the structure of an abelian group. If G is infinite, then G can be given the structure of an elementary abelian 2-group.*

We shall supply a second proof for the main assertion of Proposition 8.1, which is perhaps somewhat more illuminating. Suppose that $|G| = \mathfrak{m} \geq \aleph_0$, and let

$$\Sigma = \bigoplus_{\nu \in N} \langle \zeta_\nu \rangle, \quad (|N| = \mathfrak{m})$$

be the discrete direct sum of $\mathfrak{m}$ copies $\langle \zeta_\nu \rangle$ of the cyclic group of order 2.[10] Certainly, $\mathfrak{n} := |\Sigma| \geq \mathfrak{m}$, since the map $N \to \Sigma$ given by $\nu \mapsto \zeta_\nu$ is injective. Each element $\sigma \in \Sigma$ can be uniquely written in the form

$$\sigma = \zeta_{\nu_{\sigma,1}} + \zeta_{\nu_{\sigma,2}} + \cdots + \zeta_{\nu_{\sigma,k(\sigma)}} \quad (\nu_{\sigma,i} \in N,\ k(\sigma) \in \mathbb{N}).$$

Mapping σ to the finite subset $\{\nu_{\sigma,1}, \ldots, \nu_{\sigma,k(\sigma)}\}$ of N yields a bijection between Σ and the set $\mathscr{B}_{\text{fin}}(N)$ of finite subsets of N. Denote by $\binom{N}{k}$ the set of all k-element subsets of N. Writing down a k-element subset as a string (that is, as a k-tuple), we obtain an embedding $\binom{N}{k} \hookrightarrow N^k$, hence $\left|\binom{N}{k}\right| \leq |N|^k = |N|$, since $|N| = \mathfrak{m} \geq \aleph_0$. For each $k \in \mathbb{N}$ choose an injective map $\psi_k : \binom{N}{k} \to N$, and define an embedding of

$$\mathscr{B}_{\text{fin}}(N) = \bigsqcup_{k \in \mathbb{N}} \binom{N}{k}$$

into $\mathbb{N} \times N$ by sending $\{\nu_1, \ldots, \nu_k\}$ to $(k, \psi_k(\{\nu_1, \ldots, \nu_k\}))$. It follows that

$$\mathfrak{n} = |\Sigma| = |\mathscr{B}_{\text{fin}}(N)| \leq \aleph_0 \cdot |N| = \aleph_0 \cdot \mathfrak{m} = \mathfrak{m},$$

so $\mathfrak{n} = \mathfrak{m}$, and G can be given the structure of an elementary abelian 2-group via a bijection $\psi : \Sigma \to G$.

We remark that the Axiom of Choice (or one of its equivalent forms) is still needed in this last argument, since the equation $\aleph_0 \cdot \mathfrak{m} = \mathfrak{m}$ for infinite cardinals $\mathfrak{m}$, used above in a crucial way, cannot be established without it.

§8.2. Ring structures on non-empty sets

We now turn to similar questions and results for rings.

A ring $(R, +, \cdot)$ is called a *Boolean ring*, if every element of R is idempotent; that is, if the equation $a^2 = a$ holds for all $a \in R$. We begin with two easy observations concerning Boolean rings.

[10] We note that discrete direct sums, as they are called here, are sometimes also referred to as a *co-product*, a *finite support product*, or simply as a *sum* in the literature.

Lemma 8.2. *A Boolean ring* $(R,+,\cdot)$ *is commutative, and satisfies* $2a=0$ *for all* $a\in R$.

Proof. For an arbitrary element $a\in R$, we have

$$\begin{aligned}2a=a+a=(a+a)^2=(a+a)(a+a)&=a^2+a^2+a^2+a^2\\&=a+a+a+a=4a.\end{aligned}$$

Subtracting $2a$ from both sides gives $2a=0$, as desired. Hence, the additive group of a Boolean ring is an elementary abelian 2-group. Next, let $a,b\in R$ be arbitrary elements. Then we have

$$a+b=(a+b)^2=(a+b)(a+b)=a^2+ab+ba+b^2=a+ab+ba+b.$$

Subtracting $a+b$ from both sides yields $ab+ba=0$, or $ab=-ba=ba$, since $2ba=0$ by what we have shown above. Hence, R is commutative as claimed. □

We have the following analogue of Proposition 8.1 concerning ring structures on non-empty sets.

Proposition 8.3. *Every non-empty set R can be given the structure of a commutative ring. Moreover, if R is infinite, then R can be given the structure of a Boolean ring with identity element.*

Proof. If $|R|=n<\aleph_0$, then R may be given the structure of the residue class ring $\mathbb{Z}/n\mathbb{Z}$. Next, suppose that R is infinite. We first give R the structure of an (additive) elementary abelian 2-group (cf. Proposition 8.1). Then R has the form

$$R=\bigoplus_{\nu\in N}\langle\zeta_\nu\rangle\quad(2\zeta_\nu=0,\,\zeta_\nu\neq 0)\tag{8.22}$$

for some infinite index set N. On each direct summand in (8.22), we now introduce the structure of a ring isomorphic to $\mathbb{Z}/2\mathbb{Z}$, and form the ring-theoretic discrete direct sum. Then R is a discrete direct sum of copies of the prime field of characteristic 2 and, for an arbitrary element

$$x=\zeta_{\nu_1}+\zeta_{\nu_2}+\cdots+\zeta_{\nu_r}\in R,$$

we have

$$\begin{aligned}x^2&=(\zeta_{\nu_1}+\cdots+\zeta_{\nu_r})^2\\&=\zeta_{\nu_1}^2+\cdots+\zeta_{\nu_r}^2+2\sum_{1\le i<j\le r}\zeta_{\nu_i}\zeta_{\nu_j}\\&=\zeta_{\nu_1}^2+\cdots+\zeta_{\nu_r}^2\\&=\zeta_{\nu_1}+\cdots+\zeta_{\nu_r}\\&=x,\end{aligned}$$

so that R is indeed a Boolean ring, as desired. Moreover, every infinite Boolean ring can be embedded into a Boolean ring with identity element of the same cardinality (see Exercise 3 below), whence our result. □

As is well known, defining

$$a \cup b := a + b - ab,$$
$$a \cap b := ab$$

for the elements of a Boolean ring R with identity element, we obtain a Boolean algebra (that is, a lattice which is distributive and complemented). Moreover, any finite chain with n elements is a distributive lattice with n elements; cf. [29, §§1–4] or Chapter I in [13] for background, proofs, and definitions. From Proposition 8.3, we thus immediately obtain the following.

Corollary 8.4. *Every non-empty set L can be given the structure of a distributive lattice. Moreover, if L is infinite, then L can be given the structure of a Boolean algebra.*

Remark 8.5. In this section, we have shown, using the axiom of choice (or one of its equivalent forms), that every non-empty set can be given one of a number of specific algebraic structures. The question arises, how closely these results are related to the axiom of choice itself. Are they equivalents? We refer to [41] for more information on this particular aspect.

Exercises for § 8.

1. Prove that the relation $\preceq$ defined on the set $\mathscr{H}$ is an order relation.
2. Show that the operation $+_{\tilde{H}}$ defined on $\tilde{H}$ in (8.21) is well defined and gives $\tilde{H}$ the structure of an abelian 2-group.
3. Let R be a Boolean ring, and let $F = GF(2)$ be the prime field of characteristic 2 with elements $\overline{0}, \overline{1}$. Consider the set $R^* = R \times F$ of all ordered pairs $(r, \overline{x})$ with $r \in R$ and $\overline{x} \in F$ and, for $r, s \in R$ and $\overline{x}, \overline{y} \in F$, set

$$(r,\overline{x}) + (s,\overline{y}) := (r+s, \overline{x}+\overline{y}),$$
$$(r,\overline{x}) \cdot (s,\overline{y}) := (rs + \overline{x}s + \overline{y}r, \overline{x}\overline{y}),$$

where

$$\overline{x}r := \begin{cases} 0, & \overline{x} = \overline{0}, \\ r, & \overline{x} = \overline{1}. \end{cases}$$

Show the following:
(i) $(R^*, +, \cdot)$ is a Boolean ring;
(ii) the element $(0, \overline{1}) \in R^*$ is an identity element in R^*;
(iii) the map $\varphi : R \to R^*$ sending $r \in R$ to $(r, \overline{0}) \in R^*$ is an injective ring homomorphism.

Conclude that every Boolean ring can be embedded into a Boolean ring with identity element.

4. It follows in particular from Proposition 8.1, that every non-empty set can be given the structure of a semigroup. However, this fact can be shown without transfinite methods, for instance as follows. For a non-empty set G, define a binary operation $\circ$ by

$$a \circ b := a \quad (a, b \in G).$$

Show that $\circ$ is associative; that is, that $(G, \circ)$ is a semigroup.

§9. Orderable abelian groups and fields

We turn to the question, whether a given abelian group, or a given field, can be made into an ordered abelian group, respectively an ordered field; that is, whether or not each of these structures supports an order relation compatible with its operation(s).

§9.1. Orderable abelian groups

Let G be an additive abelian group. By a *positive cone* in G we mean a subset P of G such that

(i) P is closed under addition,

(ii) $G = P \sqcup (-P) \sqcup \{0\}$.

An abelian group is termed *orderable*, if it contains a positive cone. By an *ordered* abelian group we mean a pair (G,P) consisting of an abelian group G and a positive cone P of G. If P is a positive cone in G, then

$$a \leq b :\Longleftrightarrow b - a \in P \cup \{0\} \quad (a,b \in G) \tag{9.23}$$

defines a total order on G, which is compatible with addition in the sense that

$$a \leq b \Longrightarrow a + c \leq b + c \quad (a,b,c \in G), \tag{9.24}$$

and we have

$$P = \{a \in G : a > 0\}. \tag{9.25}$$

Conversely, given a total order $\leq$ on G satisfying (9.24), then Equation (9.25) defines a positive cone in G; see Exercise 1.

We have the following characterisation of orderable abelian groups due to F. W. Levi. Call a subset H of G a *quasi-cone*, if H is closed under addition and does not contain 0; for instance, the empty set is a quasi-cone in G.

Proposition 9.1. (Levi [58]) *An abelian group G is orderable if, and only if, it is torsion-free. If G is torsion-free and $H \subseteq G$ is a quasi-cone, then G has a positive cone P with $P \supseteq H$.*

Proof. Let (G,P) be an ordered abelian group, and let $a \in G$ be a non-zero element. Then either $a \in P$ or $-a \in P$. If $a \in P$, then we see inductively that $na \in P$ for every $n \in \mathbb{N} - \{0\}$. Similarly, if $-a \in P$, then $-(na) \in P$ for all $n \in \mathbb{N} - \{0\}$; in both cases, the order $O(a)$ of a is infinite; that is, G is torsion-free.

Now let G be a torsion-free abelian group, and let $H \subseteq G$ be a quasi-cone, and let

$$\mathscr{Q} := \{H' \subseteq G : H' \text{ is a quasi-cone, and } H' \supseteq H\} \tag{9.26}$$

be the set of all quasi-cones in G containing H, ordered by inclusion. Then $H \in \mathcal{Q}$, so $\mathcal{Q} \neq \emptyset$, and $\mathcal{Q}$ is inductively ordered; cf. Exercise 2. By Zorn's Lemma, $\mathcal{Q}, \subseteq)$ has a maximal element P. We want to show that $P \subseteq G$ is a positive cone in G. Let $a \in G$ be a non-zero element. The equation $a + (-a) = 0$ shows that at most one of $a, -a$ can belong to P, that is, $P \cap (-P) = \emptyset$. Suppose that none of $a, -a$ belongs to P. Then the sets

$$P' := \{x + ka : x \in P, k \in \mathbb{N} - \{0\}\} \cup P \cup \{ka : k \in \mathbb{N} - \{0\}\}$$
$$P'' := \{y - \ell a : y \in P, \ell \in \mathbb{N} - \{0\}\} \cup P \cup \{-\ell a : \ell \in \mathbb{N} - \{0\}\}$$

both are closed with respect to addition and properly contain P, thus $0 \in P', P''$ by maximality of P. Consequently, there are elements $u, v \in P$ and positive integers m, n such that

$$u + ma = 0 \text{ and } v - na = 0,$$

hence

$$nu + mv = 0, \tag{9.27}$$

a contradiction, since the left-hand side of (9.27) is in P, while the right-hand side is not. It follows that

$$G = P \sqcup (-P) \sqcup \{0\},$$

so that P is a positive cone in G containing H; in particular, G is orderable.

□

§9.2. Orderable fields

We now turn to order relations compatible with a given field structure. The main result of this section is Proposition 9.2, a famous result by Emil Artin and Otto Schreier, which characterises orderability of fields in terms of the concept of "formal reality". This theorem will also play a role in §14, which describes Artin's solution of Hilbert's 17th problem.

Let K be a field. A subset P of K is called a *positive cone* of K, if

(i) P is closed under addition and multiplication,

(ii) $K = P \sqcup (-P) \sqcup \{0\}$.

A field K is called *orderable*, if it has at least one positive cone. By an *ordered field* we shall mean a pair (K, P) where K is a field and P is a positive cone of K. If (K, P) is an ordered field, then setting

$$a \leq b :\Longleftrightarrow b - a \in P \cup \{0\} \quad (a, b \in K)$$

defines a total order on K, which is compatible with the field operations in the sense that, for all $a, b, c \in K$,

$$a \leq b \implies a + c \leq b + c, \tag{9.28}$$
$$a \leq b \text{ and } c \geq 0 \implies ac \leq bc. \tag{9.29}$$

Again, we have

$$P = \{a \in K : a > 0\}, \tag{9.30}$$

and conversely, if K is a field totally ordered by $\leq$ such that (9.28) and (9.29) hold, then (9.30) defines a positive cone of K.

By Proposition 9.1, an ordered field (K,P) has characteristic 0, hence contains a copy of the rationals as prime field; moreover, we have $1 \in P$, thus all positive rationals are in P.

Call a field K *formally real*, if -1 cannot be expressed as a sum of squares. This is easily seen to be equivalent to the condition that

$$a_1^2 + a_2^2 + \cdots + a_n^2 = 0 \Longrightarrow a_1 = a_2 = \cdots = a_n = 0, \quad (a_1, \ldots, a_n \in K); \tag{9.31}$$

cf. Exercise 3 below. The following result, which goes back to Artin and Schreier, provides a purely algebraic characterisation of orderable fields.

Proposition 9.2. (Artin and Schreier [10]) *A field K is orderable if, and only if, K is formally real.*

Proof. An ordered field is necessarily formally real, since, by (9.28) and (9.29), every square and thus also every sum of squares is ≥ 0, while $-1 < 0$.

Conversely, suppose that the field K is formally real, and let

$$S := \left\{ s = \sum_{i=1}^{n} a_i^2 : n \in \mathbb{N} - \{0\}, a_1, \ldots, a_n \in K, s \neq 0 \right\}$$

be the set of all non-zero sums of squares in K. Since K is formally real,

$$S = \left\{ s = \sum_{i=1}^{n} a_i^2 : n \in \mathbb{N} - \{0\}, a_1, \ldots, a_n \in K - \{0\} \right\}.$$

Let

$$\mathscr{Q} := \{A \subseteq K : A \text{ satisfies (i)}, 0 \notin A, S \subseteq A\}.$$

We have $\mathscr{Q} \neq \emptyset$ since $S \in \mathscr{Q}$, and $\mathscr{Q}$ is clearly inductively ordered under inclusion, hence, by Zorn's Lemma, has a maximal element P; see Exercises 4 and 5 below. We want to show that P also satisfies Property (ii) of a positive cone in a field, implying that P is a positive cone for K, and that K is orderable, as claimed. Suppose this is false; that is, there exists some $w \in K - \{0\}$ such that $w \notin P$ and $-w \notin P$. Let

$$P^* := \{s + wt : s \in P, t \in P \cup \{0\}\}.$$

Given two elements $s_1 + wt_1, s_2 + wt_2 \in P^*$, then

$$(s_1 + wt_1) + (s_2 + wt_2) = (s_1 + s_2) + w(t_1 + t_2) \in P^*$$

and

$$(s_1 + wt_1)(s_2 + wt_2) = (s_1 s_2 + t_1 t_2 w^2) + w(s_1 t_2 + s_2 t_1) \in P^*$$

since P is closed under addition and multiplication and contains non-zero squares. Hence, P^* satisfies Condition (i). Also, $S \subseteq P \subseteq P^*$, so $P^* \supseteq S$. Next, note that $2w \neq 0$, for otherwise K would have characteristic 2, implying $1^2 = -1$, contradicting our hypothesis that K is formally real. Since $1 \in S \subseteq P$ and $w, -w \notin P$ by assumption, we have $-w \neq 1$, i.e., $w + 1 \neq 0$, and so

$$w = \Big(\frac{2w}{w+1}\Big)^2 + w\Big(\frac{w-1}{w+1}\Big)^2 \in P^*.$$

Since $w \notin P$, P is properly contained in P^*, hence $0 \in P^*$ by maximality of P; that is, we can find $a, b \in P$ such that $a + wb = 0$. The last equation implies that

$$-w = \frac{a}{b} = \big(\frac{1}{b}\big)^2 ab \in P,$$

contradicting our assumption that $-w \notin P$. Hence, P satisfies Condition (ii) as well, and K is orderable as claimed. □

More on formally real fields and Artin's and Schreier's abstract approach to real algebra may be found in §14.1, the principal aim of §14 being a description of Artin's solution to Hilbert's 17th problem.

Exercises for §9.

1. Let (G, P) be an ordered abelian group.
   (i) Show that the binary relation $\leq$ defined in (9.23) is a total order on G satisfying Equations (9.24) and (9.25).
   (ii) Given a total order $\leq$ on the abelian group G satisfying Condition (9.24), show that Equation (9.25) defines a positive cone on G.
2. Show that the set $\mathscr{Q}$ defined in (9.26), when ordered by inclusion, is inductively ordered.
3. Show that a field K is formally real if, and only if, it satisfies Condition (9.31).
4. Show that the set S defined in the proof of Proposition 9.2 is closed under the field operations of addition and multiplication.
5. Show that the set $\mathscr{Q}$ defined in the proof of Proposition 9.2 is inductively ordered under inclusion.
6. Without appeal to Proposition 9.2, show that a formally real field has characteristic zero.
7. Call an ordered field K *archimedean*[11] if, given any element $a \in K$, there exists some $n \in \mathbb{N}$ such that $n > a$. Moreover, a subfield $K' \leq K$ is termed *dense in* K, if between any two elements of K there exists an element of

[11]The "Axiom of Archimedes" in geometry says the following: Starting from a given point P ("origin"), a given line segment PQ ("unity segment") may always be laid off in the direction PR of a point R on the line $\overline{PQ}$ defined by P and Q, such that the last end point lies beyond the given point R.

K'; that is, for $a, b \in K$ with $a < b$ say, we have $a \leq \alpha \leq b$ with some $\alpha \in K'$. Show the following.

Proposition 9.3. *An ordered field K is archimedean if, and only if, the prime field $\mathbb{Q}$ of K is dense in K.*

8. (i) Show that the ring $\mathbb{Z}$ of integers allows precisely one ordering compatible with its arithmetic.

(ii) Let K be the field of fractions of an integral domain R, and suppose that R carries an ordering compatible with its arithmetic. Show that K has precisely one compatible ordering which extends the ordering of R.[12]

(iii) Deduce that the field $\mathbb{Q}$ of rational numbers carries precisely one ordering compatible with its arithmetic.

9. Let K be an ordered field, and let $K[t]$ be the ring of polynomials over K. Call a polynomial $f(t) \in K[t]$ positive, if $f(t) \neq 0$ and the leading coefficient of $f(t)$ is positive in K. Show that the ordering of the quotient field $K(t)$ extending the ordering of $K[t]$ just defined is non-archimedean. Here, $K(t)$ results from $K[t]$ by the field of fractions construction alluded to in Footnote no. 10.

[12] For the field of fractions construction see, for instance, Section 6.2 in [23].

§10. Subdirect decomposition of algebras

The main result of the present section is Theorem 10.18, a famous result by Garrett Birkhoff, who was one of the pioneers of universal algebra. As a warm-up, we start out in Part (A) by considering the subdirect decomposition of rings, a very special (but important) case of Birkhoff's result. Part (B) then offers a quick introduction to some of the most basic concepts and ideas of Universal Algebra, culminating in the proof of Birkhoff's subdirect decomposition theorem mentioned above. The reader who wants to know more about this part of mathematics which, among other things, has important connections to Logic and Model Theory, is pointed to the book [33] by George Grätzer, which provides an excellent and comprehensive introduction, as well as (in the appendices of the second edition) an up-to-date guide to recent developments.

(A) **The Case of Rings.** Let $\mathscr{R} = \{R_\nu\}_{\nu\in N}$ be a family of rings, and let R be the set of all choice functions over $\mathscr{R}$. Thus, explicitly, an element $r \in R$ is a map $r : N \to \bigcup_\nu R_\nu$ such that $r(\nu) \in R_\nu$ for each $\nu \in N$. For $r \in R$ and $\nu \in N$, we write r_ν for the image of ν under the choice function r, and define addition and multiplication in R via

$$\left.\begin{array}{ll}(r+s)_\nu & := r_\nu + s_\nu \\ (rs)_\nu & := r_\nu s_\nu\end{array}\right\} \quad (r,s \in R),$$

where the right-hand side is computed in the ring R_ν. In this way, R becomes a ring, called the *complete direct sum* of the family of rings $\mathscr{R} = \{R_\nu\}_{\nu\in N}$, denoted by

$$R = \bigoplus_{\nu\in N}^{*} R_\nu .$$

For a fixed index ν, mapping an element $r^{(\nu)} \in R_\nu$ to the function

$$\bar{r}^{(\nu)} : N \longrightarrow \bigcup_{\mu\in N} R_\mu$$

given by

$$\bar{r}^{(\nu)}_\mu = \begin{cases} r^{(\nu)}, & \mu = \nu \\ 0 \in R_\mu, & \mu \neq \nu \end{cases} \quad (\mu \in N)$$

yields an embedding of rings $\iota_\nu : R_\nu \to R$, so that R_ν may be viewed as a subring of R. In fact, viewed in this way, R_ν is a (two-sided) ideal of R. However, in general, R is not generated as an ideal by the collection of ideals R_ν, since, for $|N| \geq \aleph_0$, not every element of R can be written as a finite sum of elements in $\bigcup_{\nu\in N} R_\nu$. The ideal

$$R_0 := \bigoplus_{\nu\in N} R_\nu = \{r \in R : r_\nu = 0 \text{ for all but finitely many } \nu\}$$

generated in R by the ideals R_ν is called the *discrete direct sum* of the rings R_ν. The rings $R_\nu\,(\nu \in N)$ are called the *components* or *summands* of R, respectively, R_0. Indeed, for each fixed $\nu \in N$, we have

$$R \cong R_\nu \oplus \bigoplus_{\substack{\mu\in N\\ \mu\neq\nu}}^{*} R_\mu \quad \text{and} \quad R_0 \cong R_\nu \oplus \bigoplus_{\substack{\mu\in N\\ \mu\neq\nu}} R_\mu,$$

i.e., each R_ν is a direct summand of R, respectively R_0.

If r is an element of R, respectively R_0, then $r_\nu\,(\nu \in N)$ is called the ν-*th component* of r (with respect to the complete, respectively discrete, direct sum). The element, whose ν-th component is the zero element of R_ν for every $\nu \in N$, is the zero element of R (and hence of R_0). For each fixed $\nu \in N$, mapping $r \mapsto r_\nu$ defines a surjective ring homomorphism $\pi_\nu : R \to R_\nu$, called the ν-*th projection* of R. If N is finite, $N = \{1,2,\ldots,n\}$ say, then $R = R_0$; that is, the complete and discrete sum of the rings $R_1,\ldots,R_n$ coincide.

Let R be the complete direct sum of the family of rings $\{R_\nu\}_{\nu\in N}$, and let $S \leq R$ be a subring. If $\pi_\nu(S) = R_\nu$ for every $\nu \in N$, then S is called a *subdirect sum* of the rings R_ν. In particular, both R and R_0 are subdirect sums of the R_ν, as is any subring of R containing R_0. Hence, a subdirect sum of rings R_ν is not uniquely determined by the components R_ν. If a ring S is isomorphic to a subdirect sum of rings R_ν, then we say that S can be *represented* as a subdirect sum of the rings R_ν. The R_ν are then called the *components* of S, and the map π_ν is called the *canonical projection* of S onto R_ν. The following result is of fundamental importance in ring theory.

Proposition 10.1. *A ring S can be represented as a subdirect sum of the rings $R_\nu\,(\nu \in N)$ if, and only if, for every $\nu \in N$ there exists an ideal A_ν in S such that*

$$S/A_\nu \cong R_\nu \tag{10.32}$$

and

$$\bigcap_{\nu\in N} A_\nu = 0. \tag{10.33}$$

Proof. Let ψ be an isomorphism of S onto a subdirect sum S' of the rings R_ν, and let A'_ν be the kernel of the surjective map $\pi_\nu|_{S'} : S' \to R_\nu$. Then $S'/A'_\nu \cong R_\nu$ and $\bigcap_{\nu\in N} A'_\nu = 0$. Defining $A_\nu = \psi^{-1}(A'_\nu)$ gives a family of ideals in S such that (10.32) and (10.33) hold.

Conversely, suppose that S contains ideals $A_\nu\,(\nu \in N)$ such that (10.32) and (10.33) are satisfied, and choose isomorphisms $\psi_\nu : S/A_\nu \to R_\nu$. Define a map $\psi : S \to R$, where $R = \bigoplus^{*}_{\nu\in N} R_\nu$ is the complete direct sum of the rings R_ν, via

$$\big(\psi(s)\big)_\nu := \psi_\nu(s + A_\nu), \quad (s \in S,\ \nu \in N).$$

For $\nu \in N$ and $s_1, s_2 \in S$,

$$\begin{aligned}(\psi(s_1+s_2))_\nu &= \psi_\nu((s_1+s_2)+A_\nu)\\ &= \psi_\nu(s_1+A_\nu)+\psi_\nu(s_2+A_\nu)\\ &= (\psi(s_1)+\psi(s_2))_\nu,\end{aligned}$$

so that

$$\psi(s_1+s_2)=\psi(s_1)+\psi(s_2).$$

Similarly, we have, for $\nu \in N$ and $s_1, s_2 \in S$,

$$\begin{aligned}(\psi(s_1 s_2))_\nu &= \psi_\nu(s_1 s_2+A_\nu)\\ &= \psi_\nu(s_1+A_\nu)\cdot\psi_\nu(s_2+A_\nu)\\ &= (\psi(s_1)\psi(s_2))_\nu,\end{aligned}$$

so that

$$\psi(s_1 s_2)=\psi(s_1)\psi(s_2),$$

showing that ψ is a homomorphism of rings. Next, suppose that $\psi(s_1) = \psi(s_2)$. Then

$$\psi_\nu(s_1+A_\nu)=\psi_\nu(s_2+A_\nu)\quad(\nu \in N),$$

thus $s_1 - s_2 \in A_\nu$ for all $\nu \in N$, implying that $s_1 = s_2$ by (10.33). Hence, ψ is an embedding of the ring S into R. Finally, since each ψ_ν is surjective,

$$\pi_\nu(\psi(S)) = \{\psi_\nu(s+A_\nu) : s \in S\} = R_\nu \quad (\nu \in N),$$

so $\psi(S)$ is a subdirect sum of the rings R_ν. □

As a first application of Proposition 10.1, we find that the ring $\mathbb{Z}$ of integers is represented as a subdirect sum of the family of finite prime fields $(K_p)_{p\text{ prime}}$; see Exercise 1 below.

Every ring can be represented as a subdirect sum of some family of rings. Indeed, given a ring R and a positive integer n, the set S of n-tuples $(r, \ldots, r)$ with $r \in R$ is a subring of the direct sum $R_1 \oplus \cdots \oplus R_n$ of n copies $R_1, \ldots, R_n$ of R, $R \cong S$, and $\pi_\nu(S) = R_\nu$ for $\nu = 1, \ldots, n$, that is, S is a subdirect sum of the rings $R_1, \ldots, R_n$. Call a representation of a ring as a subdirect sum *trivial* if at least one of the canonical projections is an isomorphism. By what we have just seen, every ring has trivial representations as a subdirect sum. A ring all of whose representations as subdirect sum are trivial, is called *subdirectly irreducible*. We have the following characterisation of subdirectly irreducible rings; cf. Exercise 3 below.

Proposition 10.2. *A ring S is subdirectly irreducible if, and only if, the intersection of all non-zero ideals in S is non-zero.*

The importance of subdirectly irreducible rings is underlined by the following observation of Birkhoff.

Proposition 10.3. (Birkhoff [12]) *Every ring is isomorphic to a subdirect sum of subdirectly irreducible rings.*

Proof. If $R = 0$, then R itself is subdirectly irreducible, so we can assume that $R \neq 0$. Given a non-zero element $a \in R$, consider the set $\mathcal{Q}_a$ of all ideals $J \leq R$ not containing a. We have $\mathcal{Q}_a \neq \emptyset$, since the zero ideal is in $\mathcal{Q}_a$, and $\mathcal{Q}_a$ is inductively ordered under inclusion; cf. Exercise 4. By Zorn's Lemma, the ordered set $(\mathcal{Q}_a, \subseteq)$ has a maximal element, J_a say. In this way, we associate to every element $a \neq 0$ of R an ideal J_a such that $a \notin J_a$ and, for every ideal $J \leq R$, we have

$$J_a \subset J \Longrightarrow a \in J.$$

By construction,

$$\bigcap_{a \in R - \{0\}} J_a = 0,$$

so, by Proposition 10.1, R is isomorphic to a subdirect sum of the rings R/J_a for $a \in R - \{0\}$. It remains to show that each quotient ring R/J_a is subdirectly irreducible. Denote by $^- : R \to R/J_a$ the canonical projection associated with the ideal J_a, and let $\bar{J}$ be an arbitrary non-zero ideal of R/J_a. Then the full pre-image of $\bar{J}$ is an ideal J in R such that $J \supset J_a$, so $a \in J$, and hence $\bar{a} \in \bar{J}$; that is, every non-zero ideal of R/J_a contains the element $\bar{a} \neq 0$ of R/J_a. It follows that

$$\bigcap_{0 \neq \bar{J} \leq R/J_a} \bar{J} \neq 0,$$

so R/J_a is subdirectly irreducible by Proposition 10.2. □

(B) **A Generalisation.** Proposition 10.3 is a very special (though important) case of a famous theorem by Garrett Birkhoff, which we shall discuss next, starting with some definitions. Along the way, the reader will meet some of the (rather abstract) concepts and arguments typical of universal algebra (as this part of mathematics is usually referred to).

1. *Finitary operations and (universal) algebras.* Given a set A and a non-negative integer n, a map $f : A^n \to A$ is called an *n-ary operation on A*. In particular, a 0-ary (nullary) operation may be thought of as a constant (a distinguished element of A).

By an *algebra*, we mean a pair $\mathfrak{A} = (A; F)$, where A is a non-empty set, and F is a (possibly empty) family of operations on A; that is, $F = (f_i)_{i \in I}$ and $f_i : A^{n_i} \to A$. The family of non-negative integers $(n_i)_{i \in I}$ is called the *type* of the algebra $\mathfrak{A}$, and is denoted by $\tau(\mathfrak{A})$. An algebra $\mathfrak{A}$ is termed *unary* if $\tau(\mathfrak{A}) = (1, 1, \ldots, 1)$; that is, if all its operations are maps on A. In this language, a group G, for instance, is an algebra of type $\tau(G) = (2, 0)$ (one binary operation $\cdot$ and one distinguished element $1 \in G$, the neutral element of multiplication). Similarly, a ring R is an algebra of type $\tau(R) = (2,2,0)$, while a ring S with identity element is of type $\tau(S) = (2,2,0,0)$. The mono-unary algebras $\mathfrak{A}$ occurring in Novotný's Theorem 7.27 are of type $\tau(\mathfrak{A}) = (1)$.

2. *Subalgebras and homomorphisms.* Let $\mathfrak{A} = (A;F)$ be an algebra, and let $B \subseteq A$ be a non-empty subset. Then B is called a *subalgebra* of $\mathfrak{A}$ if it is closed under all operations of $\mathfrak{A}$; that is, if

$$f_i(b_0,b_1,\ldots,b_{n_i-1}) \in B, \quad (i \in I,\, (b_0,b_1,\ldots,b_{n_i-1}) \in B^{n_i}).$$

With slight abuse of notation, we denote this subalgebra of $\mathfrak{A}$ as $\mathfrak{B} = (B;F)$. Clearly, the intersection of a family of subalgebras $\{\mathfrak{B}_j\}_{j\in J}$ of $\mathfrak{A} = (A;F)$ is again a subalgebra, provided that $\bigcap_{j\in J} B_j \neq \emptyset$, where $\mathfrak{B}_j = (B_j;F)$. Thus, it makes sense to define the subalgebra $(\langle H\rangle;F)$ of $\mathfrak{A}$ *generated by a non-empty subset* $H \subseteq A$ as the smallest subalgebra of $\mathfrak{A}$ containing H; cf. Exercise 5.

If $\mathfrak{A} = (A;F)$ and $\mathfrak{B} = (B;G)$ are algebras of the same type $\tau = (n_i)_{i\in I}$, then a map $\varphi : A \to B$ is called a *homomorphism* from $\mathfrak{A}$ to $\mathfrak{B}$, denoted by $\varphi : \mathfrak{A} \to \mathfrak{B}$, if

$$\varphi(f_i(a_0,a_1,\ldots,a_{n_i-1})) = g_i(\varphi(a_0),\varphi(a_1),\ldots,\varphi(a_{n_i-1})),$$
$$(i \in I,\, (a_0,a_1,\ldots,a_{n_i-1}) \in A^{n_i}).$$

We note that, if a nullary operation f_i of the algebra $\mathfrak{A}$ picks out the element a from the underlying set A, while the corresponding nullary operation g_i from $\mathfrak{B}$ picks out the element $b \in B$, then a homomorphism $\varphi : \mathfrak{A} \to \mathfrak{B}$ satisfies $\varphi(a) = b$. In particular, if R,S are rings with identity element, then an algebra homomorphism $\varphi : R \to S$ satisfies the rules

$$\varphi(r_1 + r_2) = \varphi(r_1) + \varphi(r_2), \tag{10.34}$$

$$\varphi(r_1 \cdot r_2) = \varphi(r_1) \cdot \varphi(r_2), \tag{10.35}$$

$$\varphi(1) = 1. \tag{10.36}$$

(10.34)–(10.36) are precisely the defining conditions of a homomorphism in the theory of rings with identity element.

We shall call a homomorphism $\varphi : \mathfrak{A} \to \mathfrak{B}$ a *monomorphism*, if the underlying map $\varphi : A \to B$ is injective. Similarly, $\varphi : \mathfrak{A} \to \mathfrak{B}$ is termed an *epimorphism* if $\varphi : A \to B$ is surjective, and an *isomorphism*, if $\varphi : A \to B$ is a bijection; cf. Exercise 6. Clearly, if $\mathfrak{B} = (B;F)$ is a subalgebra of $\mathfrak{A} = (A;F)$, then the canonical embedding $1_A^B : B \to A$ defines a monomorphism $1_{\mathfrak{A}}^{\mathfrak{B}} : \mathfrak{B} \to \mathfrak{A}$.

3. *Congruence relations, quotients, and projections.* Let $\mathfrak{A} = (A;F)$ be an algebra, and let θ be a binary relation on A, with $(a,b) \in \theta$ written as

$a \equiv b(\theta)$. Then θ is called a *congruence relation* on $\mathfrak{A}$, if it is an equivalence relation on A satisfying, for each $i \in I$ and all tuples $(a_0, a_1, \ldots, a_{n_i-1})$, $(b_0, b_1, \ldots, b_{n_i-1}) \in A^{n_i}$, the *substitution property*

$$a_\nu \equiv b_\nu(\theta)(0 \leq \nu < n_i) \implies f_i(a_0, a_1, \ldots, a_{n_i-1}) \equiv f_i(b_0, b_1, \ldots, b_{n_i-1})(\theta). \tag{10.37}$$

Given an algebra $\mathfrak{A} = (A; F)$ and a congruence relation θ on $\mathfrak{A}$, we may define a new algebra $\mathfrak{A}/\theta = (A/\theta; F)$ as follows: the set A/θ is the natural quotient set

$$A/\theta = \{[a]_\theta : a \in A\},$$

that is, the set of equivalence classes of A modulo θ, with operations defined by

$$f_i([a_0]_\theta, [a_1]_\theta, \ldots, [a_{n_i-1}]_\theta) := [f_i(a_0, a_1, \ldots, a_{n_i-1})]_\theta, \quad (i \in I); \tag{10.38}$$

see Exercise 7. In particular, the operation f_i, considered as operation on $\mathfrak{A}/\theta$ is again n_i-ary, thus $\tau(\mathfrak{A}/\theta) = \tau(\mathfrak{A})$.

Next, consider the map $\varphi_\theta : A \to A/\theta$ sending an element $a \in A$ to its equivalence class $[a]_\theta$ modulo θ. Given an index $i \in I$ and elements $a_0, a_1, \ldots, a_{n_i-1} \in A$, we have, by (10.38), that

$$\begin{aligned} \varphi_\theta(f_i(a_0, a_1, \ldots, a_{n_i-1})) &= [f_i(a_0, a_1, \ldots, a_{n_i-1})]_\theta \\ &= f_i([a_0]_\theta, [a_1]_\theta, \ldots, [a_{n_i-1}]_\theta) \\ &= f_i(\varphi_\theta(a_0), \varphi_\theta(a_1), \ldots, \varphi_\theta(a_{n_i-1})), \end{aligned}$$

which shows that $\varphi_\theta : \mathfrak{A} \to \mathfrak{A}/\theta$ is a homomorphism (more precisely, an epimorphism) of algebras. We call φ_θ the *canonical projection* associated with the congruence relation θ on the algebra $\mathfrak{A}$.

At this point, it might be profitable for the reader to recall that a group epimomorphism $\varphi : G \to H$ induces an isomorphism $G/N \cong H$, where $N = \ker(\varphi)$,[13] as well as an analogous statement for rings, ring epimorphisms, and ideals. Our next result provides a common generalisation of these facts.

Proposition 10.4. *Let $\varphi : \mathfrak{A} \to \mathfrak{B}$ be an epimorphism of universal algebras, and let $\theta = \theta_\varphi$ be the congruence relation on $\mathfrak{A}$ induced by φ; that is,*

$$x \equiv y\,(\theta) :\iff \varphi(x) = \varphi(y), \quad (x, y \in A).$$

Then we have $\mathfrak{A}/\theta \cong \mathfrak{B}$ via the isomorphism sending $[x]_\theta$ to $\varphi(x)$.

Proof. Let $\mathfrak{A} = (A; F)$ and $\mathfrak{B} = (B; G)$, where $\tau(\mathfrak{A}) = (n_i)_{i \in I} = \tau(\mathfrak{B})$. By definition, the set A/θ underlying the algebra $\mathfrak{A}/\theta$ is the collection of all equivalence classes $[x]_\theta$ of A modulo the congruence θ, and the operations of $\mathfrak{A}/\theta$ are given by Equation (10.38). Define a map $\hat{\varphi} : A/\theta \to B$ by setting

$$\hat{\varphi}([x]_\theta) := \varphi(x), \quad (x \in A).$$

[13] This fact is often referred to as the *first isomorphism theorem* for groups.

In view of the definition of θ, the map $\hat{\varphi}$ is well defined, and it is surjective since φ is. Also, if

$$\hat{\varphi}([x]_\theta) = \hat{\varphi}([y]_\theta),$$

then we have $\varphi(x) = \varphi(y)$, so that $x \equiv y\ (\theta)$, and hence $[x]_\theta = [y]_\theta$. This shows that $\hat{\varphi}$ is injective as well, thus a bijection. Moreover, for $i \in I$, we have

$$\begin{aligned} \hat{\varphi}(f_i([x_0]_\theta, [x_1]_\theta, \ldots, [x_{n_i-1}]_\theta)) &= \hat{\varphi}([f_i(x_0, x_1, \ldots, x_{n_i-1})]_\theta) \\ &= \varphi(f_i(x_0, x_1, \ldots, x_{n_i-1})) \\ &= g_i(\varphi(x_0), \varphi(x_1), \ldots, \varphi(x_{n_i-1})) \\ &= g_i(\hat{\varphi}([x_0]_\theta, \hat{\varphi}([x_1]_\theta), \ldots, \hat{\varphi}([x_{n_i-1}]_\theta)). \end{aligned} \tag{10.39}$$

Here, we have used the definition (10.38) of the operations on $\mathfrak{A}/\theta$ in the first step, the definition of the map $\hat{\varphi}$ in Steps 2 and 4, and the fact that, by hypothesis, φ is an algebra homomorphism in Step 3. The computation (10.39) shows that $\hat{\varphi}$ is a homomorphism of algebras, thus an isomorphism. □

4. *Direct products of algebras.* For $\nu \in N$, let $\mathfrak{A}_\nu = (A_\nu; F_\nu)$, assuming that these algebras $\mathfrak{A}_\nu$ are all of the same type $\tau = (n_i)_{i\in I}$, where $F_\nu = (f_i^{(\nu)})_{i\in I}$. We form the Cartesian product $P = \prod_{\nu\in N} A_\nu$, whose elements are the choice functions $p : N \to \bigcup_{\nu\in N} A_\nu$, where $p(\nu) \in A_\nu$ for each $\nu \in N$. We may then define operations f_i on P by setting

$$f_i(p_0, p_1, \ldots, p_{n_i-1})(\nu) := f_i^{(\nu)}(p_0(\nu), p_1(\nu), \ldots, p_{n_i-1}(\nu)), \ (i \in I,\ \nu \in N). \tag{10.40}$$

In this way, we obtain an algebra $\mathfrak{P} = (P; F)$, which is again of type τ. This algebra is called the *direct product* of the algebras $\mathfrak{A}_\nu$, and is denoted by $\mathfrak{P} = \prod_{\nu\in N} \mathfrak{A}_\nu$. The direct product comes equipped with a family of epimorphisms

$$\pi_\mu : \prod_{\nu\in N} \mathfrak{A}_\nu \longrightarrow \mathfrak{A}_\mu, \quad (\mu \in N)$$

given by $\pi_\mu(p) := p(\mu)$; see Exercise 10. These epimorphisms π_μ are called the *canonical projections* associated with the direct product $\mathfrak{P}$. Denote by θ_μ the congruence relation induced by the homomorphism π_μ; that is

$$p \equiv q\,(\theta_\mu) :\Longleftrightarrow \pi_\mu(p) = \pi_\mu(q), \quad (p, q \in \mathfrak{P});$$

cf. Exercise 8. By Proposition 10.4, we have

$$\Big(\prod_{\nu\in N} \mathfrak{A}_\nu\Big)/\theta_\mu \cong \mathfrak{A}_\mu, \quad (\mu \in N). \tag{10.41}$$

5. *The structure lattice of a universal algebra.* Given an algebra $\mathfrak{A} = (A; F)$, we denote by $\Theta(\mathfrak{A})$ the poset whose underlying set consists of all

congruence relations on $\mathfrak{A}$, subject to the partial order relation $\leq$ defined by

$$\theta \leq \theta' :\Longleftrightarrow [x \equiv y(\theta) \Rightarrow x \equiv y(\theta')]$$
$$\Longleftrightarrow \theta \subseteq \theta' \text{ as subsets of } A^2.$$

The relations $\mathbf{0}$ and $\mathbf{1}$ given by

$$x \equiv y(\mathbf{0}) :\Longleftrightarrow x = y \text{ (equality relation)}$$
$$x \equiv y(\mathbf{1}) \text{ for all } x, y \in A \text{ (all relation)}$$

are congruences such that

$$\mathbf{0} \leq \theta \leq \mathbf{1}, \quad (\theta \in \Theta(\mathfrak{A})).$$

Definition 10.5. By a *translation* of an algebra $\mathfrak{A} = (A : F)$ we mean a unary operation $g_{i,\mathbf{c},k}$ on A of the form

$$g_{i,\mathbf{c},k}(x) = f_i(c_1, \ldots, c_{k-1}, x, c_{k+1}, \ldots c_{n_i}), \quad (i \in I,\ 1 \leq k \leq n_i,\ x \in A),$$

for a suitable tuple of constants $\mathbf{c} = (c_1, \ldots, c_{k-1}, c_{k+1}, \ldots, c_{n_i})$.

The following observation will be useful in the proof of the fundamental Proposition 10.8 below.

Lemma 10.6. *Let $\mathfrak{A} = (A; F)$ be an algebra. Then an equivalence relation on A is a congruence relation on $\mathfrak{A}$ if, and only if, it satisfies the substitution property for every translation of $\mathfrak{A}$.*

Proof. This is left to the reader; cf. Exercise 12. □

Definition 10.7. Let L be a lattice, and let $K \subseteq L$. Then K is called a *closed sublattice* of L, if $\inf_L(X)$ and $\sup_L(X)$ are contained in K for all $X \subseteq K$.

Proposition 10.8. *The congruence relations on an algebra $\mathfrak{A} = (A; F)$ form a closed sublattice $\Theta(\mathfrak{A})$ of the complete lattice $\mathfrak{E}(A)$ of all equivalence relations on the set A underlying the algebra $\mathfrak{A}$.*

Proof. Let $\mathscr{F} = (\theta_\beta)_{\beta \in B}$ be any family of congruence relations on $\mathfrak{A}$, and set

$$\theta := \bigcap_{\beta \in B} \theta_\beta \quad \text{and} \quad \hat{\theta} := \left\langle \bigcup_{\beta \in B} \theta_\beta \right\rangle,$$

where $\langle\rangle$ denotes transitive closure. The equivalence relations θ and $\hat{\theta}$ are the infimum and supremum of the θ_β, respectively, in the complete lattice $\mathfrak{E}(A)$; cf. Exercise 10.13. We need to show that θ and $\hat{\theta}$ are congruence relations on $\mathfrak{A}$.

By Lemma 10.5, it suffices to check the substitution property for unary operations $g_\gamma(x)$. If $x \equiv y\ (\theta)$, then

$$x \equiv y\ (\theta_\beta), \quad (\beta \in B)$$

by definition of θ. As each θ_β has the substitution property, this implies

$$g_\gamma(x) \equiv g_\gamma(y)\ (\theta_\beta), \quad (\beta \in B),$$

which, in turn, gives $g_\gamma(x) \equiv g_\gamma(y)\ (\theta)$, showing that θ satisfies the substitution property for translations. Thus, by Lemma 10.5, θ is a congruence relation, as required.

We now turn to the relation $\hat{\theta}$. If $x \equiv y\ (\hat{\theta})$, we have a chain of elements $x = z_0, z_1, \ldots, z_m = y$ in A, and corresponding indices $\beta(j) \in B$, such that $z_{j-1} \equiv z_j\ (\theta_{\beta(j)})$ for $j = 1, 2, \ldots, m$. By the substitution property of the congruence relations $\theta_{\beta(j)}$, we infer that, for any translation g_γ,

$$g_\gamma(z_{j-1}) \equiv g_\gamma(z_j)\ (\theta_{\beta(j)}), \quad (1 \le j \le m). \tag{10.42}$$

However, (10.42), applied to the chain

$$g_\gamma(x) = g_\gamma(z_0), g_\gamma(z_1), \ldots, g_\gamma(z_m) = g_\gamma(y)$$

yields $g_\gamma(x) \equiv g_\gamma(y)\ (\hat{\theta})$, so that $\hat{\theta}$ has the substitution property for each translation g_γ; thus is a congruence relation by Lemma 10.5. □

Definition 10.9. The complete lattice $\Theta(\mathfrak{A})$ is termed the *structure lattice* of the algebra $\mathfrak{A}$.

6. *Lifting congruence relations.*

Let $\mathfrak{A} = (A; F)$ be a universal algebra, let θ be a (fixed) congruence relation, and let $\overline{\mathfrak{A}} = \mathfrak{A}/\theta$ be the quotient algebra associated with the pair $(\mathfrak{A}, \theta)$. Our aim is to obtain a complete description of the structure lattice $\Theta(\overline{\mathfrak{A}})$ in terms of the lattice $\Theta(\mathfrak{A})$. To this end, we introduce two easy constructions.

(a) Let $\hat{\theta}$ be a congruence relation on $\mathfrak{A}$ such that $\hat{\theta} \ge \theta$. Then we define a binary relation $\hat{\theta}/\theta$ on $\overline{\mathfrak{A}}$ by setting

$$[a]_\theta \equiv [b]_\theta\ (\hat{\theta}/\theta) :\Longleftrightarrow a \equiv b\ (\hat{\theta}), \quad (a, b \in A).$$

Lemma 10.10. *$\hat{\theta}/\theta$ is a well-defined congruence relation on the quotient algebra $\overline{\mathfrak{A}}$.*

The proof is left to the reader; cf. Exercise 10.14.

(b) Given a congruence $\bar{\theta}$ on the quotient algebra $\mathfrak{A}/\theta$, we define a binary relation $\hat{\hat{\theta}}$ on $\mathfrak{A}$ via

$$a \equiv b\ (\hat{\hat{\theta}}) :\Longleftrightarrow [a]_\theta \equiv [b]_\theta\ (\bar{\theta}).$$

Lemma 10.11. *$\hat{\bar{\theta}}$ is a congruence relation on $\mathfrak{A}$, and $\hat{\bar{\theta}} \geq \theta$.*

Proof. Clearly, $\hat{\bar{\theta}}$ is an equivalence relation on A, since $\bar{\theta}$ is an equivalence relation on A/θ by hypothesis. Next, assume that

$$a_\nu \equiv b_\nu \; (\hat{\bar{\theta}}), \quad (i \in I, 0 \leq \nu < n_i).$$

Then, by definition of $\hat{\bar{\theta}}$, we have

$$[a_\nu]_\theta \equiv [b_\nu]_\theta \; (\bar{\theta}), \quad (i \in I, 0 \leq \nu < n_i);$$

thus,

$$f_i([a_0]_\theta, [a_1]_\theta, \dots, [a_{n_i-1}]_\theta) \equiv f_i([b_0]_\theta, [b_1]_\theta, \dots, [b_{n_i-1}]_\theta) \; (\bar{\theta}),$$

since $\bar{\theta}$ satisfies the substitution rule by hypothesis. By definition of the algebra $\mathfrak{A}/\theta$, the last displayed congruence may be rewritten in the form

$$[f_i(a_0, a_1, \dots, a_{n_i-1})]_\theta \equiv [f_i(b_0, b_1, \dots, b_{n_i-1})]_\theta \; (\bar{\theta}),$$

and the definition of $\hat{\bar{\theta}}$ gives

$$f_i(a_0, a_1, \dots, a_{n_i-1}) \equiv f_i(b_0, b_1, \dots, b_{n_i-1}) \; (\hat{\bar{\theta}}),$$

as desired. Hence, $\hat{\bar{\theta}}$ is a congruence relation on $\mathfrak{A}$. Furthermore, if $a \equiv b \; (\theta)$, then $[a]_\theta = [b]_\theta$, which implies $[a]_\theta \equiv [b]_\theta \; (\bar{\theta})$ by reflexivity of $\bar{\theta}$, and thus $a \equiv b \; (\hat{\bar{\theta}})$ by definition of $\hat{\bar{\theta}}$. This shows that $\theta \leq \hat{\bar{\theta}}$, as required. □

We are now in position to describe, up to isomorphism, the structure lattice of a quotient algebra.

Proposition 10.12. *Let $\mathfrak{A}$ be an algebra, let θ be a congruence relation on $\mathfrak{A}$, and set*

$$\theta^+ := \{\hat{\theta} \in \Theta(\mathfrak{A}) : \hat{\theta} \geq \theta\}.$$

Then we have

$$\Theta(\mathfrak{A}/\theta) \cong (\theta^+, \leq). \tag{10.43}$$

In the notation of Lemmas 10.10 *and* 10.11, *the isomorphism* (10.43) *is given by $\bar{\theta} \mapsto \hat{\bar{\theta}}$, while its inverse is effected by the assignment $\hat{\theta} \mapsto \hat{\theta}/\theta$.*

Proof. Let

$$\Theta(\mathfrak{A}/\theta) \xrightarrow{\;\varphi\;} (\theta^+, \leq) \xrightarrow{\;\psi\;} \Theta(\mathfrak{A}/\theta)$$

be the maps involved in (10.43). Then we need to establish the following four facts:

(i) $\psi \circ \varphi = 1_{\Theta(\mathfrak{A}/\theta)}$,

(ii) $\varphi \circ \psi = 1_{\theta^+}$,

(iii) $\bar{\theta}_1 \leq \bar{\theta}_2 \implies \varphi(\bar{\theta}_1) \leq \varphi(\bar{\theta}_2)$,

(iv) $\hat{\theta}_1 \leq \hat{\theta}_2 \Longrightarrow \psi(\hat{\theta}_1) \leq \psi(\hat{\theta}_1)$.

For $a, b \in A$ and $\hat{\theta} \in \theta^+$, we have

$$a \equiv b\ (\widehat{\hat{\theta}/\theta}) \Longleftrightarrow [a]_\theta \equiv [b]_\theta\ (\hat{\theta}/\theta) \Longleftrightarrow a \equiv b\ (\hat{\theta}),$$

so that

$$\varphi(\psi(\hat{\theta})) = \widehat{\hat{\theta}/\theta} = \hat{\theta} = 1_{\theta^+}(\hat{\theta}), \quad (\hat{\theta} \in \theta^+),$$

whence item (ii). Similarly, item (i) follows since, for $a, b \in A$ and $\bar{\theta} \in \Theta(\mathfrak{A}/\theta)$, we have

$$[a]_\theta \equiv [b]_\theta\ (\bar{\theta}) \Longleftrightarrow a \equiv b\ (\hat{\bar{\theta}}) \Longleftrightarrow [a]_\theta \equiv [b]_\theta\ (\hat{\bar{\theta}}/\theta),$$

so that

$$\psi(\varphi(\bar{\theta})) = \hat{\bar{\theta}}/\theta = \bar{\theta} = 1_{\Theta(\mathfrak{A}/\theta)}(\bar{\theta}), \quad (\bar{\theta} \in \Theta(\mathfrak{A}/\theta)).$$

Next, for $a, b \in A$ and $\bar{\theta}_1, \bar{\theta}_2 \in \Theta(\mathfrak{A}/\theta)$ such that $\bar{\theta}_1 \leq \bar{\theta}_2$, we get

$$a \equiv b\ (\hat{\bar{\theta}}_1) \Longleftrightarrow [a]_\theta \equiv [b]_\theta\ (\bar{\theta}_1) \Longrightarrow [a]_\theta \equiv [b]_\theta\ (\bar{\theta}_2) \Longleftrightarrow a \equiv b\ (\hat{\bar{\theta}}_2),$$

so that

$$\varphi(\bar{\theta}_1) = \hat{\bar{\theta}}_1 \leq \hat{\bar{\theta}}_2 = \varphi(\bar{\theta}_2)$$

as desired, whence item (iii). Finally, for $a, b \in A$ and $\hat{\theta}_1, \hat{\theta}_2 \in \theta^+$ such that $\hat{\theta}_1 \leq \hat{\theta}_2$, we find that

$$[a]_\theta \equiv [b]_\theta\ (\hat{\theta}_1/\theta) \Longleftrightarrow a \equiv b\ (\hat{\theta}_1) \Longrightarrow a \equiv b\ (\hat{\theta}_2) \Longleftrightarrow [a]_\theta \equiv [b]_\theta\ (\hat{\theta}_2/\theta),$$

which implies

$$\psi(\hat{\theta}_1) = \hat{\theta}_1/\theta \leq \hat{\theta}_2/\theta = \psi(\hat{\theta}_2)$$

as required, whence (iv). This finishes the proof of Proposition 10.12. □

7. *Subdirect products.*

Definition 10.10. (i) A subalgebra $\mathfrak{C} = (S; F)$ of a direct product $\mathfrak{A} = \prod_{\nu \in N} \mathfrak{A}_\nu$, where $\mathfrak{A}_\nu = (A_\nu, F_\nu)$, is called a *subdirect product* of the algebras $\mathfrak{A}_\nu$ if, given an index $\nu \in N$ and any element $x_\nu \in A_\nu$, there exists some element $c \in S$ having x_ν for its component in $\mathfrak{A}_\nu$; that is if, and only if, the restrictions

$$\pi_\nu\big|_{\mathfrak{C}} : \mathfrak{C} \longrightarrow \mathfrak{A}_\nu, \quad (\nu \in N)$$

of the canonical projections $\pi_\nu : \mathfrak{A} \to \mathfrak{A}_\nu$ are all surjective.

(ii) An epimorphism $\varphi : \mathfrak{B} \to \mathfrak{C}$ of an algebra $\mathfrak{B}$ onto a subdirect product of algebras $\mathfrak{A}_\nu$ is termed a *representation* of $\mathfrak{B}$ as a subdirect product of the $\mathfrak{A}_\nu$.

For subdirect representations of an algebra we have the following basic result.

Proposition 10.11. *If* $\varphi : \mathfrak{A} \to \mathfrak{C}$ *is a representation of the algebra* $\mathfrak{A}$ *as a subdirect product of algebras* $\mathfrak{A}_\nu$, *then we have* $\mathfrak{C} \cong \mathfrak{A}/\inf_\nu \theta_\nu$, *where* θ_ν *is the congruence relation on* $\mathfrak{A}$ *induced by the epimorphism* $\pi_\nu\big|_{\mathfrak{C}} \circ \varphi : \mathfrak{A} \to \mathfrak{A}_\nu$. *Conversely, each choice of congruence relations* θ_ν *on the algebra* $\mathfrak{A}$ *yields a representation of* $\mathfrak{A}$ *as a subdirect product* $\mathfrak{C} = \mathfrak{A}/\inf_\nu \theta_\nu$ *of the algebras* $\mathfrak{A}_\nu = \mathfrak{A}/\theta_\nu$.

Proof. For each index ν, the epimorphism

$$\pi_\nu\big|_{\mathfrak{C}} \circ \varphi : \mathfrak{A} \longrightarrow \mathfrak{A}_\nu$$

defines a congruence θ_ν on $\mathfrak{A}$ via

$$x \equiv y\ (\theta_\nu) :\Longleftrightarrow \pi_\nu(\varphi(x)) = \pi_\nu(\varphi(y)),$$

such that $\mathfrak{A}/\theta_\nu \cong \mathfrak{A}_\nu$ (see the proof of (10.41)). Moreover, two elements of $\mathfrak{A}$ are mapped onto the same element of $\mathfrak{C}$ via φ if, and only if, they are congruent modulo every congruence θ_ν; that is, if, and only if, $x \equiv y\ (\theta)$, where $\theta = \inf_{\nu \in N} \theta_\nu$. Thus, $\mathfrak{C} \cong \mathfrak{A}/\theta$, as claimed.

Conversely, given a family $(\theta_\nu)_{\nu \in N}$ of congruence relations on $\mathfrak{A}$, consider the natural epimorphisms

$$\varphi_\nu : \mathfrak{A} \longrightarrow \mathfrak{A}_\nu := \mathfrak{A}/\theta_\nu, \quad (\nu \in N).$$

Then we have $\tau(\mathfrak{A}_\nu) = \tau(\mathfrak{A})$, so that we may form the direct product $\prod_\nu \mathfrak{A}_\nu$, and the morphisms φ_ν combine to define a homomorphism

$$\varphi : \mathfrak{A} \longrightarrow \prod_{\nu \in N} \mathfrak{A}_\nu.$$

If $\mathfrak{C} := \mathrm{image}(\varphi)$, then the co-restriction $\varphi' : \mathfrak{A} \to \mathfrak{C}$ is an epimorphism, and $\mathfrak{C}$ is a subdirect product of the algebras $\mathfrak{A}_\nu$. □

Corollary 10.12. *The isomorphic representations of an algebra* $\mathfrak{A}$ *as a subdirect product correspond bijectively to the families* $\mathscr{F}$ *of congruence relations on* $\mathfrak{A}$ *such that* $\inf \mathscr{F} = \mathbf{0}$, *where* $\mathbf{0}$ *denotes the equality relation.*

8. *The subdirect decomposition theorem.*

Definition 10.13. An algebra $\mathfrak{A}$ is termed *subdirectly irreducible* if, in any isomorphic representation of $\mathfrak{A}$ as a subdirect product of algebras $\mathfrak{A}_\nu$, one of the associated natural epimorphisms $\mathfrak{A} \to \mathfrak{A}_\nu$ is an isomorphism.

We have the following useful criterion for subdirect irreducibility, which may be read off the structure lattice of the algebra in question.

Lemma 10.14. *Let* $\mathfrak{A}$ *be an algebra, and let* $P = P(\mathfrak{A})$ *denote the poset of all congruence relations* θ *on* $\mathfrak{A}$, *such that* $\theta > \mathbf{0}$. *Then* $\mathfrak{A}$ *is subdirectly irreducible if, and only if,* P *has a least element* θ_ℓ.

Proof. If θ_ℓ exists, then $\theta > \mathbf{0}$ implies $\theta \geq \theta_\ell$. Thus, $\inf_{\theta \in P} \theta = \mathbf{0} < \theta_\ell$ implies that some $\theta \in P$ is zero, which contradicts the definition of $P(\mathfrak{A})$. Hence, we must have

$$\inf_{\theta \in P(\mathfrak{A})} \theta = \theta_\ell > \mathbf{0}.$$

Consequently, each family $\mathscr{F}$ of congruence relations on $\mathfrak{A}$, which corresponds to an isomorphic representation of $\mathfrak{A}$ as a a subdirect product satisfies $\mathbf{0} \in \mathscr{F}$ which, by Corollary 10.12 and Definition 10.13, ensures that $\mathfrak{A}$ is subdirectly irreducible. Conversely, if θ_ℓ does not exist, then $\inf_{\theta \in P} \theta = \mathbf{0}$; thus, $\mathfrak{A}$ is not subdirectly irreducible. □

Lemma 10.15. *Let $\mathfrak{A} = (A; F)$ be an algebra, and let $a, b \in A$ be distinct elements. Then there exists a congruence relation $\theta = \theta(a,b)$ on $\mathfrak{A}$, such that $a \not\equiv b\ (\theta)$, and such that $\theta(a,b)$ is maximal with respect to this property.*

Proof. The proof by means of Zorn's Lemma is straightforward, and is left to the reader; cf. Exercise 16. □

Definition 10.16. A congruence relation θ on an algebra $\mathfrak{A}$ is called *strictly irreducible*, if the set Θ of congruence relations $\mu > \theta$ on $\mathfrak{A}$ has a least element.

Lemma 10.17. *Any congruence relation $\theta(a,b)$ as in Lemma* 10.15 *is strictly irreducible.*

Proof. If ϕ is a congruence relation on $\mathfrak{A}$ such that $\phi > \theta$, then $a \equiv b\ (\phi)$ by maximality of θ. Thus, if λ denotes the intersection of all congruence relations μ on $\mathfrak{A}$ such that $\mu \geq \theta$ and $a \equiv b\ (\mu)$, then we have

$$\varphi \geq \lambda > \theta.$$

This construction exhibits a congruence relation λ with the desired properties, whence our result. □

Summarising the preceding discussion, we have arrived at the following celebrated result, first obtained by Garrett Birkhoff in [13].

Theorem 10.18. *Any universal algebra is a subdirect product of subdirectly irreducible algebras.*

Proof. Let $\mathfrak{A} = (A; F)$ be the algebra in question, and consider the family of congruence relations

$$\mathscr{F} = \{\theta(a,b) : a, b \in A,\ a \neq b\}$$

on $\mathfrak{A}$, where $\theta(a,b)$ is as described in Lemma 10.15. Set

$$\hat{\theta} := \inf \mathscr{F} = \inf_{a \neq b} \theta(a,b),$$

and suppose that $x \equiv y\ (\hat{\theta})$ for some pair $x, y \in A$ such that $x \neq y$. Since $\hat{\theta} \leq \theta(x,y)$, it follows that $x \equiv y\ (\theta(x,y))$ while, at the same time, $x \not\equiv y\ (\theta(x,y))$ by definition of the congruence $\theta(x,y)$; see Lemma 10.15. This contradiction shows that $\hat{\theta} = \mathbf{0}$, the equality relation. By Corollary 10.12 it follows that the algebra $\mathfrak{A}$ has an isomorphic representation as a subdirect product of the quotient algebras $\mathfrak{A}/\theta(a,b)$. Furthermore, by Lemma 10.17, each relation $\theta(a,b)$ is strictly irreducible; that is, the set

$$\Theta_{a,b} = \{\hat{\mu} \in \Theta(\mathfrak{A}) : \hat{\mu} > \theta(a,b)\} = \theta(a,b)^{+} \setminus \{\theta(a,b)\}$$

has a least element. Invoking Proposition 10.12, we see that this property of $\Theta_{a,b}$ is equivalent to the set

$$\Theta(\mathfrak{A}/\theta(a,b)) \setminus \{\mathbf{0}\}$$

having a least element which, in turn, means that the algebra $\mathfrak{A}/\theta(a,b)$ is subdirectly irreducible by Lemma 10.14. The result follows. □

Exercises for § 10.

1. Show that the ring $\mathbb{Z}$ of integers can be represented as a subdirect sum of the family of finite prime fields $(K_p)_{p \text{ prime}}$.
2. Show that simple rings and skew fields are subdirectly irreducible. (A ring R is called *simple*, if $R \neq 0$, and the only non-zero (2-sided) ideal in R is the full ring itself.)
3. Give a proof of Proposition 10.2.
4. Show that the set $\mathscr{Q}_a$ introduced in the proof of Proposition 10.3 is inductively ordered under inclusion.
5. Explain in detail how to define the subalgebra $(\langle H \rangle; F)$ of an algebra $\mathfrak{A} = (A; F)$ generated by a non-empty subset H of A.
6. Consult your favourite book on category theory for the definition of the terms monomorphism, epimorphism, and isomorphism (see, for instance, [43], [66], or [70]), and justify the alternative definitions of these concepts given in the last paragraph of subsection B2.
7. Show that the operations f_i of a quotient algebra $\mathfrak{A}/\theta$ are well defined by (10.38); that is, the result of each operation does not depend on the representatives chosen.
8. Let $\varphi : \mathfrak{A} \to \mathfrak{B}$ be a homomorphism of algebras. Show that, if $x\theta y$ means $\varphi(x) = \varphi(y)$, then θ is a congruence relation on $\mathfrak{A}$. Conclude that *the epimorphic images of a universal algebra $\mathfrak{A}$ are, up to isomorphism, precisely the algebras $\mathfrak{A}/\theta$ defined by a congruence θ on $\mathfrak{A}$.*
9. Let $\varphi : \mathfrak{A} \to \mathfrak{B}$ be a homomorphism of algebras, where $\mathfrak{A} = (A; F)$, $\mathfrak{B} = (B; G)$, and $\tau(\mathfrak{A}) = (n_i)_{i \in I} = \tau(\mathfrak{B})$. Prove that $(\varphi(A); G)$ is a subalgebra of $\mathfrak{B}$.

10. Let $\varphi : \mathfrak{A} \to \mathfrak{B}$ and $\psi : \mathfrak{B} \to \mathfrak{C}$ be homomorphisms of algebras, say, $\mathfrak{A} = (A;F)$, $\mathfrak{B} = (B;G)$, and $\mathfrak{C} = (C;H)$, where

$$\tau(\mathfrak{A}) = \tau(\mathfrak{B}) = \tau(\mathfrak{C}) = (n_i)_{i \in I}.$$

Show that the map $\psi \circ \varphi : A \to C$ defines a homomorphism of the algebra $\mathfrak{A}$ into the algebra $\mathfrak{C}$.

11. Show that the maps $\pi_\mu : P \to A_\mu$ associated with the direct product of algebras $\mathfrak{P} = \prod_{\nu \in N} \mathfrak{A}_\nu$ define epimorphisms $\pi_\mu : \mathfrak{P} \to \mathfrak{A}_\mu$.

12. Prove Lemma 10.5.

13. For a set A, denote by $\mathfrak{E}(A)$ the collection of all equivalence relations on A. Show that the poset $(\mathfrak{E}(A), \subseteq)$ is a complete lattice.

14. Prove Lemma 10.10.

15. Let $\mathfrak{A} = (A;F)$ be an algebra, let $\mathfrak{B} = (B;F)$ be a subalgebra of $\mathfrak{A}$, and let θ be a congruence relation on $\mathfrak{A}$. Prove that θ induces a congruence relation on $\mathfrak{B}$.

16. Let $\mathfrak{A} = (A;F)$ be an algebra, and let $a, b \in A$ be distinct elements. Show that there exists a congruence relation $\theta = \theta(a,b)$ on $\mathfrak{A}$ such that $a \not\equiv b\ (\theta)$, and such that $\theta(a,b)$ is maximal with respect to this property.

§ 11. Dependence relations, rank functions, and closure operators

§11.1. Linear dependence in vector spaces

Let V be a (left) vector space over the skew field K. We say that a vector $v \in V$ is *linearly dependent* on a set $\{x_1, x_2, \ldots, x_m\} \subseteq V$, if

$$v = \alpha_1 x_1 + \alpha_2 x_2 + \cdots + \alpha_m x_m \tag{11.44}$$

holds for a suitable choice of scalars $\alpha_1, \alpha_2, \ldots, \alpha_m \in K$. Equivalently, v is linearly dependent on $\{x_1, x_2, \ldots, x_m\}$, if we have a relation

$$\beta v + \beta_1 x_1 + \cdots + \beta_m x_m = 0, \quad (\beta \neq 0). \tag{11.45}$$

Concerning this correspondence between a vector and a (finite) set of vectors, we may observe the following:

(LD1) Every element x_μ is linearly dependent on the set $\{x_1, \ldots, x_m\}$.

(LD2) If y is linearly dependent on $\{x_1, \ldots, x_m\}$, but not on $\{x_1, \ldots, x_{m-1}\}$, then x_m is linearly dependent on $\{x_1, \ldots, x_{m-1}, y\}$.

(LD3) If z is linearly dependent on $\{y_1, \ldots, y_n\}$, and if every y_ν is linearly dependent on $\{x_1, \ldots, x_m\}$, then z is linearly dependent on $\{x_1, \ldots, x_m\}$.

A set $\{x_1, \ldots, x_m\} \subseteq V$ is termed *linearly independent* if none of its elements is linearly dependent on the others.

In [81], a book first published in 1930 under the title "Moderne Algebra", van der Waerden, in Chapter 4, §20, states (LD1)–(LD3), and, viewing them as axioms, derives from these the Steinitz exchange lemma, and thus invariance of (linear) dimension.[14] His reason for choosing this abstract approach lies in the fact that there is another dependence concept treated in his book, namely that of algebraic dependence in field theory, which can be shown to satisfy the same axioms, thus also satisfies the conclusions derived from these; cf. [81, § 74]. In fact, dependence relations satisfying an analogue of the above axioms, or some equivalent system of axioms, also occur in various other parts of mathematics; for instance in graph theory.[15] Thus, matroid theory, the systematic study of this type of dependence on finite sets was born with Whitney's fundamental paper [85] of 1935, and has developed ever since; see, for instance, Oxley's book [64] for an overview. The present section treats some basic aspects of matroid theory for sets of arbitrary cardinality.

---

[14] To be precise, Axioms (LD1)–(LD3) first occur in the second edition of van der Waerden's book, published in 1937.

[15] If $E = E(G)$ is the edge set of a finite graph G, and if $C = C(G)$ is the collection of edge sets of cycles in G, then $(E(G), C(G))$ is a matroid; an edge set X of G is termed independent if it does not contain the edge set of a cycle.

§11.2. $\mathscr{D}$-sets

When considering the axioms (D1)–(D4) of an abstract dependence relation below, it will help if the reader keeps in mind the properties (LD1)–(LD3) of linear dependence; see also Exercises 1–3.

Let S be a set, and let $\mathscr{D} \subseteq S \times \mathscr{B}(S)$ be a correspondence between elements and subsets of S. If $(x,A) \in \mathscr{D}$ for some $x \in S$ and some $A \subseteq S$, we write $\mathscr{D}(x,A)$, while $\overline{\mathscr{D}}(x,A)$ means that $(x,A) \notin \mathscr{D}$. The pair $(S,\mathscr{D})$ is called a *$\mathscr{D}$-set*, $\mathscr{D}$ itself a *dependence relation on* S, if the following conditions are satisfied for all elements $a,x \in S$, and all subsets $A,B \subseteq S$:

(D1) If $x \in A$, then $\mathscr{D}(x,A)$.

(D2) If $\mathscr{D}(x,A)$ and $\overline{\mathscr{D}}(x,A\setminus\{a\})$ for some $a \in A$, then $\mathscr{D}(a,(A\setminus\{a\}\cup\{x\})$.

(D3) If $\mathscr{D}(x,A)$, and if $\mathscr{D}(a,B)$ for all $a \in A$, then $\mathscr{D}(x,B)$.

(D4) If $\mathscr{D}(x,A)$, then there exists a finite subset A' of A, such that $\mathscr{D}(x,A')$.

If $\mathscr{D}(x,A)$ holds for some $x \in S$ and $A \subseteq S$, then we say that the element x *depends on* (or *is dependent on*) A. Also, if every element of $A \subseteq S$ depends on the subset B of S, we say that A depends (or is dependent) on B, and express this as $\mathscr{D}(A,B)$ for short. Two subsets $A,B \subseteq S$ are called *$\mathscr{D}$-equivalent*, if both $\mathscr{D}(A,B)$ and $\mathscr{D}(B,A)$ hold. A subset $A \subseteq S$ is termed *$\mathscr{D}$-dependent*, if there exists an element $a \in A$, such that $\mathscr{D}(a,A\setminus\{a\})$; otherwise A is called *$\mathscr{D}$-independent*. By Condition (D4), a set $A \subseteq S$ is $\mathscr{D}$-independent if, and only if, each of its finite subsets is $\mathscr{D}$-independent.

To give a first example of a $\mathscr{D}$-set, let S be an arbitrary set, and define a correspondence $\mathscr{D} \subseteq S \times \mathscr{B}(S)$ by

$$\mathscr{D}(x,A) :\Longleftrightarrow x \in A, \quad (x \in S, A \subseteq S). \tag{11.46}$$

One checks easily that $\mathscr{D}$ is a dependence relation, so that $(S,\mathscr{D})$ is a $\mathscr{D}$-set, and that every subset A of S is $\mathscr{D}$-independent; cf. Exercise 4. A second example of a $\mathscr{D}$-set is discussed in Exercise 6; indeed, linear dependence (the topic of this exercise) is a main motivation for introducing and naming dependence relations.

§11.3. Invariance of independent sets

We have the following fundamental result concerning $\mathscr{D}$-independent sets of a $\mathscr{D}$-set.

Proposition 11.1. *Let $(S,\mathscr{D})$ be a $\mathscr{D}$-set, and let $A,B \subseteq S$ be two $\mathscr{D}$-independent and $\mathscr{D}$-equivalent subsets of S. Then we have $|A| = |B|$.*

Proof. Let A and B be two $\mathscr{D}$-independent and $\mathscr{D}$-equivalent subsets of S. By symmetry of our hypotheses, it suffices to show that A is equipotent with a subset of B; the result will then follow from Corollary 3.7.

Let $C := A \cap B$ (the case where $C = \emptyset$ is allowed here), and consider subsets A' of A and B' of B, such that $C \subseteq A' \cap B'$. We form triples (A', B', φ'), where $\varphi' : A' \to B'$ is a bijection inducing the identity on C, and such that the set

$$K' := B' \cup (A \setminus A')$$

is $\mathscr{D}$-independent. Let $\mathscr{Q}$ be the set of all such triples (A', B', φ'). We have $\mathscr{Q} \neq \emptyset$, since $(C, C, 1_C) \in \mathscr{Q}$. Define an order relation $\preceq$ on $\mathscr{Q}$ via

$$(A', B', \varphi') \preceq (A'', B'', \varphi'') :\Longleftrightarrow A' \subseteq A'',\ B' \subseteq B'', \text{ and } \varphi'' \text{ extends } \varphi'.$$

We claim that $(\mathscr{Q}, \preceq)$ is inductively ordered. Let $\mathscr{C} = \{(A'_\nu, B'_\nu, \varphi'_\nu)\}_{\nu \in N}$ be a non-empty chain in $\mathscr{Q}$, and set $A_0 := \bigcup_\nu A'_\nu$ and $B_0 := \bigcup_\nu B'_\nu$. Then the set

$$K_0 := B_0 \cup (A \setminus A_0)$$

is $\mathscr{D}$-independent, since (by the chain condition) each of its finite subsets is contained in some set K' corresponding to a suitable triple $(A'_\mu, B'_\mu, \varphi'_\mu) \in \mathscr{C}$, which is $\mathscr{D}$-independent by definition of these triples; cf. Exercise 5. Also, again making use of the chain condition, we see that setting

$$\varphi_0(x) := \varphi'_\nu(x), \quad (x \in A'_\nu)$$

gives a well-defined bijection $\varphi_0 : A_0 \to B_0$ inducing the identity on C. It follows that the triple (A_0, B_0, φ_0) lies in $\mathscr{Q}$, and is an upper bound for $\mathscr{C}$ by construction. Hence, $(\mathscr{Q}, \preceq)$ is inductively ordered as claimed. By Zorn's Lemma, $\mathscr{Q}$ has a maximal element (A^*, B^*, φ^*), and we aim to show that $A^* = A$, implying $|A| \leq |B|$, as desired.

Suppose for a contradiction that $A^* \subset A$. Then there exists an element $a \in A$ such that $a \notin A^*$; in particular,

$$a \in K^* := B^* \cup (A \setminus A^*),$$

where K^* is $\mathscr{D}$-independent. To this element a we want to assign an element $b \in B$ such that $\overline{\mathscr{D}}(b, K^* \setminus \{a\})$. If we had $\mathscr{D}(B, K^* \setminus \{a\})$, then making use of the fact that $\mathscr{D}(a, B)$ (by $\mathscr{D}$-equivalence of the sets A and B) plus Condition (D3), we would find that $\mathscr{D}(a, K^* \setminus \{a\})$, contradicting the $\mathscr{D}$-independence of the set K^*. Hence, there exists some element $b \in B$ such that $\overline{\mathscr{D}}(b, K^* \setminus \{a\})$, implying $b \notin B^*$ by Condition (D1) and the fact that $b \neq a$ (since otherwise $a \in C \subseteq A^*$ contradicting our choice of the element a). To complete our argument, we need to show that the set

$$K^{**} := B^* \cup \{b\} \cup (A \setminus (A^* \cup \{a\})) = (K^* \setminus \{a\}) \cup \{b\}$$

is $\mathscr{D}$-independent. Suppose otherwise. Then there exists some element $k \in K^{**}$ with $\mathscr{D}(k, K^{**} \setminus \{k\})$. If $k = b$, then we have $\mathscr{D}(b, K^* \setminus \{a\})$, contradicting our choice of the element b. If, on the other hand, $k \neq b$, then we have $k \in K^{**} \setminus \{b\} = K^* \setminus \{a\}$, and $\mathscr{D}(k, K^{**} \setminus \{k, b\})$ would imply that the set $K^* \setminus \{a\}$ is $\mathscr{D}$-dependent, contradicting the fact that the larger set K^* is $\mathscr{D}$-independent, since it belongs to the triple $(A^*, B^*, \varphi^*) \in \mathscr{Q}$. Hence, we must have $\overline{\mathscr{D}}(k, K^{**} \setminus \{k, b\})$, and Condition (D2) implies that $\mathscr{D}(b, K^* \setminus \{a\})$, again contradicting our choice of the element b. Hence, K^{**} is $\mathscr{D}$-independent, as desired. Setting $A^{**} := A^* \cup \{a\}$ and $B^{**} := B^* \cup \{b\}$, and denoting by $\varphi^{**} : A^{**} \to B^{**}$ the extension of the map $\varphi^* : A^* \to B^*$ sending a to b, we now obtain a triple $(A^{**}, B^{**}, \varphi^{**}) \in \mathscr{Q}$, such that

$$(A^*, B^*, \varphi^*) \prec (A^{**}, B^{**}, \varphi^{**}),$$

contradicting the maximality of the triple (A^*, B^*, φ^*). This last contradiction shows that indeed $A^* = A$, and the proof is complete. □

A $\mathscr{D}$-independent set $M \subseteq S$ of a $\mathscr{D}$-set S is called *maximal,* or, more precisely, *maximal $\mathscr{D}$-independent* if, for each element $s \in S \setminus M$, the set $M \cup \{s\}$ is $\mathscr{D}$-dependent. We have the following useful characterisation of maximal $\mathscr{D}$-independent sets.

Lemma 11.2. *A $\mathscr{D}$-independent subset M of a $\mathscr{D}$-set S is maximal if, and only if, the relation $\mathscr{D}(s, M)$ holds for each element $s \in S$.*

Proof. Let $M \subseteq S$ be a maximal $\mathscr{D}$-independent subset of the $\mathscr{D}$-set S, and let $s \in S$ be arbitrary. If $s \in M$, then we have $\mathscr{D}(s, M)$ by Condition (D1). If, on the other hand, $s \notin M$, then the set $M \cup \{s\}$ is $\mathscr{D}$-dependent by maximality of M; that is, there exists some $b \in M \cup \{s\}$ such that $\mathscr{D}(b, (M \cup \{s\}) \setminus \{b\})$. If $b = s$, then $\mathscr{D}(s, M)$ follows from the last relation. Otherwise, $b \in M$, and $\mathscr{D}$-independence of M gives $\overline{\mathscr{D}}(b, M \setminus \{b\})$. Applying Condition (D2) with $x = b$, $A = (M \setminus \{b\}) \cup \{s\}$, and $a = s$ now gives $\mathscr{D}(s, M)$, as desired. This proves the forward implication, while the converse is immediate. □

Combining Proposition 11.1 with Lemma 11.2, we readily obtain the following.

Corollary 11.3. *Any two maximal $\mathscr{D}$-independent subsets of a $\mathscr{D}$-set are $\mathscr{D}$-equivalent and equipotent.*

The question arises whether a given $\mathscr{D}$-set necessarily contains maximal $\mathscr{D}$-independent subsets. In fact, our next result proves something stronger.

Proposition 11.4. *Let S be a $\mathscr{D}$-set, and let $A \subseteq S$ be $\mathscr{D}$-independent. Then there exists a maximal $\mathscr{D}$-independent subset M of S such that $A \subseteq M$.*

Proof. Let $\mathscr{M}$ be the set of all $\mathscr{D}$-independent subsets of S containing the given set A, ordered by inclusion. We have $\mathscr{M} \neq \emptyset$ since $A \in \mathscr{M}$. We claim that $(\mathscr{M}, \subseteq)$ is inductively ordered. Let $\mathscr{C} = \{M_i\}_{i \in I}$ be a non-empty chain in $\mathscr{M}$, and let $M^* := \bigcup_i M_i$. Then M^* is a subset of S containing A (since $\mathscr{C} \neq \emptyset$), and M^* is $\mathscr{D}$-independent by Condition (D4). Indeed, if we had $\mathscr{D}(a, M^* \setminus \{a\})$ for some $a \in M^*$, then Condition (D4) would guarantee existence of a finite subset $M' \subseteq M^* \setminus \{a\}$, such that

$$\mathscr{D}(a, M') = \mathscr{D}(a, (M' \cup \{a\}) \setminus \{a\});$$

in particular, $M' \cup \{a\}$ is $\mathscr{D}$-dependent. However, by the chain condition and the finiteness of M', $M' \cup \{a\} \subseteq M_i$ for some index $i \in I$, implying that M_i is $\mathscr{D}$-dependent, contradicting the definition of $\mathscr{M}$. Hence, $M^* \in \mathscr{M}$, and M^* is an upper bound for $\mathscr{C}$, as desired. By Zorn's Lemma, $\mathscr{M}$ contains a maximal element $\tilde{M}$, and $\tilde{M}$ is a maximal $\mathscr{D}$-independent subset of S containing the given $\mathscr{D}$-independent set A. □

§11.4. Two applications

We pause to describe two classical examples of $\mathscr{D}$-sets.

Invariance of dimension

By Exercise 9, a vector space V over the skew-field K becomes a $\mathscr{D}$-set by putting

$$\mathscr{D}(x, A) :\Longleftrightarrow x \in \langle A \rangle, \quad (x \in V, A \subseteq V),$$

where $\langle A \rangle$ denotes the linear span of the subset A of V. In this setting, the $\mathscr{D}$-independent sets are the linearly independent subsets of V, thus the maximal $\mathscr{D}$-independent subsets of V are precisely the bases of V. Consequently, Corollary 11.3 yields the following (see also Exercise 10 in §5).

Corollary 11.5. *Any two bases of a vector space over a skew-field are equipotent. This cardinal number, the size of any basis of V, is called the dimension of V, denoted* $\dim_K(V)$.

A cardinal invariant of a field extension

Let $L \supseteq K$ be a field extension. Moreover, let $a \in L$ be an element, and let $H \subseteq L$ be a subset. We call *a algebraically dependent on H over K*, if a is algebraic over $K[H]$, the subfield of L obtained by adjoining the elements of H to K. More explicitly, this means that a satisfies a polynomial equation

$$\alpha_0(H) + \alpha_1(H)a + \cdots + \alpha_g(H)a^g = 0,$$

where the coefficients $\alpha_0(H), \ldots, \alpha_g(H)$ are polynomials in the elements of H with coefficients in K, and not all $\alpha_i(H)$ are equal to zero. Define a correspondence $\mathscr{D} \subseteq L \times \mathscr{B}(L)$ by

$$\mathscr{D}(a, H) :\Longleftrightarrow a \text{ is algebraically dependent on } H \text{ over } K, \quad (a \in L, H \subseteq L). \tag{11.47}$$

This correspondence $\mathscr{D}$ is known to satisfy Axioms (D1)–(D4) of a dependence relation; cf., for instance, [81, §74] (see also Exercise 10 below). A subset $H \subseteq L$ is termed *algebraically dependent (over K)*, if there exists some $h \in H$ such that $\mathscr{D}(h, H \setminus \{h\})$; otherwise it is called *algebraically independent (over K)*. Two subsets $H_1, H_2 \subseteq L$ are called *algebraically equivalent*, if $\mathscr{D}(H_1, H_2)$ and $\mathscr{D}(H_2, H_1)$ both hold; that is, if each element of H_1 is algebraically dependent on H_2, and vice versa. By Corollary 11.3, we get the following result.

Corollary 11.6. (Steinitz [75]) *If $L \supseteq K$ is a field extension, then any two subsets of L, which are algebraically equivalent and independent over K, are equipotent.*

§11.5. Rank functions

Let S be a set. Following Whitney [85], we define a *rank function over S* to be a map $r : \mathscr{B}_{\text{fin}}(S) \to \mathbb{N}$, assigning a non-negative integer $r(A)$ to each finite subset A of S, subject to the following axioms:

(R1) $r(\emptyset) = 0$;

(R2) $r(A \cup \{x\}) = r(A) + \delta$, where $\delta \in \{0, 1\}$;

(R3) $r(A) = r(A \cup \{x\}) = r(A \cup \{y\})$ implies $r(A) = r(A \cup \{x, y\})$.

Call a finite subset A of S *r-independent*, if $r(A) = |A|$; an arbitrary subset $X \subseteq S$ is termed r-independent, if each of its finite subsets is r-independent.

We begin by drawing some easy conclusions from Axioms (R1)–(R3).

Lemma 11.7. *Let $r : \mathscr{B}_{\text{fin}}(S) \to \mathbb{N}$ be a rank function over the set S.*

(a) *If $r(A) = r(A \cup \{x_1\}) = \cdots = r(A \cup \{x_n\})$ for a finite subset A of S and elements $x_1, \ldots x_n \in S$, then $r(A) = r(A \cup \{x_1, \ldots, x_n\})$.*

(b) *If $r(A) = r(A \cup \{x\})$ for some finite subset $A \subseteq S$ and some element $x \in S$, then each finite subset Y of S satisfies the relation $r(A \cup Y) = r(A \cup Y \cup \{x\})$.*

Proof. (a) This straightforward argument is left to the reader; see Exercise 4.

(b) It clearly suffices to prove our claim in the case when $Y = \{y\}$ for some $y \in S$. There are two cases according to whether $r(A \cup \{y\})$ equals $r(A)$ or $r(A) + 1$; cf. Axiom (R2). If $r(A \cup \{y\}) = r(A)$, the Axiom (R3), in conjunction with our hypothesis, yields $r(A \cup \{y\}) = r(A \cup \{x, y\})$, as desired. If, on the other hand, $r(A \cup \{y\}) = r(A) + 1$, then (R2) gives

$$r(A \cup \{x, y\}) \leq r(A \cup \{x\}) + 1 = r(A) + 1 = r(A \cup \{y\}),$$

as well as

$$r(A \cup \{x, y\}) \geq r(A \cup \{y\}),$$

whence our result. □

We are now in a position to establish the main result of this section.

Proposition 11.8. (Rado [67]) *Let S be an arbitrary set, and let $r \colon \mathscr{B}_{\mathrm{fin}}(S) \to \mathbb{N}$ be a rank function over S. Then any two maximal r-independent subsets of S are equipotent.*

Proof. Define a correspondence $\mathscr{D} \subseteq S \times \mathscr{B}(S)$ via

$\mathscr{D}(x,A) :\Longleftrightarrow$ there exists a finite subset $A' \subseteq A$, such that $r(A' \cup \{x\}) = r(A')$.

We claim that $\mathscr{D}$ satisfies Axioms (D1)–(D4) of a dependence relation. If $x \in A$, then setting $A' := \{x\}$, we see that (D1) holds true. Also, Axiom (D4) holds by definition of $\mathscr{D}$. Next, suppose that $\mathscr{D}(x,A)$, and that $\overline{\mathscr{D}}(x, A \setminus \{a\})$ for some $a \in A$. By definition of $\mathscr{D}$, there exists some finite subset A' of A, such that $r(A' \cup \{x\}) = r(A')$. Also, by definition of $\mathscr{D}$, our hypotheses, and Axiom (R2), we have

$$r(A' \setminus \{a\}) < r((A' \setminus \{a\}) \cup \{x\}).$$

Applying (R2), we get

$$r(A') \leq r((A' \setminus \{a\}) \cup \{x\}) \leq r(A' \cup \{x\}) = r(A'),$$

so that

$$r((A' \setminus \{a\}) \cup \{x\}) = r(A' \cup \{x\}),$$

implying $\mathscr{D}(a, (A \setminus \{a\}) \cup \{x\})$, whence (D2). We now turn to Axiom (D3). Suppose that $\mathscr{D}(x,A)$ and that $\mathscr{D}(A,B)$ for some $x \in S$ and subsets $A, B \subseteq S$. We may assume without loss of generality that A and B are finite, say $A = \{a_1, \dots, a_n\}$. By Part (b) of Lemma 11.7, if X is a finite subset of S, then $X' \subseteq X$ and $r(X') = r(X' \cup \{y\})$ implies $r(X) = r(X \cup \{y\})$. We shall use this fact without further reference in the remainder of this proof. By hypothesis,

$$r(B) = r(B \cup \{a_\nu\}), \quad \nu \in [n].$$

By Part (a) of Lemma 11.7, it follows that

$$r(B) = r(A \cup B). \tag{11.48}$$

Also, since A and B are finite,

$$r(A \cup B \cup \{x\}) = r(A \cup B). \tag{11.49}$$

Furthermore, by Axiom (R2),

$$r(B) \leq r(B \cup \{x\}) \leq r(A \cup B \cup \{x\}). \tag{11.50}$$

Combining (11.48)–(11.50) yields

$$r(B) = r(B \cup \{x\});$$

thus $\mathscr{D}(x,B)$ as desired, whence (D3). Finally, we observe that a subset $A \subseteq S$ is r-independent if, and only if, it is $\mathscr{D}$-independent; cf. Exercise 8. Our claim thus follows from Corollary 11.3. □

As we have seen in the proof of Proposition 11.8, any set S with a rank function $r : \mathscr{B}_{\text{fin}} \to \mathbb{N}$ gives rise to a $\mathscr{D}$-set $(S, \mathscr{D})$ with a suitably defined dependence relation $\mathscr{D}$. Moreover, for this construction, the r-independent and $\mathscr{D}$-independent subsets of S coincide (see Exercise 8 below). Conversely, if $(S, \mathscr{D})$ is a $\mathscr{D}$-set, then we define $r(A)$ for $A \in \mathscr{B}_{\text{fin}}(S)$ as the cardinality of a maximal $\mathscr{D}$-independent subset of A. By Corollary 11.3, $r(A)$ is well defined. Also, with this definition of r, Axioms (R1)–(R3) hold, so that r is a rank function on S; cf. Exercise 9. It follows that the system of axioms (D1)–(D4) is equivalent to the system of axioms (R1)–(R3); $\mathscr{D}$-sets and sets with a rank function are just two conceptually different ways of describing the same structural object. In this context, we shall need one further observation. Let r be a rank function over the set S, let $\mathscr{D}_r$ be the dependence relation on S associated with r as in the proof of Proposition 11.8, and let $r' = r_{\mathscr{D}_r}$ be the rank function over S associated with $\mathscr{D}_r$. We claim that $r = r'$. Indeed, for $A \in \mathscr{B}_{\text{fin}}(S)$,

$$r'(A) = r_{\mathscr{D}_r}(A) = |A_0|,$$

where A_0 is a maximal $\mathscr{D}_r$-independent subset of A. However, since the $\mathscr{D}_r$-independent subsets of A coincide with the r-independent subsets of A, A_0 is a maximal r-independent subset of A; in particular, $r(A_0) = |A_0|$. Also, for any $x \in A \setminus A_0$, we have $r(A_0 \cup \{x\}) = r(A_0)$, since $A_0 \cup \{x\}$ is r-dependent by maximality of A_0. Hence, by Part (a) of Lemma 11.7, $r(A_0) = r(A)$, so that $r'(A) = r(A)$ for each finite subset A of S, as claimed.

§11.6. Connection with matroids

Let S be a set, and let $r : \mathscr{B}_{\text{fin}}(S) \to \mathbb{N}$ be a rank function over S as defined in §11.4. It is an immediate consequence of Axioms (R1) and (R2) that

$$(R1') \qquad 0 \le r(A) \le |A|, \quad A \in \mathscr{B}_{\text{fin}}(S).$$

Also, again by Axioms (R1) and (R2), we have

$$(R2') \qquad r(A) \le r(B), \quad (A, B \in \mathscr{B}_{\text{fin}}(S), A \subseteq B).$$

The following observation is more interesting.

Lemma 11.9. *Let r be a rank function over the set S. Then we have*

$$(R3') \qquad r(A \cup B) + r(A \cap B) \le r(A) + r(B), \quad (A, B \in \mathscr{B}_{\text{fin}}(S)).$$

Proof. Let $\mathscr{D}$ be the dependence relation on S associated with the rank function r as in the proof of Proposition 11.8; that is,

$$\mathscr{D}(x, A) \iff \text{there exists } A' \subseteq A,\ |A'| < \infty, \text{ with } r(A' \cup \{x\}) = r(A').$$

As observed earlier, we have $r(A) = |A_0|$, where A_0 is a maximal $\mathscr{D}$-independent subset of A.

Let M be a maximal $\mathscr{D}$-independent subset of $A \cap B$, so that

$$r(A \cap B) = |M|.$$

Then M is a $\mathscr{D}$-independent subset of $A \cup B$ so, by Proposition 11.4, is contained in some maximal $\mathscr{D}$-independent subset M' of $A \cup B$. Furthermore, $M' \cap A$ and $M' \cap B$ are $\mathscr{D}$-independent subsets of A and B, respectively, thus

$$|M' \cap A| \le r(A) \quad \text{and} \quad |M' \cap B| \le r(B).$$

Hence,

$$\begin{aligned} r(A) + r(B) &\ge |M' \cap A| + |M' \cap B| \\ &= |(M' \cap A) \cup (M' \cup B)| + |(M' \cap A) \cap (M' \cap B)| \\ &= |M' \cap (A \cup B)| + |M' \cap (A \cap B)|. \end{aligned}$$

Also, by definition of the sets M and M', we have

$$M' \cap (A \cup B) = M'$$

and

$$M' \cap (A \cap B) = M.$$

Consequently, we get

$$r(A) + r(B) \ge |M'| + |M| = r(A \cup B) + r(A \cap B),$$

as claimed. □

Indeed, more is true.

Proposition 11.10. *Let S be a set, and let $r : \mathscr{B}_{\text{fin}}(S) \to \mathbb{N}$ be a function assigning non-negative integers to finite subsets of S. Then r satisfies Axioms* (R1)–(R3) *of a rank function over S if, and only if, r satisfies Conditions* (R1')–(R3').

Proof. By our previous observations, including Lemma 11.9, a rank function r over the set S satisfies (R1')–(R3'). Conversely, suppose that r satisfies Conditions (R1')–(R3'). Then, by (R1'),

$$0 \le r(\emptyset) \le |\emptyset| = 0,$$

whence (R1). Also, for $x \in S$ and $A \in \mathscr{B}_{\text{fin}}(S)$, we have, by (R2') and (R3'),

$$r(A) \le r(A \cup \{x\}) \le r(A) + r(\{x\}) - r(A \cap \{x\}) \le r(A) + 1.$$

Hence, Axiom (R2) holds as well. Finally, suppose that, for some elements $x, y \in S$ and some finite subset $A \in \mathscr{B}_{\text{fin}}(S)$, we have

$$r(A) = r(A \cup \{x\}) = r(A \cup \{y\}).$$

In showing that $r(A\cup\{x,y\})=r(A)$, we may suppose without loss of generality that $x,y\notin A$, since otherwise our claim holds trivially. Then, by (R3') plus our hypothesis,

$$r(A\cup\{x,y\})+r((A\cup\{x\})\cap(A\cup\{y\}))\leq r(A\cup\{x\})+r(A\cup\{y\}) = 2r(A),$$

implying the desired equality since, by (R2'), we have $r(A)\leq r(A\cup\{x,y\})$ as well. Hence, Axiom (R3) holds as well, and the proof is complete. □

A finite set S equipped with a function $r:\mathscr{B}(S)\to\mathbb{N}$ satisfying Conditions (R1')–(R3') is called a *matroid*; cf., for instance, [64, Cor. 1.3.4]. Hence, $\mathscr{D}$-sets and rank functions on sets are simply equivalent generalisations of the matroid concept to arbitrary sets.

§11.7. Closure operators

The purpose of this section is to discuss yet another technical manifestation of dependence relations, namely a particular type of closure operator.

Definition 11.11. (a) Let S be a set. By a *closure operator* on S we mean a map $^{-}:\mathscr{B}(S)\longrightarrow\mathscr{B}(S)$ satisfying the following axioms:

(C1) $X\subseteq\overline{X},\quad X\subseteq S$;

(C2) $X\subseteq Y\Longrightarrow\overline{X}\subseteq\overline{Y},\quad X,Y\subseteq S$;

(C3) $\overline{\overline{X}}=X$ for all $X\subseteq S$.

(b) A closure operator $^{-}:\mathscr{B}(S)\to\mathscr{B}(S)$ on the set S is called *finitary* (or *of finite type)*, if, for each subset $X\subseteq S$ and every element $a\in S$,

$$a\in\overline{X}\Longrightarrow a\in\overline{X'}\text{ for some finite subset }X'\text{ of }X. \tag{11.51}$$

(c) We say that the closure operator $^{-}:\mathscr{B}(S)\to\mathscr{B}(S)$ satisfies the *exchange property*, if, for each $X\subseteq S$ and any elements $y,z\in S$,

$$y\notin\overline{X}\text{ and }y\in\overline{X\cup\{z\}}\Longrightarrow z\in\overline{X\cup\{y\}}. \tag{11.52}$$

Our next result describes the precise connection between closure operators and dependence relations on a set.

Proposition 11.12. (a) *Let* $^{-}:\mathscr{B}(S)\to\mathscr{B}(S)$ *be a finitary closure operator on the set S satisfying the exchange property* (11.52). *Set*

$$\mathscr{D}(x,A):\Longleftrightarrow x\in\overline{A},\quad(x\in S,A\subseteq S).$$

Then $(S,\mathscr{D})$ *is a* $\mathscr{D}$*-set.*

(b) *Let* $(S,\mathscr{D})$ *be a* $\mathscr{D}$*-set. Define a map* $^{-}:\mathscr{B}(S)\to\mathscr{B}(S)$ *via*

$$\overline{A}:=\{x\in S:\mathscr{D}(x,A)\},\quad A\subseteq S.$$

Then $^-$ *is a finitary closure operator on S satisfying the exchange property* (11.52).

Proof. (a) We need to verify Axioms (D1)–(D4).

First, if $x \in S$, $A \subseteq S$, and $x \in A$, then $x \in \overline{A}$ by Axiom (C1), so that $\mathscr{D}(x,A)$ holds, whence (D1).

Next, suppose that $\mathscr{D}(x,A)$ and $\overline{\mathscr{D}}(x,A \setminus \{a\}$, so that $x \in \overline{A}$ and $x \notin \overline{A \setminus \{a\}}$. Setting $X := A \setminus \{a\}$, $y := x$, and $z := a$ in (11.52), we find that

$$a \in \overline{(A \setminus \{a\}) \cup \{x\}},$$

so that $\mathscr{D}(a,(A \setminus \{a\}) \cup \{x\})$ holds, as desired. Hence, Axiom (D2) holds as well.

Third, suppose that $\mathscr{D}(x,A)$ and $\mathscr{D}(A,B)$, so that $x \in \overline{A}$ and $A \subseteq \overline{B}$. By Axioms (C2) and (C3) of a closure operator, we get $\overline{A} \subseteq \overline{B}$, implying $x \in \overline{B}$; that is, $\mathscr{D}(x,B)$, as required, which verifies (D3).

Finally, Axiom (D4) follows immediately from the fact that the closure operator $^-$ on S is assumed to be of finite type. All in all, we have thus shown that $\mathscr{D}$ is a dependence relation of the set S, as claimed.

(b) Here, we need to verify Axioms (C1)–(C3), as well as properties (11.51) and (11.52).

First, let $A \subseteq S$, and let $x \in A$. By Axiom (D1), we have $\mathscr{D}(x,A)$, so that $x \in \overline{A}$, proving (C1).

Next, let $A,B \subseteq S$, and suppose that $x \in \overline{A}$ and that $A \subseteq B$. Then $\mathscr{D}(x,A)$ and $\mathscr{D}(A,B)$, where we have used (C1) to deduce that $A \subseteq \overline{B}$. By Axiom (D3), we get that $\mathscr{D}(x,B)$, so that $x \in \overline{B}$, showing that $\overline{A} \subseteq \overline{B}$, whence (C2).

Third, let $A \subseteq S$ be any subset. By (C1), we have $\overline{A} \subseteq \overline{\overline{A}}$. In order to see the reverse inclusion, let $x \in \overline{\overline{A}}$ be an arbitrary element, so that $\mathscr{D}(x,\overline{A})$ by definition of $\overline{\overline{A}}$. Also, again by definition of the closure, we have $\mathscr{D}(\overline{A},A)$, implying $\mathscr{D}(x,A)$ by Axiom (D3). Hence, $x \in \overline{A}$, as desired, proving (C3).

Fourth, let $A \subseteq S$ be an arbitrary subset of S, and let $x \in \overline{A}$, so that $\mathscr{D}(x,A)$. By Axiom (D4), there exists a finite subset $A' \subseteq A$ such that $\mathscr{D}(x,A')$, implying $x \in \overline{A'}$ by definition of closure. Hence, $^-$ is of finite type, as claimed.

Finally, we need to verify that $^-$ satisfies the exchange property (11.52). Let $X \subseteq S$, and let $y,z \in S$ be elements such that $y \notin \overline{X}$ and $y \in \overline{X \cup \{z\}}$, so that $\overline{\mathscr{D}}(y,X)$ and $\mathscr{D}(y,X \cup \{z\})$. Setting $A := X \cup \{z\}$, $x := y$, and $a := z$ in Axiom (D2), we find that $\mathscr{D}(z,X \cup \{y\})$, so that $z \in \overline{X \cup \{y\}}$, as desired.

Hence, $^{-}$ is a closure operator of finite type satisfying the exchange property, as claimed. □

Exercises for § 11.

1. Prove (LD1)–(LD3).
2. Call two finite subsets $X = \{x_1, \dots, x_r\}$ and $Y = \{y_1, \dots, y_s\}$ of a vector space V *linearly equivalent,* if each y_σ is linearly dependent on X, and every x_ρ is linearly dependent on Y. Show that linear equivalence is an equivalence relation on the collection $\mathscr{B}_{\text{fin}}(V)$ of finite subsets of V.
3. In the context of, and on the basis of, Properties (LD1)–(LD3), show the following:
   (i) If $z \in V$ is linearly dependent on a finite set $Y = \{y_1, y_2, \dots, y_m\}$, and if $Z \in \mathscr{B}_{\text{fin}}(V)$ is such that $Y \subseteq Z$, then z is linearly dependent on Z.
   (ii) If the set $X = \{x_1, \dots, x_{n-1}\}$ is linearly independent, while $X \cup \{x_n\}$ is not, then x_n is linearly dependent on X.
   (iii) Given a set $X \in \mathscr{B}_{\text{fin}}(V)$, there exists a (possibly empty) subset $X' \subseteq X$, such that each element $x \in X$ is linearly dependent on X'.
   (iv) The *Steinitz exchange lemma* for finite subsets of a vector space states the following: *if* $Y = \{y_1, \dots, y_s\}$ *is linearly independent, and if each* y_σ *is linearly dependent on* $X = \{x_1, \dots, x_r\}$*, then there exists a subset* $\{x_{i_1}, \dots, x_{i_s}\} \subseteq X$ *of cardinality* s*, which may be exchanged against* Y*, such that the set resulting from* X *in this way is linearly equivalent with* X*. In particular, we have* $s \le r$*.*
   Prove the Steinitz exchange lemma by induction on s.
4. Show that the pair $(S, \mathscr{D})$ with $\mathscr{D}$ as in (11.46) is a $\mathscr{D}$-set, and that every subset A of S is $\mathscr{D}$-independent.
5. Show that every subset of a $\mathscr{D}$-independent set is again $\mathscr{D}$-independent.
6. Let $S = V$ be a vector space over the skew-field K. Define a correspondence $\mathscr{D} \subseteq V \times \mathscr{B}(V)$ via
   $$\mathscr{D}(x,A) :\Longleftrightarrow x \in \langle A \rangle, \quad (x \in V, A \subseteq V),$$
   where $\langle A \rangle$ denotes the linear span of the subset A of V. Show that $(V, \mathscr{D})$ satisfies Conditions (i)–(iv) of a $\mathscr{D}$-set.
7. Prove Part (a) of Lemma 11.7.
8. In the context of the proof of Proposition 11.8, show that a subset $A \subseteq S$ is r-independent if, and only if, it is $\mathscr{D}$-independent.
9. Let $(S, \mathscr{D})$ be a $\mathscr{D}$-set. Define a function $r : \mathscr{B}_{\text{fin}}(S) \to \mathbb{N}$ by setting $r(A) = |A'|$, where A' is a maximal $\mathscr{D}$-independent subset of A. Show that, with this definition, r satisfies Axioms (R1)–(R3), hence is a rank function over S.
10. Show that the correspondence $\mathscr{D}$ defined in (11.47) satisfies Axioms (D1)–(D4).

11. Call an element x of a group G *distinguished*, if its normal closure $\langle\!\langle x \rangle\!\rangle_G$ in G is a minimal normal subgroup of G. Let S be the set of all distinguished elements of G, and define a correspondence $\mathscr{D} \subseteq S \times \mathscr{B}(S)$ by

$$\mathscr{D}(x,A) :\Longleftrightarrow x \in \langle\!\langle A \rangle\!\rangle_G, \quad (x \in S, A \subseteq S),$$

where $\langle\!\langle A \rangle\!\rangle_G$ denotes the normal subgroup generated by the set A in G. Show that $(S, \mathscr{D})$ is a $\mathscr{D}$-set.

12. Let G be a group decomposed as a discrete direct product

$$G = \prod_{\mu \in M} G_\mu = \prod_{\nu \in N} H_\nu,$$

where the factors G_μ and H_ν are non-trivial simple groups. Making use of the result in Exercise 8, show that the index sets M and N are equipotent.

§12. Semi-simple and injective modules

Semi-simplicity is a concept widely employed in mathematics: semi-simple objects may be decomposed, in a suitable sense, into a sum of simple objects (i.e., those not containing any non-trivial proper subobjects). Injectivity, on the other hand, is a more subtle concept. Roughly speaking, for modules, it means sharing certain desirable properties with the $\mathbb{Z}$-module $\mathbb{Q}$ of rationals; for instance, if $\mathbb{Q}$ is a submodule of a module M, then it is automatically a direct summand of M. Moreover, it turns out that every module may be embedded into an injective module.

§12.1. Some preliminaries on modules

We begin with general facts and definitions. Transfinite aspects of module theory will be treated in Sections §12.2 and §12.3.

Modules are a generalisation of vector spaces, where the (skew) field of scalars is replaced by a ring. More explicitly, let M be an additive abelian group, let Λ be a ring, and suppose there is given a map $\Lambda \times M \to M$, sending a pair $(\lambda, x) \in \Lambda \times M$ to $\lambda m \in M$, such that the following axioms hold:

(M1) $\lambda(\mathbf{x}+\mathbf{y}) = \lambda\mathbf{x}+\lambda\mathbf{y}, \quad (\lambda \in \Lambda; \mathbf{x}, \mathbf{y} \in M)$,

(M2) $(\lambda+\mu)\mathbf{x} = \lambda\mathbf{x}+\mu\mathbf{x}, \quad (\lambda, \mu \in \Lambda; \mathbf{x} \in M)$,

(M3) $(\lambda\mu)\mathbf{x} = \lambda(\mu\mathbf{x}), \quad (\lambda, \mu \in \Lambda; \mathbf{x} \in M)$.

Then M is called a (left) Λ-*module*. If the ring Λ possesses an identity element 1, and if also

(M4) $1 \cdot \mathbf{x} = \mathbf{x}, \quad (\mathbf{x} \in M)$,

then M is called a *unital* Λ-*module*. In this chapter all rings will have an identity element, and all modules will be assumed to be unital.

A subset $L \subseteq M$ of a Λ-module M is called a Λ-*submodule* of M, denoted as $L \leq M$, if L is closed under addition and multiplication by elements of Λ, and forms a Λ-module in its own right with respect to the restrictions of addition and scalar multiplication. It is easy to see that a non-empty subset $L \subseteq M$ of a Λ-module M is a Λ-submodule if, and only if, it satisfies

(S1) $\mathbf{x}-\mathbf{y} \in L, \quad (\mathbf{x}, \mathbf{y} \in L)$,

(S2) $\lambda\mathbf{x} \in L, \quad (\mathbf{x} \in L, \lambda \in \Lambda)$;

cf. Exercise 1. The subsets $\{\mathbf{0}\}$ and M of a Λ-module M are Λ-submodules of M, called the *trivial* submodules of M. If a Λ-module M possesses exactly two distinct Λ-submodules, the M is called *simple* (note that, according to this definition, a simple module M satisfies $M \neq \{\mathbf{0}\}$).

A submodule $L \neq \{\mathbf{0}\}$ of a Λ-module M is called *minimal*, if every Λ-submodule K of M with $\{\mathbf{0}\} \leq K \leq L$ satisfies $K = \{\mathbf{0}\}$ or $K = L$. A minimal submodule of M is necessarily simple.

Let $S \subseteq M$ be a subset of the Λ-module M. Then we denote by $\langle S \rangle$ the Λ-submodule of M *generated by* S; that is, the intersection of all Λ-submodules of M containing S; cf. Exercise 2. It is not hard to show that (in view of the fact that M is unital),

$$\langle S \rangle = \big\{\lambda_1 x_1 + \cdots + \lambda_r \mathbf{x_r} : \lambda_1, \ldots, \lambda_r \in \Lambda, x_1, \ldots, \mathbf{x_r} \in S,\ r \in \mathbb{N}\big\}; \quad (12.53)$$

see Exercise 3. If $S \subseteq M$ consists only of one element, $\mathbf{x_0}$ say, then

$$\langle S \rangle = \big\{\lambda \mathbf{x_0} : \lambda \in \Lambda\big\}$$

is a *cyclic* Λ-module. The set

$$\mathscr{A}(M) := \big\{\lambda \in \Lambda : \lambda \mathbf{x} = \mathbf{0} \text{ for all } \mathbf{x} \in M\big\}$$

is called the *annihilator* of the Λ-module M. $\mathscr{A}(M)$ is a two-sided ideal of the ring Λ; cf. Exercise 4. If $\mathbf{x} \in M$, then the annihilator $\mathscr{A}(\langle \mathbf{x} \rangle)$ is called the *order* of the element $\mathbf{x}$.

Just as for vector spaces, the structure-preserving maps in the case of Λ-modules are the linear maps. Explicitly, if M and N are two Λ-modules, a map $\varphi : M \to N$ is called *linear* (more precisely Λ-*linear*), if it satisfies

(L1) $\varphi(\mathbf{x}+\mathbf{y}) = \varphi(\mathbf{x}) + \varphi(\mathbf{y}), \quad (\mathbf{x}, \mathbf{y} \in M)$,

(L2) $\varphi(\lambda \mathbf{x}) = \lambda \varphi(\mathbf{x}), \quad (\lambda \in \Lambda, \mathbf{x} \in M)$.

It is easy to see that the set

$$\operatorname{image}(\varphi) := \big\{\mathbf{y}' \in N : \exists \mathbf{x} \in M \text{ such that } \varphi(\mathbf{x}) = \mathbf{y}'\big\}$$

is a Λ-submodule of N, called the *image* of M under φ, or simply the image of φ. Similarly, the set

$$\ker(\varphi) := \big\{\mathbf{x} \in M : \varphi(\mathbf{x}) = \mathbf{0}\big\}$$

is a Λ-subspace of M, called the *kernel* of the linear map φ; see Exercise 5.

A sequence

$$M_1 \xrightarrow{\varphi_1} M_2 \xrightarrow{\varphi_2} M_3 \xrightarrow{\varphi_3} \cdots \xrightarrow{\varphi_{n-1}} M_n$$

of Λ-modules $M_1, \ldots, M_n$ and Λ-linear maps $\varphi_1, \ldots, \varphi_{n-1}$ is called *exact*, if

$$\operatorname{image}(\varphi_i) = \ker(\varphi_{i+1}), \quad 1 \leq i < n-1.$$

In particular, the sequence

$$M \xrightarrow{\varphi} N \longrightarrow \{\mathbf{0}\}$$

is exact, if, and only if, image$(\varphi) = N$; that is, if, and only if, φ is surjective (or a Λ-epimorphism). In this situation, we have a Λ-isomorphism

$$M/\ker(\varphi) \cong N \tag{12.54}$$

(meaning existence of a bijective Λ-linear map $M/\ker(\varphi) \to N$). Similarly, if the sequence

$$\{\mathbf{0}\} \longrightarrow M \xrightarrow{\varphi} N$$

is exact, then $\ker(\varphi) = \{\mathbf{0}\}$, so that φ is injective (or a Λ-monomorphism); see Exercises 6–9 for more details.

§12.2. Semi-simple modules

In what follows, let Λ be a ring (with identity element), and let M be a (unital) Λ-module.

Definition 12.1. A Λ-module M is called *semi-simple*, if M can be decomposed into the (discrete) direct sum of simple Λ-modules.

We have the following important characterisation of semi-simple Λ-modules.

Proposition 12.2. *A Λ-module M is semi-simple if, and only if, every submodule of M is a direct summand of M.*

Both, Definition 12.1 and Proposition 12.2, provide strong motivation to consider semi-simple modules. Our next result in particular deals with the forward implication in Proposition 12.2.

Lemma 12.3. *Let M be a module which is the (discrete) sum of simple modules S_i for $i \in I$, and let L be any submodule of M. Then there exists a subset $J \subseteq I$, such that*

$$M = L \oplus \left(\bigoplus_{j \in J} S_j \right). \tag{12.55}$$

In particular, M is semi-simple.

Proof. Let $\mathscr{F}$ be the family of all subsets J of I for which the sum

$$L + \sum_{j \in J} S_j \tag{12.56}$$

is direct. Since $\emptyset \in \mathscr{F}$, we have $\mathscr{F} \neq \emptyset$. The sum (12.56) fails to be direct if, and only if, there exists a finite subset $J_0 \subseteq J$ and vectors $\mathbf{u} \in L$ and $\mathbf{x_j} \in S_j$ for $j \in J_0$, such that $\mathbf{x_j} \neq \mathbf{0}$ and $\mathbf{u} + \sum_{j \in J_0} \mathbf{x_j} = \mathbf{0}$. It follows in particular from this that $\mathscr{F}$ is inductively ordered by inclusion. Consequently, by Zorn's Lemma, $\mathscr{F}$ has a maximal element J. By definition of $\mathscr{F}$, the sum $N := L + \sum_{j \in J} S_j$ is direct. For each index $i \in I$, the intersection $N \cap S_i$ is either S_i, or equals $\{\mathbf{0}\}$, since S_i is simple. However, if $N \cap S_i = \{\mathbf{0}\}$, then the sum $L + \sum_{j\in J} S_j + S_i$ is still direct, so that $J \subset J \cup \{i\} \in \mathscr{F}$, contradicting the maximality of J. Hence, N contains the submodules S_i for all $i \in I$, implying $N = M$, whence (12.55). The particular statement follows now by choosing $L = S_i$ in the main part of the lemma, for some index $i \in I$. □

Proof of Proposition 12.2. Suppose that each submodule of M is a direct summand; we want to show that M is semi-simple. Let us suppose for the moment (to be proved below) that every non-trivial cyclic submodule $\langle x \rangle$ of M, $\mathbf{0} \neq x \in M$, can be shown to contain a minimal (thus simple) submodule. Then, in particular, the set $\mathscr{R}$ of semi-simple submodules of M is non-empty. Moreover, $\mathscr{R}$ is seen to be inductively ordered by inclusion; see Exercise 11. By Zorn's Lemma, $\mathscr{R}$ contains a maximal element M_0. Thus, M_0 is semi-simple and a direct summand of M,

$$M = M_0 \oplus K,$$

say. Moreover, M_0 must contain every minimal submodule of M: if K_0 is a minimal submodule of M not contained in M_0, then $K_0 \cap M_0 = \{\mathbf{0}\}$ since K_0 is simple, so that $M_0 \oplus K_0$ would be a semi-simple submodule of M strictly containing M_0, contradicting the maximality of M_0. This, in turn, forces $K = \{\mathbf{0}\}$, thus $M = M_0$ is semi-simple as desired.

It remains to show that every non-trivial cyclic submodule of M contains a minimal submodule. To see this, we first note that every submodule of M inherits the direct summand property of M (that each submodule is a direct summand): if $L \leq M$ and $L' \leq L$, then L' is a submodule of M, thus $M = L' \oplus K$ for some submodule K of M, so that $L = L' \oplus (L \cap K)$. Now let $\mathbf{0} \neq x \in M$ be any non-zero vector. If $\langle x \rangle$ does not contain a minimal submodule, then $\langle x \rangle$ decomposes as the direct sum of infinitely many non-trivial submodules, implying that $\langle x \rangle$ is not finitely generated, which is absurd. □

Corollary 12.4. *Each submodule and every quotient module of a left semi-simple module is itself semi-simple.*

Proof. Let M be a semi-simple left R-module, and let $N \leq M$ be a submodule. If P is any submodule of N, then P is a submodule of M and M is semi-simple by hypothesis thus, by Proposition 12.2, P is a direct summand of M, so that $M = P \oplus P'$. Hence, intersecting with N, we find that $N = P \oplus (P' \cap N)$, which shows that P is a direct summand of N. By Proposition 12.2, N is a semi-simple left R-module, as claimed.

Next, let M/N be a quotient of M. By Proposition 12.2, N is a direct summand of M, so that $M = N \oplus N'$ for some submodule N' of M. By what we have shown in the last paragraph, N' is semi-simple, and $M/N \cong N'$. □

In what follows, we want to show that the decomposition of a semi-simple module into a (discrete) direct sum of simple modules is essentially unique. The following observation will be helpful.

Lemma 12.5. (Schur [71]) *Let M_1 and M_2 be simple (left) Λ-modules, and let $\varphi : M_1 \to M_2$ be a linear map. Then φ is either the zero map, or an isomorphism.*

Proof. By Exercise 5 below, the kernel $\ker(\varphi)$ of φ is a submodule of M_1, while $\mathrm{image}(\varphi)$ is a submodule of M_2. As M_1 is simple by hypothesis, we have $\ker(\varphi) = \{\mathbf{0}\}$, or $\ker(\varphi) = M_1$. In the second case, φ is the zero map, while in the first case φ is injective and $\mathrm{image}(\varphi) > \{\mathbf{0}\}$. Hence, in this first case, $\mathrm{image}(\varphi) = M_2$, since M_2 is simple, so that φ is an isomorphism. □

We can now prove the following uniqueness result.

Proposition 12.6. *Let*

$$M = \bigoplus_{i\in I} S_i = \bigoplus_{j\in J} T_j$$

be two decompositions of the semi-simple Λ-module M as a discrete direct sum of simple modules S_i and T_j, respectively. Then there exists a bijection $f : I \to J$, such that $S_i \cong T_{f(i)}$ for each index $i \in I$.

Proof. Choose an index $i \in I$, and let $\mathbf{x_i}$ be a non-zero vector of S_i. Then we have a unique decomposition

$$\mathbf{x_i} = \mathbf{y_{j_1}} + \cdots + \mathbf{y_{j_r}}, \quad (\mathbf{y_{j_\rho}} \in T_{j_\rho} \setminus \{\mathbf{0}\}). \tag{12.57}$$

Since S_i is simple, we have $\langle \mathbf{x_i} \rangle = S_i$, thus

$$S_i \leq \bigoplus_{\rho=1}^{r} T_{j_\rho}.$$

Moreover, the representation (12.57) induces linear maps

$$\varphi_{j_\rho}^{(i)} : S_i \longrightarrow T_{j_\rho},\ \mathbf{x_i} \longmapsto \mathbf{y_{j_\rho}}; \quad (\rho = 1, 2, \ldots, r).$$

By Lemma 12.5, each map $\varphi_{j_\rho}^{(i)}$ is a Λ-isomorphism. Hence, the direct sum decomposition $M = \bigoplus_{j\in J} T_j$ contains summands isomorphic to S_i, and S_i is contained in the direct sum of these summands. Set

$$L_i := \bigoplus_{\substack{k\in I\\ S_k \cong S_i}} S_k \quad \text{and} \quad L_i' := \bigoplus_{\substack{\ell\in J\\ T_\ell \cong S_i}} T_\ell.$$

Then we have $L_i = L_i'$ for each $i \in I$. It remains to see that the cardinality of the set of simple direct summands of L_i is an invariant of L_i, as this implies

$$|\{k \in I : S_k \cong S_i\}| = |\{\ell \in J : T_\ell \cong S_i\}|, \quad i \in I.$$

This equation in turn shows that, for each $i \in I$, there exists a bijection

$$f_i : \{k \in I : S_k \cong S_i\} \longrightarrow \{\ell \in J : T_\ell \cong S_i\},$$

and we have $T_{f_i(k)} \cong S_k$ by construction. Letting i run through a subset I' of I, such that S_i for $i \in I'$ exactly runs through the different isomorphism types of the S_i in the decomposition of M, and setting $f := \bigoplus_{i \in I} f_i$, we obtain the desired bijection $f : I \to J$ with $S_i \cong T_{f(i)}$ for all $i \in I$.

Let

$$\tilde{L}_i := \{\mathbf{x} \in L_i : \langle \mathbf{x} \rangle \text{ is simple}\},$$

and define a correspondence $\mathscr{D} \subseteq \tilde{L}_i \times \mathscr{B}(\tilde{L}_i)$ via

$$\mathscr{D}(\mathbf{x}, \mathbf{U}) :\Longleftrightarrow \mathbf{x} \in \langle \mathbf{U} \rangle; \quad (\mathbf{x} \in \tilde{L}_i, \mathbf{U} \subseteq \tilde{L}_i). \tag{12.58}$$

It is straightforward to show that $\mathscr{D}$ is a dependence relation on the set $\tilde{L}_i$ as defined in §11; cf. Exercise 12. Now let

$$L_i = \bigoplus_{k \in \tilde{I}} S_k = \bigoplus_{\ell \in \tilde{J}} T_\ell$$

be two decompositions of L_i as (discrete) direct sum of simple modules. From each of the summands S_k and T_ℓ we choose exactly one non-zero element $\mathbf{x_k}$, respectively $\mathbf{y}_\ell$. Then $\mathbf{X} = \{\mathbf{x_k} : k \in \tilde{I}\}$ and $\mathbf{Y} = \{\mathbf{y}_\ell : \ell \in \tilde{J}\}$ are maximal $\mathscr{D}$-independent subsets of the $\mathscr{D}$-set $\tilde{L}_i$, thus $|\tilde{I}| = |\tilde{J}|$ by Corollary 11.3. □

§12.3. Injective modules

The abelian group $\mathbb{Q}$ of rational numbers, when viewed as a (left) $\mathbb{Z}$-module $\mathbb{Q} = {}_{\mathbb{Z}}\mathbb{Q}$ exhibits a number of remarkable and desirable features. For instance, one may observe the following:

(i) if $M \cong \mathbb{Q}$ is a submodule of some other $\mathbb{Z}$-module N, say, then M is automatically a direct summand of N; that is, there exists a submodule $M' \leq N$ such that $N = M \oplus M'$;

(ii) given a submodule X of a $\mathbb{Z}$-module Y, then any module homomorphism $X \to \mathbb{Q}$ can be extended to a module homomorphism $Y \to \mathbb{Q}$.

An injective module is a module $Q = {}_\Lambda Q$ over a ring Λ which shares these (and other) properties of the module ${}_{\mathbb{Z}}\mathbb{Q}$. These objects were originally introduced around 1940 by Baer [11] in the context of abelian groups: an abelian group A is injective as a $\mathbb{Z}$-module if, and only if, it is *divisible*; that is if, given an element $a \in A$ and a positive integer n, there always exists an element $x \in A$ such that $nx = a$ (Baer showed, among other things, that divisible groups satisfy the analogue of property (i)).

Let Λ be a ring with identity element, and let M be a unital Λ-module. Then M is termed *injective*, if it satisfies the following: for each Λ-module L and every Λ-linear map $\varphi : L' \to M$, where L' is a Λ-submodule of L, there exists a Λ-linear map $\hat{\varphi} : L \to M$, such that $\hat{\varphi}|_{L'} = \varphi$. Equivalently, M is injective, if every diagram of the form

$$\begin{array}{ccccc} \mathbf{0} & \longrightarrow & L' & \longrightarrow & L \\ \downarrow & & \varphi\downarrow & & \\ \mathbf{0} & \longrightarrow & M & \underset{1_M}{\longrightarrow} & M \end{array}$$

with exact top row can be extended to a commutative diagram

$$\begin{array}{ccccc} \mathbf{0} & \longrightarrow & L' & \longrightarrow & L \\ \downarrow & & \varphi\downarrow & & \downarrow\hat{\varphi} \\ \mathbf{0} & \longrightarrow & M & \underset{1_M}{\longrightarrow} & M \end{array}$$

of Λ-modules and Λ-linear maps. The reader should observe that this definition of an injective Λ-module is an obvious generalisation of property (ii) above to arbitrary rings Λ with identity element.

Injective modules have a number of remarkable properties, some of which are stated in our next result; Part (iii) of Theorem 12.7 perhaps most clearly indicates the importance of this class of modules.[16]

Theorem 12.7. (Baer [11], Cartan and Eilenberg [22]) *Let M be a Λ-module. Then the following assertions are equivalent:*

(i) *For each left ideal Λ' of Λ and every Λ-linear map $\varphi : \Lambda' \to M$, there exists some element $\mathbf{x} \in M$, such that $\varphi(\lambda) = \lambda\mathbf{x}$ for all $\lambda \in \Lambda'$.*

(ii) *M is injective.*

(iii) *If M is a submodule of the Λ-module N, then M is a direct summand of N.*

For the proof of the implication (i) $\Rightarrow$ (ii) in Theorem 12.7, we require the following auxiliary result.

Lemma 12.8. *Suppose that the Λ-module M satisfies Condition* (i) *of Theorem* 12.7, *let $\mathbf{0} \to L' \to L$ be an exact sequence of Λ-modules, and let $\varphi : L' \to M$ be Λ-linear. If $\mathbf{x} \in L$ is an arbitrary vector, then φ can be extended to a Λ-linear map $\tilde{\varphi} : \langle L', \mathbf{x} \rangle \to M$.*

---

[16] In this context, see also Remark 12.10.

Proof. If $\langle \mathbf{x} \rangle \cap L' = \{\mathbf{0}\}$, then $\langle L', \mathbf{x} \rangle = L' \oplus \langle \mathbf{x} \rangle$, and our claim clearly holds in that situation. Thus, we may suppose that $\langle \mathbf{x} \rangle \cap L' \neq \{\mathbf{0}\}$. Set

$$\Lambda' := \{\lambda \in \Lambda : \lambda \mathbf{x} \in L'\}. \tag{12.59}$$

Then Λ' is a non-zero left ideal in Λ, and the map $\varphi' : \Lambda' \to M$ given by $\varphi'(\lambda') := \varphi(\lambda' \mathbf{x})$ is Λ-linear; cf. Exercise 13. By Condition (i), there exists some vector $\mathbf{x}_0 \in M$ such that

$$\varphi'(\lambda') = \varphi(\lambda' \mathbf{x}) = \lambda' \mathbf{x}_0, \quad \lambda' \in \Lambda'. \tag{12.60}$$

We claim that the assignment

$$\mathbf{x}' + \lambda \mathbf{x} \longmapsto \varphi(\mathbf{x}') + \lambda \mathbf{x}_0, \quad (\mathbf{x}' \in L', \lambda \in \Lambda)$$

defines a Λ-linear map $\tilde{\varphi} : \langle L', \mathbf{x} \rangle \to M$, which extends φ.

To see that $\tilde{\varphi}$ is well defined, suppose that

$$\mathbf{x}_1' + \lambda_1 \mathbf{x} = \mathbf{x}_2' + \lambda_2 \mathbf{x}, \quad (\mathbf{x}_1', \mathbf{x}_2' \in L'; \lambda_1, \lambda_2 \in \Lambda). \tag{12.61}$$

Then

$$(\lambda_1 - \lambda_2)\mathbf{x} = \mathbf{x}_2' - \mathbf{x}_1' \in L',$$

thus

$$\lambda_0 := \lambda_1 - \lambda_2 \in \Lambda'. \tag{12.62}$$

Equations (12.61) and (12.62) taken together imply

$$\mathbf{x}_2' = \mathbf{x}_1' + \lambda_0 \mathbf{x},$$

hence, making use of (12.60), we get

$$\varphi(\mathbf{x}_2') = \varphi(\mathbf{x}_1' + \lambda_0 \mathbf{x}) = \varphi(\mathbf{x}_1') + \lambda_0 \mathbf{x}_0.$$

Re-inserting the definition of λ_0 and rewriting the last equation yields

$$\varphi(\mathbf{x}_2') + \lambda_2 \mathbf{x}_0 = \varphi(\mathbf{x}_1') + \lambda_1 \mathbf{x}_0,$$

as desired. Consequently, the map $\tilde{\varphi} : \langle L', \mathbf{x} \rangle \to M$ is indeed well defined, and it is easy to see that $\tilde{\varphi}$ is Λ-linear; cf. Exercise 14. □

Furthermore, for the proof of implication (iii) ⇒ (i), we shall need the following result.

Lemma 12.9. *Let Λ be a ring with identity element, and let M be a (unital) Λ-module. Then there exists a Λ-module $\bar{M}$ containing M as a submodule, such that $\bar{M}$ has the following property:*

(∗) *for each left ideal Λ' of Λ and every Λ-linear map $\varphi : \Lambda' \to M$, there exists some element $\mathbf{x}_0 \in \bar{M}$ such that $\varphi(\lambda') = \lambda' \mathbf{x}_0$ for all $\lambda' \in \Lambda'$.*

Proof. Let Φ be the set of all pairs (Λ', φ), where Λ' is a left ideal of Λ and $\varphi : \Lambda' \to M$ is Λ-linear. Let F_Φ be the free Λ-module with basis Φ (see Part (a) of Exercise 16 below). Moreover, let $\bar{M}$ be the quotient of $M \oplus F_\Phi$ by the Λ-submodule N generated by the elements

$$(\varphi(\lambda'), -\lambda'(\Lambda', \varphi)), \quad ((\Lambda', \varphi) \in \Phi, \lambda' \in \Lambda').$$

Then $\bar{M}$ is a Λ-module, and the map $\mathbf{x} \mapsto (\mathbf{x}, \mathbf{0})$ for $\mathbf{x} \in M$ induces a Λ-linear map $\iota : M \to \bar{M}$. If $\iota(\mathbf{x}) = \mathbf{0}$, then, using Λ-linearity of the map φ, we have

$$\begin{aligned}(\mathbf{x}, \mathbf{0}) &= \sum_\rho \lambda_\rho (\varphi_\rho(\lambda'_\rho), -\lambda'_\rho(\Lambda'_\rho, \varphi_\rho)) \\ &= \sum_\rho (\varphi_\rho(\lambda_\rho \lambda'_\rho), -\lambda_\rho \lambda'_\rho(\Lambda'_\rho, \varphi_\rho)) \\ &= (\sum_\rho \varphi_\rho(\lambda_\rho \lambda'_\rho), -\sum_\rho \lambda_\rho \lambda'_\rho(\Lambda'_\rho, \varphi_\rho)).\end{aligned}$$

It follows in particular that

$$\sum_\rho = \lambda_\rho \lambda'_\rho(\Lambda'_\rho, \varphi_\rho) = \mathbf{0}$$

in F_Φ, which implies $\lambda_\rho \lambda'_\rho = 0$ for all ρ, thus

$$\mathbf{x} = \sum_\rho \varphi_\rho(\lambda_\rho \lambda'_\rho) = \mathbf{0} \in M.$$

Hence, ι is an embedding of Λ-modules, so that M may be regarded as a submodule of $\bar{M}$.

Finally, let $\varphi : \Lambda' \to M$ be a Λ-linear map, where Λ' is a left ideal of Λ. Then $(\Lambda', \varphi) \in \Phi$. Let $\mathbf{x}_0$ be the image in $\bar{M}$ of the element $(\mathbf{0}, (\Lambda', \varphi)) \in M \oplus F_\Phi$. Then, for $\lambda' \in \Lambda'$, we have

$$\iota(\varphi(\lambda')) = (\varphi(\lambda'), \mathbf{0}) + N = (\mathbf{0}, \lambda'(\Lambda', \varphi)) + N = \lambda'((\mathbf{0}, (\Lambda', \varphi)) + N) = \lambda' \mathbf{x}_0.$$

Hence, identifying an element $\mathbf{x} \in M$ with $\iota(\mathbf{x}) \in \bar{M}$, we have indeed

$$\varphi(\lambda') = \lambda' \mathbf{x}_0, \quad \lambda' \in \Lambda',$$

as required. □

Proof of Theorem 12.7. (i) $\Rightarrow$ (ii). Let $\mathbf{0} \to L' \to L$ be an exact sequence of Λ-modules, and let $\varphi : L' \to M$ be Λ-linear. We want to show that φ extends to a Λ-linear map $\hat{\varphi} : L \to M$. In order to do that, consider the set

$$\mathscr{E} := \{(\tilde{L}, \tilde{\varphi}) : L' \leq \tilde{L} \leq L, \tilde{\varphi} : \tilde{L} \to M \ \Lambda\text{-linear, and } \tilde{\varphi}|_{L'} = \varphi\} \tag{12.63}$$

of partial extensions of φ. The set $\mathscr{E}$ is not empty, since $(L', \varphi) \in \mathscr{E}$. For $(\tilde{L}_1, \tilde{\varphi}_1)$, $(\tilde{L}, \tilde{\varphi}_2) \in \mathscr{E}$, define

$$(\tilde{L}_1, \tilde{\varphi}_1) = (\tilde{L}_2, \tilde{\varphi}_2) :\Longleftrightarrow \tilde{L}_1 = \tilde{L}_2 \text{ and } \tilde{\varphi}_1 = \tilde{\varphi}_2,$$

and introduce an order relation $\leq$ on $\mathscr{E}$ via

$$(\tilde{L}_1, \tilde{\varphi}_1) \leq (\tilde{L}_2, \tilde{\varphi}_2) :\Longleftrightarrow \tilde{L}_1 \subseteq \tilde{L}_2 \text{ and } \tilde{\varphi}_2|_{\tilde{L}_1} = \tilde{\varphi}_1. \tag{12.64}$$

One shows that $(\mathscr{E}, \leq)$ is inductively ordered (cf. Exercise 15), thus, by Zorn's Lemma, $\mathscr{E}$ has a maximal element $(\tilde{L}^*, \tilde{\varphi}^*)$. If we had $\tilde{L}^* < L$, then we could pick an element $\mathbf{x} \in L \setminus \tilde{L}^*$, and apply Lemma 12.8 to show existence of some $(\tilde{\tilde{L}}, \tilde{\tilde{\varphi}}) \in \mathscr{E}$, such that $(\tilde{L}^*, \tilde{\varphi}^*) < (\tilde{\tilde{L}}, \tilde{\tilde{\varphi}})$, contradicting maximality of $(\tilde{L}^*, \tilde{\varphi}^*)$. Hence, $\tilde{L}^* = L$, so that $\tilde{\varphi}^* : L \to M$ is a Λ-linear extension of φ, as desired, implying that the Λ-module M is injective.

(ii) $\Rightarrow$ (iii). Suppose that the Λ-module M is injective, and let N be a Λ-module containing M as a submodule. We want to show that M is a direct summand of N. By our hypothesis, there exists a Λ-linear map $\psi : N \to M$, such that the diagram

$$\begin{array}{ccccc}
\mathbf{0} & \longrightarrow & M & \xrightarrow{\ \iota\ } & N \\
\big\downarrow & & {\scriptstyle 1_M}\big\downarrow & & \big\downarrow{\scriptstyle \psi} \\
\mathbf{0} & \longrightarrow & M & \xrightarrow[1_M]{} & M
\end{array}$$

commutes, where $\iota : M \to N$ is the natural embedding; that is, $\psi|_M = 1_M$. If $\mathbf{y} \in N$ is an arbitrary element, and $\psi(\mathbf{y}) = \mathbf{x} \in M$, then there also exists some element $\tilde{\mathbf{x}} \in M$, such that $\psi(\tilde{\mathbf{x}}) = \mathbf{x}$. Thus,

$$\psi(\mathbf{y} - \tilde{\mathbf{x}}) = \psi(\mathbf{y}) - \psi(\tilde{\mathbf{x}}) = \mathbf{0},$$

so that

$$\tilde{\mathbf{y}} := \mathbf{y} - \tilde{\mathbf{x}} \in \ker(\psi).$$

It follows that $\mathbf{y} = \tilde{\mathbf{x}} + \tilde{\mathbf{y}}$, showing that

$$N = M + \ker(\psi).$$

Let $\mathbf{x} \in M \cap \ker(\psi)$. Then $\mathbf{x} \in M$ and $\psi(\mathbf{x}) = \mathbf{0}$, thus $\mathbf{x} = \mathbf{0}$, since $\psi|_M = 1_M$. It follows that

$$N = M \oplus \ker(\psi),$$

in particular, M is a direct summand of N, as required.

(iii) $\Rightarrow$ (i). Let M be a Λ-module satisfying Condition (iii), let Λ' be a left ideal of Λ, and let $\varphi : \Lambda' \to M$ be a Λ-linear map. By Lemma 12.9, there exists a Λ-module N containing M as a submodule as well as an element $\mathbf{x}_0 \in N$, such that $N = M + \langle \mathbf{x}_0 \rangle$ and

$$\varphi(\lambda') = \lambda' \mathbf{x}_0, \quad \lambda' \in \Lambda'. \tag{12.65}$$

By Condition (iii), M is a direct summand of N, thus

$$N = M \oplus K \tag{12.66}$$

with some Λ-submodule K of N. Consequently, each element $\mathbf{y} \in N$ may be written uniquely in the form

$$\mathbf{y} = \tilde{\mathbf{x}} + \tilde{\mathbf{y}}, \quad (\tilde{\mathbf{x}} \in M, \tilde{\mathbf{y}} \in K);$$

in particular, we have

$$\mathbf{x}_0 = \tilde{\mathbf{x}}_0 + \tilde{\mathbf{y}}_0, \quad (\tilde{\mathbf{x}}_0 \in M, \tilde{\mathbf{y}}_0 \in K),$$

where $\mathbf{x}_0$ is as in (12.65). It follows that

$$\varphi(\lambda') = \lambda'\mathbf{x}_0 = \lambda'\tilde{\mathbf{x}}_0 + \lambda'\tilde{\mathbf{y}}_0, \quad \lambda' \in \Lambda',$$

implying

$$\varphi(\lambda') - \lambda'\tilde{\mathbf{x}}_0 = \lambda'\tilde{\mathbf{y}}_0, \quad \lambda' \in \Lambda'. \tag{12.67}$$

As the left-hand side of Equation (12.67) is in M, while the right-hand side lies in K, we conclude from (12.66) that $\lambda'\tilde{\mathbf{y}}_0 = \mathbf{0}$ for all $\lambda' \in \Lambda'$, so that

$$\varphi(\lambda') = \lambda'\tilde{\mathbf{x}}_0, \quad \lambda' \in \Lambda'.$$

Hence, M satisfies Condition (i), and the proof of Theorem 12.7 is complete.

□

Remark 12.10. In the light of Theorem 12.7, Lemma 12.9 states precisely that every Λ-module can be embedded into an injective Λ-module.

Exercises for § 12.

1. Let M be a Λ-module, and let $L \subseteq M$ be a non-empty subset. Show that L is a Λ-submodule of M if, and only if, Conditions (S1) and (S2) hold.
2. Let M be a Λ-module, and let $\mathscr{S} = \{L_i\}_{i\in I}$ be a non-empty family of Λ-submodules. Show that the intersection $\bigcap \mathscr{S} = \bigcap_{i\in I} L_i$ is again a Λ-submodule of M.
3. Let M be a (unital) Λ-module, and let $S \subseteq M$ be a subset.
   (a) If $S = \emptyset$, what is $\langle S \rangle$?
   (b) Prove Equation (12.53).
4. Show that the annihilator $\mathscr{A}(M)$ of a Λ-module M is a two-sided ideal of the ring Λ.
5. Let M,N be Λ-modules, and let $\varphi : M \to N$ be a linear map. Show that $\ker(\varphi)$ is a Λ-submodule of M, and that $\text{image}(\varphi)$ is a Λ-submodule of the module N.
6. Let M,N be Λ-modules, and let $\varphi : M \to N$ be a Λ-linear map. Show that φ is injective if, and only if, $\ker(\varphi) = \{\mathbf{0}\}$.
7. Let M be a Λ-module, and let $L \leq M$ be a Λ-submodule. Define an equivalence relation $\sim$ on M by

   $$\mathbf{x} \sim \mathbf{y} :\Longleftrightarrow \mathbf{y} - \mathbf{x} \in L, \quad (\mathbf{x},\mathbf{y} \in M).$$

   Show that the set of equivalence classes $M/L = \{\mathbf{x} + L : \mathbf{x} \in M\}$ forms a Λ-module in a natural way, and that the map $\pi : M \to M/L$ given by $\mathbf{x} \mapsto \mathbf{x} + L$ is a surjective Λ-linear map.

8. Let M and N be Λ-modules, and let $\varphi : M \to N$ be a Λ-linear map.

(i) The map φ is termed a Λ-*epimorphism* if, for all Λ-linear maps ψ, χ defined on N, $\psi \circ \varphi = \chi \circ \varphi$ implies $\psi = \chi$. Show that φ is a Λ-epimorphism if, and only if, φ is surjective.

(ii) The map φ is called a Λ-*monomorphism* if, for all Λ-linear maps ψ, χ with co-domain M, $\varphi \circ \psi = \varphi \circ \chi$ implies $\psi = \chi$. Show that φ is a Λ-monomorphism if, and only if, φ is injective.

(iii) The map φ is termed a Λ-*isomorphism*, if there exists a Λ-linear map $\varphi' : N \to M$ such that $\varphi' \circ \varphi = 1_M$ and $\varphi \circ \varphi' = 1_N$. Show that φ is a Λ-isomorphism if, and only if, φ is a bijection.

9. Let M and N be Λ-modules, and let $\varphi : M \to N$ be Λ-linear. Exhibit a Λ-isomorphism showing that $M/\ker(\varphi) \cong \mathrm{image}(\varphi)$.

10. Making use of Zorn's Lemma, but without recourse to Lemma 12.3, prove the forward implication of Proposition 12.2.

11. Show that the set $\mathscr{R}$ of all semi-simple submodules of M (as introduced in the proof of Proposition 12.2) is inductively ordered by inclusion.

12. Prove that the correspondence $\mathscr{D} \subseteq \tilde{L}_i \times \mathscr{B}(\tilde{L}_i)$ as defined in (12.58) satisfies Axioms (D1)–(D4) of a dependence relation.

13. Prove that the subset Λ' of the ring Λ defined in (12.59) is a non-zero left ideal of Λ.

14. Show that the map $\tilde{\varphi}$ defined in the proof of Lemma 12.8 is Λ-linear.

15. Show that the set $\mathscr{E}$ defined in (12.63) is inductively ordered under the relation $\leq$ introduced in (12.64).

16. A family of vectors $\mathbf{x}_1, \ldots, \mathbf{x}_n$ in a Λ-module M is called *linearly independent*, if a relation

$$\lambda_1 \mathbf{x}_1 + \lambda_2 \mathbf{x}_2 + \cdots + \lambda_n \mathbf{x}_n = \mathbf{0}$$

with $\lambda_\nu \in \Lambda$ holds in M only for $\lambda_1 = \lambda_2 = \cdots = \lambda_n = 0$. More generally, a subset $X \subseteq M$ is termed linearly independent, if every finite subset of X is linearly independent. A linearly independent generating system of a Λ-module M is called a *basis* of M, and a Λ-module is called *free*, if it possesses a basis. Show the following:

(a) For each ring Λ with identity element, and every set X, there exists a free Λ-module F_X with basis X.

(b) If M is a free Λ-module with basis X, then every set-theoretic map $\varphi : X \to N$ into a Λ-module N extends uniquely to a Λ-linear map $\hat{\varphi} : M \to N$.

(c) If

$$\{\mathbf{0}\} \longrightarrow M' \stackrel{\alpha}{\longrightarrow} M \stackrel{\beta}{\longrightarrow} M'' \longrightarrow \{\mathbf{0}\} \tag{12.68}$$

is an exact sequence of Λ-modules and Λ-linear maps, in which M'' is free, then there exists a Λ-linear map $\beta^* : M'' \to M$ such that $\beta \circ \beta^* = 1_{M''}$ (we say that the sequence (12.68) splits).

(d) Let Λ be a ring with identity element, and let M be any Λ-module. Then there exists a free Λ-module F and a submodule $G \leq F$, such that $M \cong F/G$.

§ 13. The Jacobson radical of a ring

The construct figuring in the title of this section was originally introduced by Jacobson in 1945 as a technique for building up a general theory of radicals for arbitrary rings; cf. [44] as well as Chapter I of [45].[17] Via Nakayama's Lemma, which lays open the connection between the Jacobson radical $\mathfrak{J}(\Lambda)$ of a ring Λ and the finitely generated modules over Λ, $\mathfrak{J}(\Lambda)$ also connects to certain topics in algebraic geometry; see, in particular, [61] and Chapter 2 in [5]. Briefly, the Jacobson radical $\mathfrak{J}(\Lambda)$ of a ring Λ may be defined as the ideal consisting of those elements of Λ which annihilate all simple Λ-modules; that is,

$$\mathfrak{J}(\Lambda) = \{\lambda \in \Lambda : \lambda M = \mathbf{0} \text{ for all } M \text{ simple}\}.$$

If Λ is commutative, then $\mathfrak{J}(\Lambda)$ is precisely the intersection of all maximal ideals of Λ. The Jacobson radical has a number of internal characterisations (see Theorem 13.9 below), some of which are suitable for extending the notion to non-unital rings.

§13.1. Irreducible modules

Definition 13.1. Let Λ be a ring, and let M be a Λ-module. Then M is called *irreducible*, if $M \neq \{\mathbf{0}\}$, and $\Lambda\mathbf{x} = M$ for each vector $\mathbf{x} \in M \setminus \{\mathbf{0}\}$.

The following characterisation of irreducible modules will be useful in what follows.

Lemma 13.2. *Let Λ be a ring, and let M be a Λ-module. Then the following assertions are equivalent:*

(i) *M is irreducible;*

(ii) *M is simple, and $\Lambda \cdot M \neq \{\mathbf{0}\}$.*

Proof. (i) $\Rightarrow$ (ii). Suppose that M is irreducible. Then, in particular, $M \neq \{\mathbf{0}\}$, thus there exists some vector $\mathbf{x} \in M$ with $\mathbf{x} \neq \mathbf{0}$. Hence,

$$\Lambda \cdot M \supseteq \Lambda \cdot \mathbf{x} = M \neq \{\mathbf{0}\},$$

so $\Lambda \cdot M \neq \{\mathbf{0}\}$. Next, let M' be a non-trivial Λ-submodule of M. Then there exists some $\mathbf{x} \in M'$ with $\mathbf{x} \neq \mathbf{0}$. It follows that $M = \Lambda \cdot \mathbf{x} \leq M'$, so that $M' = M$. Hence, M is simple, whence (ii).

(ii) $\Rightarrow$ (i). Conversely, suppose that M is simple, and that $\Lambda \cdot M \neq \{\mathbf{0}\}$. Then certainly $M \neq \{\mathbf{0}\}$. Let $\mathbf{x} \in M \setminus \{\mathbf{0}\}$ be an arbitrary non-zero vector, and consider $\Lambda\mathbf{x}$. It is straightforward to see that $\Lambda\mathbf{x}$ is a Λ-submodule of

[17] Roughly speaking, a radical of a ring is an ideal of "undesirable" elements of the ring; cf. [2]–[4] and [51].

M. Since M is simple by hypothesis, we have $\Lambda\mathbf{x} = \{\mathbf{0}\}$ or $\Lambda\mathbf{x} = M$, and we wish to exclude the former possibility. Let

$$N := \{\mathbf{y} \in M : \Lambda\mathbf{y} = \{\mathbf{0}\}\}.$$

We check that N is a Λ-submodule of M. Since $\Lambda \cdot \mathbf{0} = \{\mathbf{0}\}$, we have $\mathbf{0} \in N$; in particular, $N \neq \emptyset$. Next, if $\mathbf{y}_1, \mathbf{y}_2 \in N$, then $\Lambda\mathbf{y}_1 = \{\mathbf{0}\} = \Lambda\mathbf{y}_2$, thus

$$\Lambda \cdot (\mathbf{y}_1 + \mathbf{y}_2) \leq \Lambda\mathbf{y}_1 + \Lambda\mathbf{y}_2 = \{\mathbf{0}\},$$

implying $\mathbf{y}_1 + \mathbf{y}_2 \in N$. Moreover, if $\lambda \in \Lambda$ and $\mathbf{y} \in N$, then $\Lambda\mathbf{y} = \{\mathbf{0}\}$, so

$$\Lambda \cdot (\lambda\mathbf{y}) \leq \Lambda\mathbf{y} = \{\mathbf{0}\}.$$

Hence, $\lambda\mathbf{y} \in N$, and N is a Λ-submodule of M, as claimed. Since $\Lambda \cdot M \neq \{\mathbf{0}\}$ by hypothesis, there must exist some vector $\mathbf{y} \in M$, such that $\Lambda\mathbf{y} \neq \{\mathbf{0}\}$, thus $\mathbf{y} \notin N$, and $N < M$. By simplicity of M, we get $N = \{\mathbf{0}\}$. It follows that $\mathbf{x} \notin N$, so $\Lambda\mathbf{x} \neq \{\mathbf{0}\}$, hence $\Lambda\mathbf{x} = M$ by simplicity of M, as desired, whence (i). □

Corollary 13.3. *Let Λ be a ring with identity element, and let M be a unital Λ-module. Then M is irreducible if, and only if, M is simple.*

Proof. If M is irreducible, then M is simple by the forward direction of Lemma 13.2. Suppose conversely that M is simple. Then M is non-trivial by definition, thus contains some non-zero vector $\mathbf{x}$. Hence,

$$\Lambda \cdot M \ni 1 \cdot \mathbf{x} = \mathbf{x} \neq \mathbf{0},$$

so that $\Lambda \cdot M \neq \{\mathbf{0}\}$. Irreducibility of M now follows from the backward implication of Lemma 13.2. □

§13.2. Modularity, regularity, and the circle operation

Let Λ be a ring. A left ideal Λ' of Λ is termed *left modular*, if Λ contains a right identity element modulo Λ'; that is, if Λ contains some element λ_0, such that

$$\lambda - \lambda\lambda_0 \in \Lambda', \quad \lambda \in \Lambda.$$

Similarly, a right ideal Λ'' of Λ is called *right modular*, if Λ contains a left identity element $\tilde{\lambda}_0$ modulo Λ'':

$$\lambda - \tilde{\lambda}_0\lambda \in \Lambda'', \quad \lambda \in \Lambda.$$

We note that a proper left (right) modular ideal of a ring Λ can always be embedded into a maximal left (right) ideal of Λ; cf. Exercise 2 below for this generalisation of Krull's theorem (Proposition 4.8). A two-sided ideal $\Lambda' \leq \Lambda$ which is both left and right modular is called a *modular ideal* of Λ. Thus, an ideal Λ' of Λ is modular if, and only if, the factor ring Λ/Λ' contains an identity element. If Λ possesses a right or left identity element, then every left ideal (respectively every right ideal) of Λ is left (respectively right) modular. A maximal left ideal (right ideal) of Λ, which is left (right) modular, is termed a *modular maximal left ideal (right ideal)* of Λ.

Definition 13.4. Let Λ be a ring, let M be a (left) Λ-module, and let $\mathbf{x} \in M$. Then the set

$$\mathrm{Ann}_\Lambda(\mathbf{x}) := \{\lambda \in \Lambda : \lambda\mathbf{x} = \mathbf{0}\}$$

is called the *annihilator* of $\mathbf{x}$.

Lemma 13.5. *Let Λ be a ring, and let M be a Λ-module.*

(a) *The annihilator* $\mathrm{Ann}_\Lambda(\mathbf{x})$ *of a vector* $\mathbf{x} \in M$ *is a left ideal of* Λ.

(b) *If M is irreducible and* $\mathbf{x} \neq \mathbf{0}$, *then* $\mathrm{Ann}_\Lambda(\mathbf{x})$ *is a modular maximal left ideal of* Λ.

Proof. (a) This is left to the reader; cf. Exercise 5.

(b) Suppose that M is irreducible Λ-module and that $\mathbf{x} \in M \setminus \{\mathbf{0}\}$. The map $\Lambda \to M$ given by $\lambda \mapsto \lambda\mathbf{x}$ is surjective and Λ-linear. Its kernel is the annihilator $\mathrm{Ann}_\Lambda(\mathbf{x})$ of $\mathbf{x}$, a left ideal of Λ by Part (a). We have $M \cong \Lambda/\mathrm{Ann}_\Lambda(\mathbf{x})$ as Λ-modules and, by Lemma 13.2, the module M is simple, implying that $\mathrm{Ann}_\Lambda(\mathbf{x})$ is a maximal left ideal of Λ. Moreover, since $\mathbf{x} \in M = \Lambda\mathbf{x}$, there exists some element $\lambda_0 \in \Lambda$ with $\lambda_0\mathbf{x} = \mathbf{x}$. Therefore,

$$\lambda\mathbf{x} = (\lambda\lambda_0)\mathbf{x}, \quad \lambda \in \Lambda,$$

so that

$$(\lambda - \lambda\lambda_0)\mathbf{x} = \mathbf{0}, \quad \lambda \in \Lambda.$$

Hence, $\lambda - \lambda\lambda_0 \in \mathrm{Ann}_\Lambda(\mathbf{x})$, showing that $\mathrm{Ann}_\Lambda(\mathbf{x})$ is a modular maximal left ideal. □

Call a left ideal Λ' of a ring Λ *quasi-modular* if, for each $\lambda \in \Lambda - \Lambda'$, there exists some $\tilde{\lambda} \in \Lambda$ with $\lambda\tilde{\lambda} \notin \Lambda'$. Clearly, every modular left ideal of a ring is quasi-modular, but the converse does not hold in general; cf. [77].

Now let $\lambda_1 \in \Lambda$ be a fixed arbitrary element of the ring Λ, and set

$$\Lambda_1' := \{\lambda - \lambda\lambda_1 : \lambda \in \Lambda\}. \tag{13.1}$$

Then Λ_1' is a modular left ideal of Λ, and we have $\Lambda_1' = \Lambda$ if, and only if, $\lambda_1 \in \Lambda_1'$; cf. Exercise 4. In this case, the element λ_1 is termed *left quasi-regular* (l.q.r.); that is, an element $\lambda_1 \in \Lambda$ is l.q.r. if, and only if, there exists some $\bar{\lambda}_1 \in \Lambda$, such that

$$\lambda_1 + \bar{\lambda}_1 - \bar{\lambda}_1\lambda_1 = 0. \tag{13.2}$$

An element $\bar{\lambda}_1 \in \Lambda$ as in (13.2) is called a *left quasi-inverse* of λ_1. Similarly, an element $\lambda_1 \in \Lambda$ is called *right quasi-regular* (r.q.r.), if there exists an element $\bar{\lambda}_1' \in \Lambda$, such that

$$\lambda_1 + \bar{\lambda}_1' - \lambda_1\bar{\lambda}_1' = 0,$$

and $\bar{\lambda}_1'$ is called a *right quasi-inverse* of λ_1. If $\lambda_1 \in \Lambda$ is both l.q.r. and r.q.r., then λ_1 is called *quasi-regular* (q.r.).

In an arbitrary ring Λ, we may define a new binary operation $\circ$ called *circle operation*, via

$$\lambda \circ \mu := \lambda + \mu - \lambda\mu, \quad (\lambda, \mu \in \Lambda).$$

Direct computation shows that $\circ$ is associative on Λ; cf. Exercise 6. Moreover, we have

$$0 \circ \lambda = \lambda = \lambda \circ 0, \quad (\lambda \in \Lambda);$$

that is, the element 0 is a neutral element for the circle operation, so that $(\Lambda, \circ)$ is a semigroup with identity element . If Λ contains an identity element 1, then we also have

$$1 \circ \lambda = 1 = \lambda \circ 1, \quad (\lambda \in \Lambda).$$

The fact that $\bar{\lambda}_1$ is a left quasi-inverse of $\lambda_1 \in \Lambda$ can now be expressed more concisely in the form $\bar{\lambda}_1 \circ \lambda_1 = 0$. Similarly, $\lambda_1 \circ \bar{\lambda}_1' = 0$ means that $\bar{\lambda}_1'$ is a right quasi-inverse of λ_1.

Lemma 13.6. *Let Λ be a ring, and let $\mathscr{C}(\Lambda)$ be the set of quasi-regular elements of Λ. Then $\mathscr{C}(\Lambda)$ is a group under the circle operation $\circ$ of Λ, called the* circle group *of Λ.*

Proof. Since 0 is a neutral element for the circle operation on Λ, we have in particular that $0 \in \mathscr{C}(\Lambda)$; thus, $\mathscr{C}(\Lambda)$ is non-empty. Next, let $\lambda_1, \lambda_2 \in \mathscr{C}(\Lambda)$ be quasi-regular elements of Λ. Then there exist elements $\bar{\lambda}_1, \bar{\lambda}_1', \bar{\lambda}_2, \bar{\lambda}_2' \in \Lambda$, such that the equations

$$\bar{\lambda}_1 \circ \lambda_1 = 0 = \lambda_1 \circ \bar{\lambda}_1',$$
$$\bar{\lambda}_2 \circ \lambda_2 = 0 = \lambda_2 \circ \bar{\lambda}_2'$$

hold true. Consequently, making use of associativity of the circle operation, we get

$$\begin{aligned}(\lambda_1 \circ \lambda_2) \circ (\bar{\lambda}_2' \circ \bar{\lambda}_1') &= \lambda_1 \circ (\lambda_2 \circ \bar{\lambda}_2') \circ \bar{\lambda}_1' \\ &= (\lambda_1 \circ 0) \circ \bar{\lambda}_1' \\ &= \lambda_1 \circ \bar{\lambda}_1' = 0\end{aligned}$$

as well as

$$\begin{aligned}(\bar{\lambda}_2 \circ \bar{\lambda}_1) \circ (\lambda_1 \circ \lambda_2) &= \bar{\lambda}_2 \circ (\bar{\lambda}_1 \circ \lambda_1) \circ \lambda_2 \\ &= (\bar{\lambda}_2 \circ 0) \circ \lambda_2 \\ &= \bar{\lambda}_2 \circ \lambda_2 = 0.\end{aligned}$$

Hence, $\lambda_1 \circ \lambda_2$ is again quasi-regular, so that $\lambda_1 \circ \lambda_2 \in \mathscr{C}(\Lambda)$, and $\mathscr{C}(\Lambda)$ is closed under the circle operation. So far, we have shown that $(\mathscr{C}(\Lambda), \circ)$ is a semigroup with identity element 0. It remains to see that each element of $\mathscr{C}(\Lambda)$ has a (two-sided) inverse with respect to the circle operation. Let $\lambda_1 \in \mathscr{C}(\Lambda)$. Then there exist elements $\bar{\lambda}_1, \bar{\lambda}_1' \in \Lambda$, such that

$$\bar{\lambda}_1 \circ \lambda_1 = 0 = \lambda_1 \circ \bar{\lambda}_1'.$$

We conclude that

$$\bar{\lambda}_1 = \bar{\lambda}_1 \circ 0 = \bar{\lambda}_1 \circ (\lambda_1 \circ \bar{\lambda}_1') = (\bar{\lambda}_1 \circ \lambda_1) \circ \bar{\lambda}_1' = 0 \circ \bar{\lambda}_1' = \bar{\lambda}_1',$$

so that $\bar{\lambda}_1 = \bar{\lambda}_1'$ is again quasi-regular, and inverse to λ_1 with respect to the circle operation. It follows that $(\mathscr{C}(\Lambda), \circ)$ is a group, as claimed. □

See Exercises 7–10 below for more on the circle group construction.

§13.3. The Jacobson radical, I: definition and first results

We come to the crucial definition of this section.

Definition 13.7. Let Λ be a ring. Then the set

$$\mathfrak{J}(\Lambda) := \{\lambda \in \Lambda : \mu\lambda \text{ l.q.r. for each } \mu \in \Lambda\}$$

is called the *Jacobson radical* (or simply *radical*) of Λ.

Our next result records some of the more immediate properties of the Jacobson radical.

Proposition 13.8. *Let Λ be a ring.*

(i) *The Jacobson radical $\mathfrak{J}(\Lambda)$ is a two-sided ideal of Λ.*

(ii) *Each element of $\mathfrak{J}(\Lambda)$ is l.q.r.*

(iii) *We have $\mathfrak{J}(\Lambda/\mathfrak{J}(\Lambda)) = \{\mathbf{0}\}$.*

Proof. (i) Let $\lambda_1, \lambda_2 \in \mathfrak{J}(\Lambda)$, and let $\mu \in \Lambda$ be arbitrary. We first want to show that $\lambda_1 - \lambda_2 \in \mathfrak{J}(\Lambda)$; that is, that $\mu(\lambda_1 - \lambda_2)$ is l.q.r. By hypothesis, there exist elements $\lambda_1', \lambda_2' \in \Lambda$, such that

$$\lambda_1' \circ (\mu\lambda_1) = 0 = \lambda_1' + \mu\lambda_1 - \lambda_1'\mu\lambda_1 \tag{13.3}$$

and

$$\lambda_2' \circ ((\lambda_1'\mu - \mu)\lambda_2) = 0 = \lambda_2' + (\lambda_1'\mu - \mu)\lambda_2 - \lambda_2'(\lambda_1'\mu - \mu)\lambda_2. \tag{13.4}$$

Thus,

$$\begin{aligned}(\lambda_2' \circ \lambda_1') \circ (\mu(\lambda_1 - \lambda_2)) &= \mu(\lambda_1 - \lambda_2) + (\lambda_1' + \lambda_2' - \lambda_2'\lambda_1') \\ &\qquad - (\lambda_1' + \lambda_2' - \lambda_2'\lambda_1')\mu(\lambda_1 - \lambda_2) \\ &= [\mu\lambda_1 + \lambda_1' - \lambda_1'\mu\lambda_1] \\ &\qquad + [(\lambda_1'\mu - \mu)\lambda_2 + \lambda_2' - \lambda_2'(\lambda_1'\mu - \mu)\lambda_2] \\ &\qquad - \lambda_2'[\mu\lambda_1 + \lambda_1' - \lambda_1'\mu\lambda_1] \\ &= [\lambda_1' \circ (\mu\lambda_1)] + [\lambda_2' \circ ((\lambda_1'\mu - \mu)\lambda_2)] \\ &\qquad - \lambda_2'[\lambda_1' \circ (\mu\lambda_1)] \\ &= 0,\end{aligned}$$

where, in the second to last line, we have applied (13.3) to the first and third bracket, and (13.4) to the second bracket. Hence, $\mu(\lambda_1 - \lambda_2)$ is again l.q.r., implying $\lambda_1 - \lambda_2 \in \mathfrak{J}(\Lambda)$, since $\mu \in \Lambda$ was chosen arbitrarily.

Next, for an arbitrary element $\nu \in \Lambda$, the element $\nu(\mu\lambda_1) = (\nu\mu)\lambda_1$ is l.q.r. (with left quasi-inverse λ_3', say), since $\lambda_1 \in \mathfrak{J}(\Lambda)$; thus, $\mu\lambda_1 \in \mathfrak{J}(\Lambda)$, showing that $\mathfrak{J}(\Lambda)$ is a left ideal of Λ.

Finally, we have (with λ_3' as introduced in the last paragraph),

$$\begin{aligned}(\mu\lambda_1\lambda_3'\nu - \mu\lambda_1\nu) \circ (\mu\lambda_1\nu) &= \mu\lambda_1\nu + (\mu\lambda_1\lambda_3'\nu - \mu\lambda_1\nu) \\ &\qquad - (\mu\lambda_1\lambda_3'\nu - \mu\lambda_1\nu)\mu\lambda_1\nu \\ &= \mu\lambda_1(\nu\mu\lambda_1 + \lambda_3' - \lambda_3'\nu\mu\lambda_1)\nu \\ &= 0,\end{aligned}$$

since $\lambda_3' \circ (\nu\mu\lambda_1) = 0$ by definition of λ_3'. It follows that $\mu(\lambda_1\nu)$ is l.q.r., implying $\lambda_1\nu \in \mathfrak{J}(\Lambda)$ for all $\nu \in \Lambda$. Hence, $\mathfrak{J}(\Lambda)$ is a two-sided ideal of Λ, as claimed.

(ii) Let $\lambda \in \mathfrak{J}(\Lambda)$ be arbitrary. Then, in particular, the element λ^2 is l.q.r., thus there exists some $\mu \in \Lambda$ such that

$$\mu \circ \lambda^2 = 0 = \mu + \lambda^2 - \mu\lambda^2.$$

Consequently,

$$(\mu - \lambda + \mu\lambda) \circ \lambda = \lambda + (\mu - \lambda + \mu\lambda) - (\mu - \lambda + \mu\lambda)\lambda = \mu + \lambda^2 - \mu\lambda^2 = 0,$$

showing that λ is l.q.r., as claimed.

(iii) Given a ring Λ, set $\overline{\Lambda} := \Lambda/\mathfrak{J}(\Lambda)$, and denote by $^{-} : \Lambda \to \overline{\Lambda}$ the corresponding canonical ring projection, sending $\mu \in \Lambda$ to $\overline{\mu} \in \overline{\Lambda}$. Let $\overline{\lambda} \in \mathfrak{J}(\overline{\Lambda})$ and $\mu \in \Lambda$ be arbitrary elements. By definition of the radical, the element $\overline{\mu}\overline{\lambda}$ is l.q.r., hence there exists some $\lambda' \in \Lambda$ such that

$$\overline{\lambda'} \circ (\overline{\mu}\overline{\lambda}) = \overline{0} = \overline{\mu}\overline{\lambda} + \overline{\lambda'} - \overline{\lambda'}\overline{\mu}\overline{\lambda},$$

so that

$$\lambda' \circ (\mu\lambda) = \mu\lambda + \lambda' - \lambda'\mu\lambda \in \mathfrak{J}(\Lambda).$$

Since, by Part (ii), every element of $\mathfrak{J}(\Lambda)$ is l.q.r., there exists some $\nu \in \Lambda$ with

$$\nu \circ (\lambda' \circ (\mu\lambda)) = 0 = (\mu\lambda + \lambda' - \lambda'\mu\lambda) + \nu - \nu(\mu\lambda + \lambda' - \lambda'\mu\lambda).$$

Consequently,

$$(\lambda' + \nu - \nu\lambda') \circ (\mu\lambda) = \mu\lambda + (\lambda' + \nu - \nu\lambda') - (\lambda' + \nu - \nu\lambda')\mu\lambda = 0.$$

As $\mu \in \Lambda$ was arbitrary, we deduce that $\lambda \in \mathfrak{J}(\Lambda)$, implying $\overline{\lambda} = \overline{0}$. Hence, $\mathfrak{J}(\overline{\Lambda}) = \{\overline{0}\}$, as claimed. □

§13.4. The Jacobson radical, II: characterisations

Let S be a subset of the ring Λ. We denote by $\langle S \rangle_\ell$ and $\langle S \rangle_r$ the left (respectively right) ideal generated by S in Λ; that is, the smallest left (respectively right) ideal of Λ containing S. If $\langle S \rangle_\ell = \Lambda$, then we call S a *left-generating system* of Λ. Furthermore, we denote by $\Phi_\ell(\Lambda)$ the set of all $\lambda \in \Lambda$, such that

$$\Lambda = \langle S \cup \{\lambda\mu\} \rangle_\ell \implies \Lambda = \langle S \rangle_\ell; \quad \mu \in \Lambda, S \subseteq \Lambda.$$

Hence, $\Phi_\ell(\Lambda)$ consists of those $\lambda \in \Lambda$ such that, for each $\mu \in \Lambda$, the element $\lambda\mu$ may be deleted from every left-generating system of Λ. The set $\Phi_r(\Lambda)$ is defined in an analogous way: $\Phi_r(\Lambda)$ consists of those elements $\lambda \in \Lambda$ such that, for each $\mu \in \Lambda$, the element $\lambda\mu$ may be deleted from every right-generating system of Λ.

Combining results of Jacobson [44], Kertész [48], and Szász [77]–[78], we obtain a number of important characterisations of the Jacobson radical of a ring.

Theorem 13.9. *Let Λ be a ring. Then the following subsets of Λ coincide:*

(i) $\mathfrak{J}(\Lambda)$,

(ii) *the set $\mathfrak{Q}_1$ of all elements $\lambda \in \Lambda$ such that the product $\mu\lambda\nu$ is l.q.r. for all $\mu, \nu \in \Lambda$,*

(iii) *the set $\mathfrak{Q}_2$ of all elements $\lambda \in \Lambda$ such that the product $\mu\lambda\nu$ is q.r. for all $\mu, \nu \in \Lambda$,*

(iv) *the set $\mathfrak{Q}_3$ of all elements $\lambda \in \Lambda$ such that*

$$\mu \in \langle \mu + \nu\lambda\mu \rangle_\ell, \quad (\mu, \nu \in \Lambda),$$

(v) $\Phi_\ell(\Lambda)$,

(vi) *the intersection $\mathfrak{J}$ of all quasi-modular maximal left ideals of Λ,*

(vii) *the intersection $\mathfrak{J}'$ of all modular maximal left ideals of Λ,*

(viii) *the set $\mathfrak{M}$ of all elements $\lambda \in \Lambda$ such that every maximal left ideal of Λ contains the right ideal $\lambda\Lambda$.*

Proof. **1.** $\mathfrak{J}(\Lambda) \subseteq \mathfrak{Q}_1$. Let $\lambda \in \mathfrak{J}(\Lambda)$. Then we have $\mu\lambda\nu \in \mathfrak{J}(\Lambda)$ for all $\mu, \nu \in \Lambda$, since $\mathfrak{J}(\Lambda)$ is a two-sided ideal of Λ by Part (i) of Proposition 13.8. By Part (ii) of Proposition 13.8, every element in the Jacobson radical of a ring is l.q.r.; thus $\mu\lambda\nu$ is l.q.r., implying $\lambda \in \mathfrak{Q}_1$ as desired, since $\mu, \nu \in \Lambda$ were arbitrary.

2. $\mathfrak{Q}_1 \subseteq \mathfrak{Q}_2$. Let $\lambda \in \mathfrak{Q}_1$, and let $\mu, \nu \in \Lambda$ be arbitrary. As the element $\mu\lambda\nu$ is l.q.r., there exists some $\rho \in \Lambda$, such that

$$\rho \circ (\mu\lambda\nu) = 0 = \mu\lambda\nu + \rho - \rho\mu\lambda\nu; \tag{13.5}$$

in particular, ρ is r.q.r. Solving Equation (13.5) for ρ, we get

$$\rho = \rho\mu\lambda\nu - \mu\lambda\nu = (\rho\mu - \mu)\lambda\nu.$$

Since $\lambda \in \mathfrak{Q}_1$, it follows that ρ is also l.q.r., thus q.r., and the product $\mu\lambda\nu$ is q.r., being the quasi-inverse of ρ by the proof of Lemma 13.6. Hence, $\lambda \in \mathfrak{Q}_2$, as required.

3. $\mathfrak{Q}_2 \subseteq \mathfrak{Q}_3$. Suppose that $\lambda \in \Lambda$ and that $\lambda \notin \mathfrak{Q}_3$. We shall show that then $\lambda \notin \mathfrak{Q}_2$. Since $\lambda \notin \mathfrak{Q}_3$, there exist elements $\mu, \nu \in \Lambda$ such that $\mu \notin \langle \mu + \nu\lambda\mu \rangle_\ell$. Given λ, μ, ν as above, consider the set $\mathscr{I}$ consisting of all left ideals Λ' of Λ, such that

$$(a) \qquad \langle \mu + \nu\lambda\mu \rangle_\ell \leq \Lambda' \quad \text{and}$$

$$(b) \qquad \mu \notin \Lambda',$$

ordered by inclusion. We have $\mathscr{I} \neq \emptyset$, since $\langle \mu + \nu\lambda\mu \rangle_\ell \in \mathscr{I}$. If $\mathscr{C} = \{\Lambda_i'\}_{i \in I}$ is a non-empty chain in $(\mathscr{I}, \subseteq)$, then one checks easily that $\Lambda_0' := \bigcup_{i \in I} \Lambda_i'$ is again a left ideal of Λ satisfying Conditions (a) and (b). Thus,

$(\mathscr{I}, \subseteq)$ is inductively ordered. Consequently, by Zorn's Lemma, there exists a left ideal Λ^* of Λ satisfying Conditions (a) and (b), and maximal with respect to these properties. Consider the Λ-module

$$M = (\langle\mu\rangle_\ell + \Lambda^*)/\Lambda^*,$$

and suppose that $N := I/\Lambda^*$ is a non-trivial Λ-submodule of M, where I is a left ideal of Λ such that $\Lambda^* < I \leq \langle\mu\rangle_\ell + \Lambda^*$. Since $\langle\mu\rangle_\ell + \Lambda^* \leq \Lambda^*$ by construction, the left ideal I satisfies Condition (a). Thus, by maximality Λ^*, I cannot satisfy Condition (b), so that $\mu \in I$, and therefore $I = \langle\mu\rangle_\ell + \Lambda^*$ and $N = M$. Hence, M is a simple Λ-module. Also, as Λ^* satisfies Conditions (a) and (b), we have $\nu\lambda\mu \notin \Lambda^*$, so that $\nu\lambda\overline{\mu} \neq \overline{\mathbf{0}}$, where $\overline{\mu} := \mu + \Lambda^*$. It follows that $\Lambda \cdot M \neq \{\mathbf{0}\}$, and we conclude from Lemma 13.2 that M is irreducible. Consequently, there exists some $\kappa \in \Lambda$ such that

$$\kappa\nu\lambda\overline{\mu} = \overline{\mu} \neq \overline{\mathbf{0}}. \tag{13.6}$$

If $\lambda \in \Omega_2$ then, in particular, the element

$$(\kappa\nu\lambda)^2 = (\kappa\nu)\lambda(\kappa\nu\lambda)$$

must be l.q.r, so that there has to exist some $\gamma \in \Lambda$, such that

$$\gamma \circ (\kappa\nu\lambda)^2 = 0 = (\kappa\nu\lambda)^2 + \gamma - \gamma(\kappa\nu\lambda)^2.$$

Hence, by (13.6),

$$\overline{\mu} = (\kappa\nu\lambda)^2\overline{\mu} = (\gamma(\kappa\nu\lambda)^2 - \gamma)\overline{\mu} = \gamma\overline{\mu} - \gamma\overline{\mu} = \overline{\mathbf{0}},$$

a contradiction, since $\mu \notin \Lambda^*$ by definition of $\mathscr{I}$. Therefore, $\lambda \notin \Omega_2$, as desired.

4. $\Omega_3 \subseteq \Phi_\ell(\Lambda)$. Let $\lambda \in \Lambda$ and $\lambda \notin \Phi_\ell(\Lambda)$. We want to show that $\lambda \notin \Omega_3$. By hypothesis, there exists a subset $S \subseteq \Lambda$ and an element $\mu \in \Lambda$, such that $\langle S \cup \{\lambda\mu\}\rangle_\ell = \Lambda$ and $\lambda\mu \notin \langle S\rangle_\ell$. Consider the family $\mathscr{I}$ of left ideal $\Lambda' \leq \Lambda$ such that

(a) $S \subseteq \Lambda'$,

(b) $\lambda\mu \notin \Lambda'$,

ordered by inclusion. We note that $\mathscr{I} \neq \emptyset$, since $\langle S\rangle_\ell \in \mathscr{I}$. Arguing in the usual way, one checks that $(\mathscr{I}, \subseteq)$ is inductively ordered; hence, by Zorn's Lemma, there exists a left ideal $\Lambda^* \leq \Lambda$ satisfying Conditions (a) and (b), and being maximal subject to these properties; in particular, $\mu \notin \Lambda^*$. If $\nu \in \Lambda$ and $\nu \notin \Lambda^*$ then, by construction, $\lambda\mu \in \langle\Lambda^* \cup \{\nu\}\rangle_\ell$, thus

$$\langle\Lambda^* \cup \{\nu\}\rangle_\ell = \Lambda.$$

Hence, Λ^* is a maximal left ideal of the ring Λ. Consider the quotient

$$M := \Lambda/\Lambda^*,$$

viewed as a Λ-module. Then M is simple, and we have $\lambda\overline{\mu} \neq \overline{\mathbf{0}}$, where $\overline{\mu} := \mu + \Lambda^*$; thus $\Lambda \cdot M \neq \{\mathbf{0}\}$, and M is irreducible by Lemma 13.2. Consequently, there exists some $\kappa \in \Lambda$, such that

$$\kappa\lambda\overline{\mu} = -\overline{\mu} \neq \overline{\mathbf{0}}.$$

If $\lambda \in \mathfrak{Q}_3$, then there exist an integer n and some $\varepsilon \in \Lambda$, such that

$$\mu = n(\mu + \kappa\lambda\mu) + \varepsilon(\mu + \kappa\lambda\mu);$$

cf. Exercise 14. It follows that

$$\overline{\mu} = n\overline{\mu} - n\overline{\mu} + \varepsilon\overline{\mu} - \varepsilon\overline{\mu} = \overline{\mathbf{0}},$$

a contradiction. Hence, $\lambda \notin \mathfrak{Q}_3$, as desired.

5. $\Phi_\ell(\Lambda) \subseteq \mathfrak{J}$. Let $\lambda \in \Lambda$ and $\lambda \notin \mathfrak{J}$. We shall show that $\lambda \notin \Phi_\ell(\Lambda)$. Since $\lambda \notin \mathfrak{J}$, there exists a quasi-modular maximal left ideal Λ' in Λ with $\lambda \notin \Lambda'$ and $\lambda\tilde{\lambda} \notin \Lambda'$ for some suitable $\tilde{\lambda} \in \Lambda$. By maximality of Λ',

$$\langle \Lambda' \cup \{\lambda\tilde{\lambda}\}\rangle_\ell = \Lambda.$$

However, $\langle \Lambda' \rangle_\ell = \Lambda' < \Lambda$, thus $\lambda \notin \Phi_\ell(\Lambda)$, as required.

6. $\mathfrak{J} \subseteq \mathfrak{J}'$. This is clear, since every modular maximal left ideal of Λ is, in particular, a quasi-modular maximal left ideal.

7. $\mathfrak{J}' \subseteq \mathfrak{M}$. Suppose that $\lambda \in \Lambda$ and that $\lambda \notin \mathfrak{M}$. Then there exists a maximal left ideal Λ^* of Λ, such that $\lambda\mu \notin \Lambda^*$ for some suitable element $\mu \in \Lambda$. As before, one concludes by means of Lemma 13.2 that the Λ-module $M := \Lambda/\Lambda^*$ is irreducible. By Part (b) of Lemma 13.5, the annihilator

$$\mathrm{Ann}_\Lambda(\overline{\mu}) = \{\nu \in \Lambda : \nu\mu \in \Lambda^*\}, \quad \overline{\mu} := \mu + \Lambda^*,$$

is a modular maximal left ideal of Λ, and we have $\lambda \notin \mathrm{Ann}_\Lambda(\overline{\mu})$, showing that $\lambda \notin \mathfrak{J}'$.

8. $\mathfrak{M} \subseteq \mathfrak{J}(\Lambda)$. Let $\lambda \in \Lambda$ be such that $\lambda \notin \mathfrak{J}(\Lambda)$. Then there exists some $\mu \in \Lambda$, such that $\mu\lambda$ is not l.q.r. Let

$$\Lambda_0' := \{\nu - \nu\mu\lambda : \nu \in \Lambda\}.$$

Then Λ_0' is a left ideal of Λ with $\mu\lambda \notin \Lambda_0'$. Indeed, if $\mu\lambda \in \Lambda_0'$, then there exists some $\nu \in \Lambda$, such that

$$-\mu\lambda = \nu - \nu\mu\lambda,$$

or $\nu \circ (\mu\lambda) = 0$, so that ν would be a left quasi-inverse for $\mu\lambda$. Consider the family $\mathscr{I}$ consisting of all left ideals Λ' of Λ, such that (a) $\Lambda_0' \leq \Lambda'$ and (b) $\mu\lambda \notin \Lambda'$, ordered by inclusion. It is easily shown that $(\mathscr{I}, \subseteq)$ is inductively ordered; thus, by Zorn's Lemma, there exists a left ideal Λ^* of Λ satisfying Conditions (a) and (b), and maximal with respect to these properties. We claim that Λ^* is a maximal left ideal of Λ. Indeed, suppose that $\lambda^* \in \Lambda$, and $\lambda^* \notin \Lambda^*$. Then

$$\mu\lambda \in \langle \Lambda^* \cup \{\lambda^*\}\rangle_\ell.$$

For an arbitrary element $\nu \in \Lambda$, this yields

$$\nu = (\nu - \nu\mu\lambda) + \nu\mu\lambda \in \langle \Lambda^* \cup \{\lambda^*\}\rangle_\ell,$$

implying $\langle \Lambda^* \cup \{\lambda^*\}\rangle_\ell = \Lambda$. Hence, Λ^* is a maximal left ideal of Λ, as claimed. Moreover, we have

$$\mu\lambda - (\mu\lambda)^2 \in \Lambda_0' \leq \Lambda^*,$$

while $\mu\lambda \notin \Lambda^*$. It follows that $(\mu\lambda)^2 \notin \Lambda^*$. Now suppose that $\lambda\Lambda \leq \Lambda^*$. Then, since $\lambda\mu\lambda \in \lambda\Lambda$, we would also have $\lambda\mu\lambda \in \Lambda^*$, implying $(\mu\lambda)^2 \in \Lambda^*$ since Λ^* is a left ideal. However, as we have just seen, this is not the case; hence, $\lambda\Lambda \not\leq \Lambda^*$, so that $\lambda \notin \mathfrak{M}$. The proof of Theorem 13.9 is thus complete. □

We conclude this section with two consequences of Theorem 13.9. First, we give several characterisations of radical rings; that is, rings Λ coinciding with their Jacobson radical $\mathfrak{J}(\Lambda)$.

Corollary 13.10. *Let Λ be a ring. Then the following assertions are equivalent:*

(i) $\mathfrak{J}(\Lambda) = \Lambda$;

(ii) *every element of Λ is l.q.r;*

(iii) *Λ does not possess a quasi-modular maximal left ideal;*

(iv) *Λ does not possess a modular maximal left ideal.*

Proof. (i) $\Leftrightarrow$ (ii). If $\mathfrak{J}(\Lambda) = \Lambda$, then each element of Λ is l.q.r. by Part (ii) of Proposition 13.8. If, conversely, every element of Λ is l.q.r., then we have $\mathfrak{Q}_1 = \Lambda$, implying $\mathfrak{J}(\Lambda) = \Lambda$ by Theorem 13.9.

(i) $\Leftrightarrow$ (iii). If $\mathfrak{J}(\Lambda) = \Lambda$, then the intersection $\mathfrak{J}$ of all quasi-modular maximal left ideals equals Λ by Theorem 13.9, which happens if, and only if, this intersection is an empty one.

(i) $\Leftrightarrow$ (iv). By Theorem 13.9, we have $\mathfrak{J}(\Lambda) = \Lambda$ if, and only if, the intersection $\mathfrak{J}'$ of all modular maximal left ideals equals Λ, which happens if, and only if, there are no modular maximal left ideals in Λ. □

The final result of this section provides another characterisation of the Jacobson radical of a ring Λ in the case when Λ is commutative.

Corollary 13.11. *Let Λ be a commutative ring. Then the Jacobson radical of Λ equals the intersection of those maximal ideals Λ' of Λ, which satisfy the condition*

$$\lambda^2 \in \Lambda' \Longrightarrow \lambda \in \Lambda', \quad (\lambda \in \Lambda). \tag{13.7}$$

Proof. By Theorem 13.9, it suffices to show that a maximal ideal Λ' of a commutative ring Λ is quasi-modular if, and only if, Λ' satisfies Condition (13.7). If Λ' is a maximal ideal of Λ satisfying (13.7), then Λ' is clearly quasi-modular.

Conversely, let Λ' be a quasi-modular maximal ideal of Λ, and let $\lambda \in \Lambda$ be such that $\lambda^2 \in \Lambda'$. If $\lambda \notin \Lambda'$, then by quasi-modularity there exists some $\mu \in \Lambda$ such that $\mu\lambda \notin \Lambda'$. By Lemma 13.2, the Λ-module $M = \Lambda/\Lambda'$ is irreducible, and we have $\mu\overline{\lambda} \neq \overline{0}$, where $\overline{\lambda} := \lambda + \Lambda'$ and $\overline{0} = \Lambda'$. By irreducibility of M, there exists some $\nu \in \Lambda$, such that $\nu\mu\overline{\lambda} = \overline{\mu}$, where $\overline{\mu} := \mu + \Lambda'$. Hence, there exists some $\lambda' \in \Lambda'$, such that

$$\mu = \nu\mu\lambda + \lambda'. \tag{13.8}$$

Multiplying both sides of (13.8) by λ, we find that

$$\mu\lambda = \nu\mu\lambda^2 + \lambda'\lambda \in \Lambda',$$

a contradiction. Hence, $\lambda \in \Lambda'$, and Λ' satisfies Condition (13.7). □

Exercises for § 13.

1. Suppose that $\Lambda_0 \leq \Lambda$ is a left modular ideal of the ring Λ, and that Λ_1 is a left ideal of Λ containing Λ_0. Deduce that Λ_1 is a left modular ideal of Λ.

2. Establish the following generalisation of Proposition 4.8.

Proposition 13.12. *Every proper right modular ideal of a ring Λ can be embedded into a maximal right ideal of Λ.*

3. Let Λ be a (multiplicative) semigroup. Show that, if Λ contains a left and a right identity element, then Λ contains a two-sided identity element. Deduce from this that a ring Λ containing both a left and a right identity element, has a two-sided identitiy element (this observation is relevant for the characterisation of modular ideals).

4. Let Λ be a ring, let $\lambda_1 \in \Lambda$ be an arbitrary element, and let Λ_1' be as defined in (13.1). Show that Λ_1' is a modular left ideal of Λ, satisfying $\Lambda_1' = \Lambda$ if, and only if, $\lambda_1 \in \Lambda_1'$.

5. Prove Part (a) of Lemma 13.5.

6. Show that the circle operation $\circ$ on a ring Λ is associative.

7. (i) Let Λ_1 and Λ_2 be rings, and let $\varphi : \Lambda_1 \to \Lambda_2$ be a ring homomorphism. Show that the restriction $\mathscr{C}(\varphi) = \varphi|_{\mathscr{C}(\Lambda_1)} : \mathscr{C}(\Lambda_1) \to \mathscr{C}(\Lambda_2)$ is well defined and a group homomorphism.

(ii) We have $\mathscr{C}(1_\Lambda) = 1_{\mathscr{C}(\Lambda)}$, where Λ is a ring, and 1_Λ is the identity map on Λ.

(iii) Let $\Lambda_1, \Lambda_2, \Lambda_3$ be rings, and let $\varphi_1 : \Lambda_1 \to \Lambda_2$ and $\varphi_2 : \Lambda_2 \to \Lambda_3$ be ring homomorphisms. Show that $\mathscr{C}(\varphi_2 \circ \varphi_1) = \mathscr{C}(\varphi_2) \circ \mathscr{C}(\varphi_1)$.

8. Let Λ be a ring with (two-sided) identity element 1. Show that

$$\mathscr{C}(\Lambda) = \{1 - \lambda : \lambda \in \mathscr{U}(\Lambda)\} \cong \mathscr{U}(\Lambda), \tag{13.9}$$

where $\mathscr{U}(\Lambda)$, the unit group of Λ, is the multiplicative group consisting of the invertible elements in Λ. In particular, for a ring Λ with identity element 1, we have $\mathscr{C}(\Lambda) = \Lambda \setminus \{1\}$ if, and only if, Λ is a skew-field.

9. For an integer $m \geq 1$, let

$$m\mathbb{Z} = \{m\zeta : \zeta \in \mathbb{Z}\}$$

be the ideal of the ring $\mathbb{Z}$ of rational integers generated by m. Note that $m\mathbb{Z}$ is a commutative ring but, for $m > 1$, has no identity element. Compute the abelian group $\mathscr{C}(m\mathbb{Z})$ for $m \geq 1$.

10. Consider the ideal

$$2\mathbb{Z}_8 = \{[0]_8, [2]_8, [4]_8, [6]_8\}$$

of the ring $\mathbb{Z}_8 = \mathbb{Z}/8\mathbb{Z}$. Show that the ring $2\mathbb{Z}_8$ coincides with its Jacobson radical, that $(2\mathbb{Z}_8, +) \cong C_4$, while $\mathscr{C}(2\mathbb{Z}_8) = (2\mathbb{Z}_8, \circ) \cong C_2 \times C_2$. This observation serves to demonstrate that, for a ring Λ which coincides with its radical, the circle group $\mathscr{C}(\Lambda) = (\Lambda, \circ)$ need not be isomorphic to the additive group $(\Lambda, +)$ of the ring.

11. Give an example of a non-commutative ring without identity element, possessing zero divisors.

12. (i) Let Λ be a ring, and let $\{\Lambda_\nu\}_{\nu \in N}$ be a family of (left/right/two-sided) ideals of Λ. Show that $\Lambda' = \bigcap_{\nu \in N} \Lambda_\nu$ is again a (left/right/two-sided) ideal of Λ.

(ii) Let Λ be a ring, and let S be any subset of Λ. Deduce from Part (i) that there exists a smallest (left/right/two-sided) ideal of Λ containing S. These ideals are denoted $\langle S \rangle_\ell$, $\langle S \rangle_r$, and $\langle S \rangle$, respectively.

13. Let Λ be a ring, and let $\Lambda_1, \Lambda_2 \leq \Lambda$ be left ideals. Show that the sum

$$\Lambda_1 + \Lambda_2 := \{\lambda_1 + \lambda_2 : \lambda_1 \in \Lambda_1, \lambda_2 \in \Lambda_2\}$$

of Λ_1 and Λ_2 is again a left ideal of Λ, containing Λ_1 and Λ_2.

14. Let Λ be a ring, and let $\lambda \in \Lambda$. Show that

$$\langle \lambda \rangle_\ell = \{n\lambda + \varepsilon\lambda : n \in \mathbb{Z}, \varepsilon \in \Lambda\}. \tag{13.10}$$

15. Let $\mathbb{Z}$ be the ring of rational integers.

(a) Show that every ideal of the ring $\mathbb{Z}$ is principal, that is, generated by one element.

(b) Prove that $\mathfrak{J}(\mathbb{Z}) = \{0\}$.

16. Compute the Jacobson radical of the ring $\mathbb{Z}_n = \mathbb{Z}/n\mathbb{Z}$ for $n \geq 2$.

§14. Artin's solution of Hilbert's 17th problem

David Hilbert (1862–1943) was born and educated in Königsberg, and took his doctorate there under the supervision of Ferdinand Lindemann[18] with a topic in invariant theory; cf. [37]. In 1892, he succeeded Hurwitz, obtaining the latter's (associate) professorship in Königsberg. In 1895 Hilbert accepted a full professorship in Göttingen, where he stayed for the rest of his life. With fundamental and influential work in more or less all disciplines of pure mathematics, he became a leading international authority, and many (later famous) mathematicians took their doctorate in Göttingen with Hilbert: O. Blumenthal, H. Weyl, R. Courant, E. Hecke, H. Steinhaus, etc. In 1900, Hilbert became president of the DMV (German Mathematical Society), and led the German delegation for the International Mathematical Congress of 1900 in Paris. His legendary talk at this meeting discussed a number of fundamental open problems, which, to a considerable degree, were going to guide mathematical research for many years to come.[19] As 17th of 23 problems, Hilbert raised the following question [38, pp. 284–285]:

$(*)$ **Hilbert's 17th Problem:** Call a rational function $F(x_1,x_2,\dots,x_n)$ of variables $x_1,x_2,\dots,x_n$ with rational coefficients *definite*, if

$$F(x_1,x_2,\dots,x_n) \geq 0$$

for each tuple $(x_1,x_2,\dots,x_n) \in \mathbb{R}^n$, where F is defined. Can a definite function F always be decomposed as the sum of squares of rational functions with rational coefficients?

In [39, Chap. VII], Hilbert again alludes to this problem, and solves it in the affirmative for $n = 1$. For $n = 2$, Hilbert [40] had previously shown that a definite function of two variables is the sum of squares of rational functions with real coefficients (see also Landau [54] and [55]). Hilbert's 17th problem was first resolved (in the affirmative) by Emil Artin [9], making use of his characterisation in [8] of the field of real algebraic numbers, and his joint work with Otto Schreier on the algebraic construction of real fields in [10]. The purpose of this section is to describe Artin's original proof. As a preparation, we begin by summarising certain facts and results concerning formally real fields, confining ourselves to the barest sketch of the necessary arguments, details being found in [8] and [10].

---

[18] Lindemann was the first to show transcendence of the circle number π. He also proved transcendence of e^{α} for algebraic $\alpha \neq 0$.

[19] For more on Hilbert's life and work, the reader is referred to Constance Reid's biography [68].

§14.1. More on formally real fields (after Artin and Schreier [10])

Recall from §9 that a field k is called formally real, if -1 is not a sum of squares in k. A formally real field necessarily has characteristic 0 since, in characteristic $p > 0$, we have

$$-1 = \underbrace{1^2 + \cdots + 1^2}_{p-1\,\text{summands}}.$$

As is shown in Proposition 9.2, a field k is orderable if, and only if, k is formally real.

Definition 14.1. A field k is called *real-closed*, if k is formally real, while no proper algebraic extension of k is formally real.

By Proposition 9.2, a real-closed field is orderable; however, we have the following more precise result.

Lemma 14.2. *A real-closed field k supports exactly one ordering compatible with its arithmetic.*

Proof. One shows that (i) every element $a \in k^\times$ is a square, or $-a$ is a square, (ii) that the two cases in (i) are mutually exclusive, and (iii) that sums of squares in k are again squares. Having shown this, a compatible order relation on k is defined by

$$a > 0 \iff a \text{ is a square and non-zero.}$$

This compatible ordering of k is the only one since, in any ordering of a field, squares are always non-negative. □

Lemma 14.3. *If k is real-closed, then every polynomial of odd degree over k has a root.*

Proof. One shows that a polynomial over k of odd degree cannot be irreducible, and inducts on the (odd) degree. □

Lemma 14.4. *A real-closed field k is not algebraically closed, but $k(i)$ is algebraically closed, where i denotes a root of the polynomial $x^2 + 1$.*

Proof. One shows that if, in an ordered field k, each positive element has a square root, and every polynomial of odd degree has a root, then $k(i)$ is algebraically closed. □

Lemma 14.5. *Let K be an algebraically closed field, and let k be a subfield of K such that $K = k(\xi)$; that is, K results from k by a simple extension. Then k is real-closed.*

Proof. If ξ is transcendental over k, then the equation $x^2 - \xi = 0$ has no root in K, contradicting our hypothesis that K be algebraically closed. Hence, K is a simple algebraic extension of k, and the rest of the proof proceeds as in [8]. □

Combining Lemmas 14.4 and 14.5, we see that the class of real-closed fields may be characterised as the class of fields of characteristic zero which become algebraically closed after some simple extension.

Lemma 14.6. *Let k be a real-closed field, and let $f(x)$ be a polynomial with coefficients in k. Suppose that $a, b \in k$ are such that $f(a) < 0$ and $f(b) > 0$. Then there exists some element $c \in k$ such that c is strictly between a and b, and $f(c) = 0$.*

Proof. Since $k(i)$ is algebraically closed by Lemma 14.4, $f(x)$ may be decomposed over k as a product of linear and irreducible quadratic factors; cf. Exercise 1 below. An irreducible quadratic polynomial $x^2 + px + q$ is always positive over k; indeed, we have

$$x^2 + px + q = \left(x + \frac{p}{2}\right)^2 + \left(q - \frac{p^2}{4}\right),$$

where the first summand is always non-negative, and the second summand is positive by irreducibility. Thus, a change of sign of f can only come from a linear factor changing sign, that is, from a root between a and b. □

Theorem 14.7. *All results of real algebra hold in a real-closed field. In particular, we have*

(1) *uniform continuity for a polynomial on a closed interval $[a, b]$;*

(2) *Rolle's theorem;*

(3) *Sturm's theorem concerning the number of zeros of a polynomial in an interval;*

(4) *the result that each rational function whose denominator does not vanish in $[a, b]$ has a global maximum and a global minimum, and these extremal values are among the values for $x = a$, b, ξ_j, where ξ_j runs through the zeros of the derivative of our given function in the interval $[a, b]$;*

(5) *the result that all zeros of the polynomial*

$$x^n + a_1 x^{n-1} + \cdots + a_n$$

are, in absolute value, less than $1 + |a_1| + \cdots + |a_n|$; see Exercise 3 below for a more general and slightly sharper result.

In view of Lemma 14.6, proofs are exactly as in the classical case; see, for example, §§35, 91, 112, and §114 in [83].

Next, we turn to the existence of real-closed extensions of certain formally real fields.

Lemma 14.8. *Let k be formally real, and let K be algebraically closed over k. Then there exists a real-closed field Ω, such that $k \le \Omega \le K$ and $K = \Omega(i)$.*

Proof. Consider K as a well-ordered set

$$1 = a_0, a_1, a_2, \ldots, a_\omega, \ldots \tag{14.11}$$

and, for each ordinal ν occurring in (14.11), define fields k_ν, k_ν^* as follows: set $k_0 = k_0^* = k$ and, if k_μ, k_μ^* are defined for $\mu < \nu$, let

$$k_\nu^* := \bigcup_{\mu<\nu} k_\mu$$

and

$$k_\nu := \begin{cases} k_\nu^*(a_\nu) & \text{if this field is formally real,} \\ k_\nu^* & \text{otherwise.} \end{cases}$$

By Satz 2 in [75, §2], all k_ν^* are fields, thus so are the k_ν, and the same argument shows that all k_ν are formally real, implying that the field

$$\Omega := \bigcup_\nu k_\nu$$

is formally real. If $a = a_\nu \in K \setminus \Omega$, then $a \notin k_\nu$, so that $k_\nu^*(a)$, and thus also $\Omega(a)$, is not formally real. Hence, Ω is real-closed. Moreover, as a simple transcendental extension of a formally real field is again formally real, K is algebraic over Ω and since, by Lemma 14.4, $\Omega(i)$ is algebraically closed, uniqueness of the algebraically closed extension of Ω forces $K = \Omega(i)$. □

We list some consequences of Lemma 14.8.

Corollary 14.9. *A formally real field k has a real-closed algebraic extension.*

Proof. In Lemma 14.8 choose K as the algebraically closed algebraic extension of k. □

Corollary 14.10. *A formally real field admits at least one order compatible with its arithmetic.*

Proof. This follows from Corollary 14.9 and Lemma 14.2. □

Corollary 14.11. *An algebraically closed field K of characteristic 0 contains a real-closed subfield Ω such that $K = \Omega(i)$.*

Proof. Take $k = \mathbb{Q}$ (i.e., the prime field of k), and apply Lemma 14.8. □

For ordered fields, Corollary 14.9 may be considerably strengthened as follows.

Lemma 14.12. *Let K be an ordered field. Then, up to equivalence, there exists precisely one real-closed algebraic extension Ω of K, whose ordering extends the ordering on K. Moreover, the group $G_K(\Omega)$ of field automorphisms of Ω fixing the elements of K is trivial.*

Proof. See Satz 8 in [10]. □

§14.2. A characterisation of sums of squares

Given an arbitrary field k, denote by $\Sigma_\square(k)$ the set of elements of k which can be expressed as a sum of squares. Clearly, $\Sigma_\square(k)$ contains the elements 0 and 1, and is closed under addition and multiplication. Moreover, $\Sigma_\square(k)$ is also closed under division since, for $a = \sum_\nu \alpha_\nu^2$ and $b = \sum_\mu \beta_\mu^2 \neq 0$, we have

$$\frac{a}{b} = \frac{a}{b^2} \cdot b = \Big(\sum_\nu \Big(\frac{\alpha_\nu}{b} \Big)^2 \Big) \cdot \Big(\sum_\mu \beta_\mu^2 \Big).$$

Following [9, Sec. 1], our aim in this subsection is to obtain an applicable description of the set $\Sigma_\square(k)$ for arbitrary fields k.

We first observe that attention may be restricted here to formally real fields k. Indeed, if we have

$$-1 = \xi_1^2 + \xi_2^2 + \cdots + \xi_n^2$$

in k, and if the characteristic of k is different from 2, then

$$a = \Big(\frac{1+a}{2} \Big)^2 + (\xi_1^2 + \xi_2^2 + \cdots + \xi_n^2) \Big(\frac{1-a}{2} \Big)^2, \quad a \in k,$$

so that $\Sigma_\square(k) = k$ in this case. Also, if k has characteristic 2, then each sum of squares in k is itself a square.

Definition 14.13. Let k be a formally real field. Then an element $\alpha \in k$ is called *totally positive*, if $\alpha \geq 0$ for each order relation on k compatible with the field arithmetic.

The desired characterisation of sums of squares in a formally real field is now the following.

Proposition 14.14. (Satz 1 in [9]) *Let k be a formally real field, and let $\alpha \in k$. Then α is a sum of squares in k if, and only if, α is totally positive.*

Proof. Let k be formally real, pick $\alpha \in k \setminus \Sigma_\square(k)$, and let K be an algebraically closed algebraic extension of k. Then, suitably modifying the proof of Lemma 14.8, we may ensure existence of a field Ω such that $k \leq \Omega \leq K$, $\alpha \notin \Sigma_{\square}(\Omega)$, and such that $\alpha \in \Sigma_{\square}(\Omega')$ for each proper algebraic extension Ω' of Ω; cf. Exercise 2. Since k, being formally real by hypothesis, has characteristic 0, so has Ω; consequently, the fact that $\Sigma_{\square}(\Omega) \neq \Omega$ ensures that Ω is formally real by the observations directly preceding Definition 14.13.

Next, we claim that $-\alpha$ is a square in Ω. Indeed, if $-\alpha$ is not a square in Ω, then $\Omega(\sqrt{-\alpha})$ is a proper algebraic extension of Ω, so that α is a sum of squares in $\Omega(\sqrt{-\alpha})$. Suppose that

$$\alpha = \sum_i \left(\xi_i\sqrt{-\alpha} + \eta_i\right)^2 = -\alpha\sum_i \xi_i^2 + \sum_i \eta_i^2 + 2\sqrt{-\alpha}\sum_i \xi_i\eta_i, \tag{14.12}$$

where $\xi_i, \eta_i \in \Omega$. In (14.12), the third term on the right-hand side must vanish, since otherwise $\sqrt{-\alpha} \in \Omega$, contradicting our assumption. It follows that

$$\alpha\Big(1 + \sum_i \xi_i^2\Big) = \sum_i \eta_i^2.$$

As Ω is formally real, we have $1 + \sum_i \xi_i^2 \neq 0$, so that we get

$$\alpha = \frac{\sum_i \eta_i^2}{1^2 + \sum_i \xi_i^2}.$$

It follows that α is itself a sum of squares in Ω, a final contradiction. Hence, $-\alpha$ is a square in Ω, as claimed.

We can now quickly finish the proof of Proposition 14.14. If Ω' is any proper algebraic extension of Ω, then $\alpha \in \Sigma_{\square}(\Omega')$, hence

$$-1 = \frac{\alpha}{-\alpha} \in \Sigma_{\square}(\Omega'),$$

which shows that Ω' is not formally real, and that Ω is real-closed. By Lemma 14.2, Ω has a (unique) ordering, and $-\alpha > 0$ in that ordering, since $-\alpha$ is a non-zero square. The ordering of Ω induces an ordering of k and, in this ordering, $\alpha < 0$. Hence, α is not totally positive. Since, on the other hand, a sum of squares is non-negative in each ordering of k, the result follows. □

We note that, as a special case, Proposition 14.14 contains a result of Hilbert and Landau concerning the decomposition of totally positive algebraic numbers as sums of squares; cf. [56].

§14.3. Decomposition of definite functions

Proposition 14.14 makes it clear that, in order to be able to resolve Hilbert's 17th problem, we need to investigate total positivity of definite functions. The key result turns out to be the following.

Theorem 14.15. (Satz 3 in [9]) *Let R be a real number field, and let $K = R(x_1, x_2, \dots, x_n)$ be the field of rational functions of n variables $x_1, x_2, \dots, x_n$ with coefficients in R. Suppose that K carries some (fixed) ordering extending the natural ordering of R. Setting $\mathbf{x} = (x_1, x_2, \dots, x_n)$, suppose further that*

$$(\varphi_1(\mathbf{x}),\, \varphi_2(\mathbf{x}),\, \dots,\, \varphi_m(\mathbf{x})) \in K^m$$

is any finite list of functions from K, with m arbitrary. Then there exists an n-tuple $\mathbf{a} = (a_1, a_2, \dots, a_n) \in \mathbb{Q}^n$ such that each $\varphi_\mu(\mathbf{x})$ is defined at $\mathbf{x} = \mathbf{a}$, and such that $\varphi_\mu(\mathbf{a})$ has the same sign as $\varphi_\mu(\mathbf{x})$ for all $\mu \in [m]$.

More explicitly, the requirement concerning signs in Theorem 14.15 means that if, in the given ordering of K, we have $\varphi_\mu(\mathbf{x}) \lesseqgtr 0$, then $\varphi_\mu(\mathbf{a}) \lesseqgtr 0$ in the natural ordering of R, with $\varphi_\mu(\mathbf{x}) = 0$, of course, implying $\varphi_\mu(\mathbf{a}) = 0$.

Assuming Theorem 14.15 to be true, we quickly reach an affirmative solution of Hilbert's 17th problem. Indeed, let R be a real number field which allows for precisely one ordering compatible with its arithmetic like, for instance, the field of rational numbers (see Exercise 8 in §9), or the field of real algebraic numbers. Then the function field $K = R(x_1, x_2, \dots, x_n)$, being formally real, possesses orderings by Corollary 14.10 (or apply Proposition 9.2), and each ordering of K automatically extends the ordering of R. Now let $\varphi(\mathbf{x}) \in K$ be any definite rational function. Then, in particular, we have $\varphi(\mathbf{a}) \geq 0$ for each n-tuple $\mathbf{a} = (a_1, a_2, \dots, a_n)$ of rational numbers such that $\varphi(\mathbf{x})$ is defined at $\mathbf{x} = \mathbf{a}$. By Theorem 14.15, we must have $\varphi(\mathbf{x}) \geq 0$ for every ordering of K since, otherwise, we would have $\varphi(\mathbf{a}) < 0$ for suitable $\mathbf{a} \in \mathbb{Q}^n$. Hence, $\varphi(\mathbf{x})$ is totally positive in K, which implies $\varphi(\mathbf{x}) \in \Sigma_\square(K)$ by Proposition 14.14.

We now focus on the proof of Theorem 14.15, proceeding by induction on n (the number of variables x_ν). For $n = 0$, there is nothing to prove since, in this case, $K = R$, so that the functions $\varphi_\mu(\mathbf{x})$ are constants. Suppose that our assertion holds for n variables. In order to establish the assertion of the theorem for $n+1$ variables, we first deduce some consequences of the induction hypothesis.

Let Ω be the real-closed algebraic extension of K, whose ordering extends that of K (this field exists by Lemma 14.12), and consider functions $f(t, \mathbf{x})$ which are polynomials in a new variable t, with coefficients in the field K. Given a list

$$\mathscr{F} = (f_1(t,\mathbf{x}),\, f_2(t,\mathbf{x}),\, \dots,\, f_k(t,\mathbf{x})) \in (K[t])^k,$$

we call a property $\mathfrak{p}$ of $\mathscr{F}$ *admissible*, if there exists a tuple

$$(\varphi_1(\mathbf{x}),\, \varphi_2(\mathbf{x}),\, \dots,\, \varphi_m(\mathbf{x})) \in K^m$$

(for some m) such that, for each list $\mathbf{a} = (a_1, a_2, \dots, a_n) \in \mathbb{Q}^n$ with $\varphi_\mu(\mathbf{a})$ defined for $\mu = 1, 2, \dots, m$ and

$$\operatorname{sign}_K(\varphi_\mu(\mathbf{x})) = \operatorname{sign}_R(\varphi_\mu(\mathbf{a})), \quad 1 \le \mu \le m, \tag{14.13}$$

the functions $f_\kappa(t, \mathbf{x})$ are well-defined for $\mathbf{x} = \mathbf{a}$, and the list

$$(f_1(t, \mathbf{a}), f_2(t, \mathbf{a}), \dots, f_k(t, \mathbf{a}))$$

has property $\mathfrak{p}$. Clearly, a property composed of admissible properties is itself admissible. Moreover, we note that, by our induction hypothesis, rational numbers $a_1, a_2, \dots, a_n$ satisfying (14.13) do exist.

We shall need two auxiliary results, which we prove next.

Lemma 14.16. *The property of a function $f(t, \mathbf{x}) \in K[t]$ to have exactly r roots as a function of t in the field Ω is admissible.*

Proof. Let

$$f(t, \mathbf{x}) = \psi_0(\mathbf{x})t^s + \psi_1(\mathbf{x})t^{s-1} + \cdots + \psi_s(\mathbf{x}),$$

and let

$$F(t) = F_{\mathbf{A}}(t) = A_0 t^s + A_1 t^{s-1} + \cdots + A_s \in \Omega[t]$$

be the general polynomial of degree s over Ω. By Parts (3) and (5) of Theorem 14.7, there exists a finite chain of polynomials $\Phi_\alpha(\mathbf{A}, t)$ with rational coefficients, and in variables $A_0, A_1, \dots, A_s, t$, such that the distribution of signs in the chains of Ω-values resulting for $t = \pm t_0$ with $t_0 := 1 + \sum_{\sigma=1}^{s} |A_\sigma / A_0|$, encodes the number of roots in Ω of the corresponding function $F_{\mathbf{A}}(t)$, where $\mathbf{A} := (A_0, A_1, \dots, A_s)$. We now form a list L of functions $\varphi(\mathbf{x}) \in K$ consisting of

(i) the coefficients $\psi_0(\mathbf{x}), \psi_1(\mathbf{x}), \dots, \psi_s(\mathbf{x})$ of $f(t, \mathbf{x})$, and

(ii) the finitely many functions $\Phi_\alpha(\psi_0(\mathbf{x}), \psi_1(\mathbf{x}), \dots, \psi_s(\mathbf{x}))$ with $t = \pm t_0$, where

$$t_0 = 1 + \sum_{\sigma=1}^{s} |\psi_\sigma(\mathbf{x}) / \psi_0(\mathbf{x})|.$$

If $\mathbf{a} = (a_1, a_2, \dots, a_n)$ is any n-tuple of rational numbers such that $\varphi(\mathbf{a})$ is defined for each $\varphi(\mathbf{x}) \in L$, and such that $\operatorname{sign}_K \varphi(\mathbf{x}) = \operatorname{sign}_R \varphi(\mathbf{a})$ (such $\mathbf{a}$ exists by our induction hypothesis), then $f(t, \mathbf{a})$ is defined and has the same number of roots in Ω as $f(t, \mathbf{x})$, since the distribution of signs in the chain of elements in K

$$\Phi_\alpha(\psi_0(\mathbf{x}),\ \psi_1(\mathbf{x}),\ \dots,\ \psi_s(\mathbf{x}))$$

is the same as that in the chain $\Phi_\alpha(\psi_0(\mathbf{a}),\ \psi_1(\mathbf{a}),\ \dots,\ \psi_s(\mathbf{a}))$ of elements in R. □

Lemma 14.17. *Suppose that the finite list of functions*

$$(f_1(t,\mathbf{x}), f_2(t,\mathbf{x}), \ldots, f_k(t,\mathbf{x})) \in (K[t])^k \tag{14.14}$$

has the property that, as a function of t, $f_\kappa(t,\mathbf{x})$ *has a certain root* α_κ *in* Ω *for* $\kappa = 1,2,\ldots,k$, *and that* $\alpha_1 < \alpha_2 < \cdots < \alpha_k$. *Then this property of* (14.14) *is admissible.*

Proof. Starting from the field K, we adjoin the roots $\alpha_1, \alpha_2, \ldots, \alpha_k \in \Omega$, as well as the square roots

$$\sqrt{\alpha_\kappa - \alpha_\lambda} \in \Omega, \quad 1 \le \lambda < \kappa \le k,$$

which exist since Ω is real-closed. The field $\tilde{K}$ resulting in this way satisfies $K \le \tilde{K} \le \Omega$ and is algebraic over K; in particular, $\tilde{K}$ is formally real, since Ω is. Also, since $\tilde{K}$ is algebraic over K, and the characteristic of K is zero, $\tilde{K}$ is a separable extension of K; see, for instance [81, § 45]. Thus, by the theorem on the existence of primitive elements, $\tilde{K}$ is a simple algebraic extension of K, say, $\tilde{K} = K(\xi)$; cf., for instance, [81, § 46]. Suppose that $F(\xi,\mathbf{x}) = 0$, where $F(t,\mathbf{x}) \in K[t]$ is irreducible over K. By the structure theorem for simple algebraic extensions, the quantities α_κ and $\sqrt{\alpha_\kappa - \alpha_\lambda}$ with $\kappa > \lambda$ may be expressed as polynomials in ξ with coefficients in K, say, $\alpha_\kappa = g_\kappa(\xi,\mathbf{x})$ for $\kappa \in [k]$, and

$$\sqrt{\alpha_\kappa - \alpha_\lambda} = h_{\kappa,\lambda}(\xi,\mathbf{x}), \quad (\kappa > \lambda).$$

Since $f_\kappa(g_\kappa(\xi,\mathbf{x}),\mathbf{x}) = 0$, the function $f_\kappa(g_\kappa(t,\mathbf{x}),\mathbf{x})$ must be divisible by $F(t,\mathbf{x})$, say

$$f_\kappa(g_\kappa(t,\mathbf{x}),\mathbf{x}) = F(t,\mathbf{x})G_\kappa(t,\mathbf{x}). \tag{14.15}$$

Similarly, the identity

$$g_\kappa(\xi,\mathbf{x}) - g_\lambda(\xi,\mathbf{x}) - (h_{\kappa,\lambda}(\xi,\mathbf{x}))^2 = 0, \quad (\kappa > \lambda)$$

implies that $F(t,\mathbf{x})$ divides the polynomial $g_\kappa(t,\mathbf{x}) - g_\lambda(t,\mathbf{x}) - (h_{\kappa,\lambda}(t,\mathbf{x}))^2$, say

$$g_\kappa(t,\mathbf{x}) - g_\lambda(t,\mathbf{x}) - (h_{\kappa,\lambda}(t,\mathbf{x}))^2 = F(t,\mathbf{x})\Phi_{\kappa,\lambda}(t,\mathbf{x}), \quad (\kappa > \lambda). \tag{14.16}$$

Moreover, we have $h_{\kappa,\lambda}(\xi,\mathbf{x}) \ne 0$ for $\kappa > \lambda$, since $\alpha_\kappa > \alpha_\lambda$. Thus, the multiplicative inverse of $h_{\kappa,\lambda}(\xi,\mathbf{x})$ exists in $\tilde{K}$, and may again be expressed as a polynomial in ξ over K, say,

$$\frac{1}{h_{\kappa,\lambda}(\xi,\mathbf{x})} = H_{\kappa,\lambda}(\xi,\mathbf{x}), \quad (\kappa > \lambda).$$

Reasoning as above, this fact implies a relation of the form

$$h_{\kappa,\lambda}(t,\mathbf{x})H_{\kappa,\lambda}(t,\mathbf{x}) - 1 = F(t,\mathbf{x})\Psi_{\kappa,\lambda}(t,\mathbf{x}), \quad (\kappa > \lambda). \tag{14.17}$$

We now form a list L of functions $\varphi(\mathbf{x}) \in K$ consisting of

(i) all coefficients of the polynomials $f_\kappa(t,\mathbf{x})$, $F(t,\mathbf{x})$, $g_\kappa(t,\mathbf{x})$, $h_{\kappa,\lambda}(t,\mathbf{x})$, $G_\kappa(t,\mathbf{x})$, $\Phi_{\kappa,\lambda}(t,\mathbf{x})$, $H_{\kappa,\lambda}(t,\mathbf{x})$, and $\Psi_{\kappa,\lambda}(t,\mathbf{x})$, and

(ii) a set of functions $\varphi(\mathbf{x})$ guaranteeing that, after specialising, $F(t,\mathbf{a})$ has the same number of roots in Ω as $F(t,\mathbf{x})$ (such a list of functions exists by Lemma 14.16).

Let $\mathbf{a} = (a_1, a_2, \dots, a_n) \in \mathbb{Q}^n$ be such that all entries $\varphi(\mathbf{x})$ of L are defined at $\mathbf{x} = \mathbf{a}$ and satisfy our sign condition; that is, $\operatorname{sign}_K \varphi(\mathbf{x}) = \operatorname{sign}_R \varphi(\mathbf{a})$. Then, in particular, $F(t,\mathbf{a})$ has a root $\bar{\xi}$ in Ω. Let $\bar{\alpha}_\kappa := g_\kappa(\bar{\xi},\mathbf{a})$. Setting $t = \bar{\xi}$ and $\mathbf{x} = \mathbf{a}$ in (14.15), we find that

$$f_\kappa(\bar{\alpha}_\kappa,\mathbf{a}) = F(\bar{\xi},\mathbf{a})\,G_\kappa(\bar{\xi},\mathbf{a}) = 0, \quad 1 \le \kappa \le k,$$

so that $\bar{\alpha}_\kappa$ is a root of the polynomial $f_\kappa(t,\mathbf{a})$ in Ω. Moreover, the same substitution applied to (14.17) yields

$$h_{\kappa,\lambda}(\bar{\xi},\mathbf{a})\,H_{\kappa,\lambda}(\bar{\xi},\mathbf{a}) = 1, \quad \kappa > \lambda,$$

which implies in particular that $h_{\kappa,\lambda}(\bar{\xi},\mathbf{a}) \neq 0$. Consequently, Equation (14.16) tells us that

$$\bar{\alpha}_\kappa - \bar{\alpha}_\lambda = (h_{\kappa,\lambda}(\bar{\xi},\mathbf{a}))^2 > 0, \quad \kappa > \lambda,$$

which completes the argument. □

After these preparations, we are now in a position to finish the proof of Theorem 14.15. We introduce a new variable $\tau = x_0$, and consider the function field

$$K(\tau) = R(\tau, x_1, \dots, x_n)$$

where, as before, $K = R(x_1, x_2, \dots, x_n)$. Suppose that $K(\tau)$ carries a fixed ordering extending the natural ordering of the real number field R, let Ω' be the real-closed algebraic extension of $K(\tau)$ whose ordering extends that of $K(\tau)$ and, as before, let Ω be the real closure of K, noting that $\Omega \subseteq \Omega'$. Consider a finite list

$$(f_1(\tau,\mathbf{x}),\, f_2(\tau,\mathbf{x}),\, \dots,\, f_k(\tau,\mathbf{x})) \in (K(\tau) \setminus \{0\})^k$$

of non-zero elements of $K(\tau)$. Then we need to show that there exists an $(n+1)$-tuple $\mathbf{a}' = (b, a_1, \dots, a_n) \in \mathbb{Q}^{n+1}$ such that each of the rational functions $f_\kappa(\tau,\mathbf{x})$ remains defined for the substitution $(\tau,\mathbf{x}) = (b,\mathbf{a})$, and such that the sign requirement

$$\operatorname{sign}_{K(\tau)} f_\kappa(\tau,\mathbf{x}) = \operatorname{sign}_R f_\kappa(b,\mathbf{a}), \quad 1 \le \kappa \le k,$$

holds true. Let

$$f_\kappa(\tau,\mathbf{x}) = \frac{g_\kappa(\tau,\mathbf{x})}{h_\kappa(\tau,\mathbf{x})}, \quad 1 \le \kappa \le k,$$

where $g_\kappa(\tau,\mathbf{x}), h_\kappa(\tau,\mathbf{x}) \in K[\tau]$, let

$$P_1(\tau,\mathbf{x}),\, P_2(\tau,\mathbf{x}),\, \dots,\, P_m(\tau,\mathbf{x}) \in K[\tau]$$

be the pairwise distinct monic irreducible polynomials in τ occurring in the decomposition of numerator and denominator of the functions $f_\kappa(\tau,\mathbf{x})$, and let

$$\psi_1(\mathbf{x}),\, \psi_2(\mathbf{x}),\, \dots,\, \psi_s(\mathbf{x}) \in K$$

be the factors independent of τ needed to form the $f_\kappa(\tau,\mathbf{x})$ from the irreducibles $P_\mu(\tau,\mathbf{x})$.

By irreducibility of the $f_\kappa(\tau,\mathbf{x})$, it suffices to prove the existence of some vector $(b,\mathbf{a}) \in \mathbb{Q}^{n+1}$ such that

(I) $P_\mu(b,\mathbf{a})\,(1 \le \mu \le m)$ and $\psi_\sigma(\mathbf{a})\,(1 \le \sigma \le s)$ are defined;

(II) $$\operatorname{sign}_{K(\tau)} P_\mu(\tau,\mathbf{x}) = \operatorname{sign}_R P_\mu(b,\mathbf{a}), \quad 1 \le \mu \le m;$$

(III) $$\operatorname{sign}_{K(\tau)} \psi_\sigma(\mathbf{x}) = \operatorname{sign}_R \psi_\sigma(\mathbf{a}), \quad 1 \le \sigma \le s.$$

Replacing the variable τ by t, consider the functions

$$P_1(t,\mathbf{x}), P_2(t,\mathbf{x}), \ldots, P_m(t,\mathbf{x}) \in K[t],$$

and let $\alpha_1 < \alpha_2 < \cdots < \alpha_\ell$ be the distinct roots of the $P_\mu(t,\mathbf{x})$ in Ω as functions of t, ordered according to magnitude, say, $P_{\lambda_i}(\alpha_i,\mathbf{x}) = 0$ for $1 \le i \le \ell$.

We now form a list L of functions $\varphi(\mathbf{x}) \in K$ consisting of

(i) the functions $\psi_1(\mathbf{x}), \psi_2(\mathbf{x}), \ldots, \psi_s(\mathbf{x})$;

(ii) for each $\mu \in [m]$ a list of functions $\varphi(\mathbf{x})$ guaranteeing that the polynomial $P_\mu(t,\mathbf{x})$ has the same number of roots in Ω as any of its specialisations $P_\mu(t,\mathbf{a})$ (these lists exist by Lemma 14.16);

(iii) a list of functions $\varphi(\mathbf{x})$, guaranteed to exist by Lemma 14.17, such that the functions $P_{\lambda_1}(t,\mathbf{x}),\ldots,P_{\lambda_\ell}(t,\mathbf{x})$ satisfy, for each specialisation $\mathbf{a}$, $P_{\lambda_i}(\bar{\alpha}_i,\mathbf{a}) = 0$ with some $\bar{\alpha}_i \in \Omega$ and $1 \le i \le \ell$, and $\bar{\alpha}_1 < \bar{\alpha}_2 < \cdots < \bar{\alpha}_\ell$.

Existence of a specialisation $\mathbf{a} \in \mathbb{Q}^n$ for the list L follows from the induction hypothesis. Since $\psi_1(\mathbf{x}),\ldots,\psi_s(\mathbf{x})$ are in L, the values $\psi_1(\mathbf{a}),\ldots,\psi_s(\mathbf{a})$ are defined for each specialisation $\mathbf{x} = \mathbf{a}$ of L, and the sign requirement (III) holds. It thus suffices to consider the functions $P_\mu(t,\mathbf{a})$ for a given specialisation $\mathbf{a}$ of L. We note that the sign of $P_\mu(t,\mathbf{x})$ in $K(t)$ and that of $P_\mu(b,\mathbf{a})$ in R may both be determined in the real closure Ω' of $K(t)$, since the ordering of Ω' extends that of $K(t)$ while, by construction, the ordering of $K(t)$ extends the (natural) ordering of R.

As $P_\mu(t,\mathbf{x})$ and its specialisation $P_\mu(t,\mathbf{a})$ have the same number of zeros in Ω, they have in fact the *same set* of zeros in Ω, since $P_\mu(\omega,\mathbf{x}) = 0$ with $\omega \in \Omega$ implies $P_\mu(\omega,\mathbf{a}) = 0$. Consequently, the total number of zeros of the functions $P_\mu(t,\mathbf{a})$ over Ω is again ℓ, so that $\bar{\alpha}_1 < \bar{\alpha}_2 < \cdots < \bar{\alpha}_\ell$ are precisely the distinct roots of the functions $P_\mu(t,\mathbf{a})$ over Ω.

Over Ω', the functions $P_\mu(t,\mathbf{x})$ decompose into linear and quadratic factors (the latter coming from non-real roots) and the same is true for the functions $P_\mu(t,\mathbf{a})$; indeed, the decomposition of $P_\mu(t,\mathbf{a})$ is induced by that of $P_\mu(t,\mathbf{x})$. A monic irreducible quadratic factor of $P_\mu(t,\mathbf{x})$ is positive in Ω' (see the proof of Lemma 14.6), so that the sign of $P_\mu(t,\mathbf{x})$ in Ω' (and thus in $K(t)$) is determined by the sign of the linear factors $t-\alpha_i$ of $P_\mu(t,\mathbf{x})$. Suppose that the variable t satisfies $\alpha_i < t \le \alpha_{i+1}$ in Ω' for some i with $1 \le i < \ell$. Now, the polynomial $P_\mu(t,\mathbf{a})$ factors over Ω' into linear and quadratic factors in exactly the same way as $P_\mu(t,\mathbf{x})$, replacing the α_i by $\bar{\alpha}_i$. Substituting for the variable t some element t_0 of Ω', the quadratic factors of $P_\mu(t,\mathbf{a})$ will give positive values since Ω' is real-closed. Hence, the sign of $P_\mu(t_0,\mathbf{a})$ for $t_0 \in \Omega'$ depends only on the magnitude of t_0 as measured against the grid $\bar{\alpha}_1 < \bar{\alpha}_2 < \cdots < \bar{\alpha}_\ell$. Consequently, if we choose a value t_0 for t with $\bar{\alpha}_i < t_0 < \bar{\alpha}_{i+1}$ then, simultaneously, all $P_\mu(t_0,\mathbf{a})$ will get the same sign as the corresponding $P_\mu(t,\mathbf{x})$. It only remains to see that we may choose such t_0 as a rational number. However, the real number field R is archimedean, and the roots $\bar{\alpha}_i$ of the polynomials $P_\mu(t,\mathbf{a}) \in R[t]$ are algebraic over R, since $P_{\lambda_i}(\bar{\alpha}_i,\mathbf{a}) = 0$. Consequently, $R(\bar{\alpha}_1,\ldots,\bar{\alpha}_\ell)$ is a finite algebraic extension of R, so that $R(\bar{\alpha}_1,\ldots,\bar{\alpha}_\ell) \le \Omega'$ is again archimedean, since we can bound the $\bar{\alpha}_i$ by elements of R in view of Theorem 14.7(5). It follows that the prime field $\mathbb{Q}$ of $R(\bar{\alpha}_1,\ldots,\bar{\alpha}_\ell)$ is dense in $R(\bar{\alpha}_1,\ldots,\bar{\alpha}_\ell)$ (see Exercise 7 in §9), so that there exists some rational number b with $\bar{\alpha}_i < b < \bar{\alpha}_{i+1}$. Also, if $t < \alpha_1$ or $t > \alpha_\ell$, then we choose a rational number b with $b < \bar{\alpha}_1$ or $b > \bar{\alpha}_\ell$, respectively. This completes the proof of Theorem 14.15.

Exercises for § 14.

1. Let $p(x)$ be a polynomial with coefficients in a real-closed field k, which is irreducible over k. Show that $p(x)$ is either linear or has degree two.

2. Let k be formally real, let $\alpha \in k \setminus \Sigma_\square(k)$, and let K be an algebraically closed algebraic extension of k. Show that there exists a field Ω between k and K, such that $\alpha \notin \Sigma_\square(\Omega)$, and $\alpha \in \Sigma_\square(\Omega')$ for each proper algebraic extension Ω' of Ω.

3. Let K be an ordered field, let

$$f(t) = t^n + a_1 t^{n-1} + \cdots + a_n \in K[t],$$

and set

$$M := \max\{1, |a_1| + \cdots + |a_n|\}.$$

Show that $f(s) > 0$ for $s > M$ and that $(-1)^n f(s) > 0$ for $s < -M$. Conclude that a root s_0 of $f(t)$ in K satisfies $|s_0| \le M$.

Appendix: Solutions to exercises

Convention. *In what follows, the notation x.y. refers to Exercise Number y in § x.*

1.1. Let $\mathfrak{X}$ be an arbitrary class. Then, by definition of the class $\mathfrak{T}$, we have $\mathfrak{X} \subseteq \mathfrak{T}$. Hence, if $\mathfrak{T}$ were a set, so would be $\mathfrak{X}$ by Axiom A4. Since $\mathfrak{X}$ was arbitrary, it would follow that every class is a set. However, we do know from Russel's Paradox that proper classes exist. This contradiction shows that $\mathfrak{T}$ is a proper class.

1.2. Let

$$\mathscr{M} := \{n \in \mathbb{N} : n \notin n\}.$$

For $n = 0$, the assertion $n \in n$ simply says that $\emptyset \in \emptyset$, which is impossible. Thus, $0 \notin 0$, and so $0 \in \mathscr{M}$. Let $n \in \mathscr{M}$, and suppose that $n+1 \notin \mathscr{M}$, so that

$$n+1 \in n+1 = \{0, 1, \ldots, n\}.$$

Let $n+1 = i$ for some i with $0 \le i \le n$. If $n+1 = 0$, then $n \in \emptyset$, which is impossible. Thus, we must have $i > 0$, and we conclude that $n = i-1 \in n$, contradicting our hypothesis that $n \in \mathscr{M}$. Hence, $n+1 \in \mathscr{M}$, and so $\mathscr{M} = \mathbb{N}$, as claimed.

1.3. Let $M_1 = \{\{a_1\}, \{a_1, b_1\}\}$ and $M_2 = \{\{a_2\}, \{a_2, b_2\}\}$. If $M_1 = M_2$, then we have

$$\{a_1\} = \{a_2\} \text{ and } \{a_1, b_1\} = \{a_2, b_2\},$$

or

$$\{a_1\} = \{a_2, b_2\} \text{ and } \{a_1, b_1\} = \{a_2\}.$$

In the first case, we conclude that $a_1 = a_2$, and subsequently that $b_1 = b_2$; in the second case, we get that $a_1 = a_2 = b_2$, and subsequently that $b_1 = a_2$, so that $a_1 = a_2 = b_1 = b_2$. Hence, $a_1 = a_2$ and $b_1 = b_2$ holds in both cases. Conversely, suppose that $a_1 = a_2$ and $b_1 = b_2$. Then $\{a_1\} = \{a_2\}$ and $\{a_1, b_1\} = \{a_2, b_2\}$, thus $M_1 = M_2$.

1.4. If $x_1 = y_2$ and $y_1 = x_2$, and $x_2 \ne y_2$, then $\{x_1, y_1\} = \{x_2, y_2\}$, but we would have $x_1 \ne x_2$ as well as $y_1 \ne y_2$. Thus, with this definition, the crucial property (1.2) of ordered pairs would be violated.

1.5. (i) For a natural number $n \ge 1$ and a set X, let us define an n-tuple $(x_1, x_2, \ldots, x_n)$ of elements in X as a map $\varphi : [n] \to X$, sending $i \in [n]$ to $x_i \in X$. Then we have

$$(x_1, x_2, \ldots, x_n) = (y_1, y_2, \ldots, y_n) \tag{14.18}$$

if, and only if,

$$x_i = \varphi(i) = \psi(i) = y_i \text{ for all } i \in [n].$$

Here, $\varphi, \psi : [n] \to X$ are the maps corresponding to the left-hand and right-hand side of (14.18), respectively.

(ii) Alternatively, let us define n-tuples recursively via

$$(x_1) = \{x_1\},$$
$$(x_1, x_2, \ldots, x_{n+1}) = ((x_1, \ldots, x_n), x_{n+1}), \quad n \ge 2.$$

Suppose that

$$(x_1,x_2,\ldots,x_n)=(y_1,y_2,\ldots,y_n).$$

If $n=1$, then

$$(x_1)=(y_1)\Longleftrightarrow\{x_1\}=\{y_1\}\Longleftrightarrow x_1=y_1,$$

as desired. Now suppose that our claim holds for some value $k\geq 1$ of the variable n. Then

$$\begin{aligned}(x_1,x_2,\ldots,x_{k+1})=(y_1,y_2,\ldots,y_{k+1}) &\Longleftrightarrow (x_1,\ldots,x_k)=(y_1,\ldots,y_k) \text{ and } x_{k+1}=y_{k+1}\\ &\Longleftrightarrow (x_1=y_1,x_2=y_2,\ldots,x_k=y_k) \text{ and } x_{k+1}\\ &\qquad\qquad\qquad\qquad\qquad\qquad\qquad\quad =y_{k+1}\\ &\Longleftrightarrow x_i=y_i \text{ for } i=1,2,\ldots,k+1,\end{aligned}$$

whence our claim.

1.6. (i) We have

$$\{2,4,6,8,\ldots,\}=2\mathbb{N}=\{X: X\in\mathbb{N} \text{ and } 2 \text{ divides } X\}.$$

(ii) Here,

$$\{1,-1,i,-i\}=\{X: X\in\mathbb{C} \text{ and } |X^2|=1\},$$

where $|\cdot|$ denotes the absolute value on the complex numbers defined by

$$|z|^2=|a+bi|^2=a^2+b^2.$$

(iii) We have

$$\{-4,-\frac{3}{2},-\frac{2}{3},-\frac{1}{4},\frac{1}{4},\frac{2}{3},\frac{3}{2},4\}=\{X: X=\frac{p}{q}\in\mathbb{Q},\ q>0, \text{ and } |p|+q=5\}.$$

1.7. (i) This is clear.

(ii) If $x\in X$, then $x\in Y$, since $X\subseteq Y$. Moreover, if $x\in Y$, then $x\in X$, since $Y\subseteq X$. Thus, we have

$$x\in X\Longleftrightarrow x\in Y,$$

and therefore $X=Y$, as claimed.

(iii) If $x\in X$, then $x\in Y$, since $X\subseteq Y$ by assumption. Moreover, if $x\in Y$, then $x\in Z$, since $Y\subseteq Z$, also by hypothesis. Hence, we have

$$x\in X\Longrightarrow x\in Z,$$

and thus $X\subseteq Z$, as desired.

1.8. The assertions that $B\subseteq C$ and $C\subseteq A$ imply $B\subseteq A$ by transitivity. Together with the first hypothesis, that $A\subseteq B$, this implies $A=B$ by anti-symmetry. It remains to see that $B=C$, which follows from $B\subseteq C$ and $C\subseteq A=B$, again by anti-symmetry.

1.9. We prove $(D_\cap)$, the proof of $(D_\cup)$ is completely analogous (indeed dual). We have

$$\begin{aligned} x \in A \cap (B \cup C) &\Longleftrightarrow x \in A \text{ and } x \in B \cup C \\ &\Longleftrightarrow x \in A \text{ and } (x \in B \text{ or } x \in C) \\ &\Longleftrightarrow (x \in A \text{ and } x \in B) \text{ or } (x \in A \text{ and } x \in C) \\ &\Longleftrightarrow x \in A \cap B \text{ or } x \in A \cap C \\ &\Longleftrightarrow x \in (A \cap B) \cup (A \cap C). \end{aligned}$$

1.10. We show that (i) $\Leftrightarrow$ (ii), and that (i) $\Leftrightarrow$ (iii).

(i) $\Rightarrow$ (ii). If $A \subseteq B$, then

$$x \in A \cap B \Longleftrightarrow x \in A \text{ and } x \in B \Longleftrightarrow x \in A,$$

thus $A \cap B = A$, whence (ii).

(ii) $\Rightarrow$ (i). Let $x \in A$. Then $x \in A \cap B$ by (ii), implying $x \in B$. Hence, we have $A \subseteq B$, whence (i).

(i) $\Rightarrow$ (iii). Suppose that $A \subseteq B$. Then

$$x \in A \cup B \Longleftrightarrow x \in A \text{ or } x \in B \Longleftrightarrow x \in B.$$

Thus, $A \cup B = B$, whence (iii).

(iii) $\Rightarrow$ (i). Let $x \in A$. Then $x \in A \cup B$, so $x \in B$ by (iii). It follows that $A \subseteq B$, that is, (i), as claimed.

1.11. (i) We have

$$A \Delta \emptyset = (A - \emptyset) \cup (\emptyset - A) = A \cup \emptyset = A.$$

(ii) By definition of the symmetric difference, and commutativity of union,

$$A \Delta B = (A - B) \cup (B - A) = (B - A) \cup (A - B) = B \Delta A.$$

(iii) We have

$$\begin{aligned} x \in (A\Delta B)\Delta C &\Longleftrightarrow (x \in A\Delta B \text{ and } x \notin C) \text{ or } (x \in C \text{ and } x \notin A\Delta B) \\ &\Longleftrightarrow (x \in A \text{ and } x \notin B \text{ and } x \notin C) \text{ or } (x \notin A \text{ and } x \in B \text{ and } x \notin C) \text{ or} \\ &\qquad (x \in A \text{ and } x \in B \text{ and } x \in C) \text{ or } (x \notin A \text{ and } x \notin B \text{ and } x \in C) \\ &\Longleftrightarrow (x \in A \text{ and } x \in B \text{ and } x \in C) \text{ or } (x \in A \text{ and } x \notin B \text{ and } x \notin C) \text{ or} \\ &\qquad (x \notin A \text{ and } x \in B \text{ and } x \notin C) \text{ or } (x \notin A \text{ and } x \notin B \text{ and } x \in C) \\ &\Longleftrightarrow (x \in A \text{ and } x \notin B\Delta C) \text{ or } (x \notin A \text{ and } x \in B\Delta C) \\ &\Longleftrightarrow x \in A\Delta(B\Delta C). \end{aligned}$$

(iv) We have

$$\begin{aligned} x \in A \cap (B\Delta C) &\iff x \in A \text{ and } x \in B\Delta C \\ &\iff x \in A \text{ and } (x \in B \cup C \text{ and } x \notin B \cap C) \\ &\iff x \in A \text{ and } x \in B \cup C \text{ and } (x \notin B \text{ or } x \notin C) \\ &\iff (x \in A \text{ and } x \in C \text{ and } x \notin B) \text{ or } (x \in A \text{ and } x \in B \text{ and } x \notin C) \\ &\iff x \in (A \cap B)\Delta(A \cap C). \end{aligned}$$

(v) We have

$$x \in A - B \Longrightarrow x \in (A-B) \cup (B-A) = A\Delta B,$$

whence the result.

(vi) If $A = B$, then $A - B = \emptyset = B - A$, thus $A\Delta B = \emptyset$. If, conversely, $A\Delta B = \emptyset$, then $A - B = \emptyset = B - A$. Hence, $A \subseteq B$ and $B \subseteq A$, implying $A = B$ by anti-symmetry.

(vii) Given sets A and B, suppose that $A\Delta C = B\Delta C$ holds for some set C. We want to show that then necessarily $A = B$. By symmetry, it suffices to show that $A \subseteq B$. Let $x \in A$, and suppose that $x \notin C$. Then $x \in A\Delta C$, thus $x \in B\Delta C$, so $x \in B$ as $x \notin C$. If, on the other hand, $x \in C$, then $x \notin A\Delta C$, thus $x \notin B\Delta C$, which forces $x \in B$ (as $x \in C$).

1.12. We have

$$\begin{aligned} (a,b) \in X \times (Y \cup Z) &\iff a \in X \text{ and } b \in Y \cup Z \\ &\iff a \in X \text{ and } (b \in Y \text{ or } b \in Z) \\ &\iff (a \in X \text{ and } b \in Y) \text{ or } (a \in X \text{ and } b \in Z) \\ &\iff (a,b) \in X \times Y \text{ or } (a,b) \in X \times Z \\ &\iff (a,b) \in (X \times Y) \cup (X \times Z). \end{aligned}$$

1.13. We have

$$x \in X \iff (x,x) \in X \times X \iff (x,x) \in Y \times Y \iff x \in Y.$$

Hence, $X = Y$, as claimed.

1.14. Fix an element $x_0 \in X$ (this we can do since the class X is guaranteed to be non-empty). Then we have

$$y \in Y \iff (x_0, y) \in X \times Y \iff (x_0, y) \in X \times Z \iff y \in Z.$$

Hence, $Y = Z$, as desired.

1.15. (a) One should define

$$m \leq n :\iff m \subseteq n \quad (m, n \in \mathbb{N}).$$

(b) Set $n_0 := f(n)$. If $k \in n+1$ is such that $f(k) \subset n_0$ (that is, $f(k) < n_0$), then $k \neq n$; thus, setting $g(k) = f(k)$ for such k ensures that the numbers $0, 1, \ldots, n_0 - 1$ are all in the range of the function g, as they are in the range of f. Next, suppose that

$k \in n+1$ is such that $f(k) \supset n_0$ (that is $f(k) > n_0$). Again, it follows that $k \neq n$, and we also have $f(k) \neq 0$. Thus, we may set $g(k) := f(k) - 1$ in this case and, since $n_0 + 1, n_0 + 2, \ldots$ are all in the range of f by hypothesis, the numbers $n_0, n_0 + 1, \ldots$ are in the range of g. It follows that g has range equal to $\mathbb{N}$, as claimed.

1.16. Denote by $[q,r]$ the least common multiple of the positive integers q and r. Then

$$M_q \cap M_r = \{n \in \mathbb{N} : n \text{ is divisible by } [q,r]\}.$$

1.17. Clearly, $X = B - A$ is a solution. Let X' be an arbitrary solution of this system of equations. Then the condition that $A \cup X = A \cup B$ forces $X \subseteq X'$, while the condition that $A \cap X = \emptyset$ implies $X' \subseteq X$. Hence, $X' = X$, so the solution X for this system of equations is unique for any given sets A, B.

1.18. We have

$$[0,1] \subseteq (-1/n, 1+1/n) \subseteq [-1/n, 1+1/n], \quad n \geq 1,$$

so that both intersections have to contain the interval $[0,1]$. If $x \in \mathbb{R}$ satisfies $x < 0$, then there exists some natural number n_0 so that $x < -1/n_0 < 0$ and, consequently, $x \notin [-1/n, 1+1/n]$ for all $n \geq n_0$. Similarly, if $x > 1$, there exists some natural number n_0 such that $1 < 1 + 1/n_0 < x$, so that, again, $x \notin [-1/n, 1+1/n]$ for all $n \geq n_0$. Hence, both intersections equal $[0,1]$, as claimed.

2.1. We have $|A \times B| = 3 \cdot 2 = 6$, hence there are $2^6 = 64$ correspondences from A to B.

2.2. We have

$$\begin{aligned} (x,y) \in (\Phi \cap \Psi)^{-1} &\iff (y,x) \in \Phi \cap \Psi \\ &\iff (y,x) \in \Phi \text{ and } (y,x) \in \Psi \\ &\iff (x,y) \in \Phi^{-1} \text{ and } (x,y) \in \Psi^{-1} \\ &\iff (x,y) \in \Phi^{-1} \cap \Psi^{-1}. \end{aligned}$$

Also,

$$\begin{aligned} (x,y) \in (\Phi \cup \Psi)^{-1} &\iff (y,x) \in \Phi \cup \Psi \\ &\iff (y,x) \in \Phi \text{ or } (y,x) \in \Psi \\ &\iff (x,y) \in \Phi^{-1} \text{ or } (x,y) \in \Psi^{-1} \\ &\iff (x,y) \in \Phi^{-1} \cup \Psi^{-1}. \end{aligned}$$

2.3. Equation (2.1). We have

$$\begin{aligned} (a,b) \in \theta \circ (\Psi \circ \Phi) &\iff (a,x) \in \Psi \circ \Phi \text{ and } (x,b) \in \theta \text{ for some } x \\ &\iff (a,y) \in \Phi,\ (y,x) \in \Psi, \text{ and } (x,b) \in \theta \text{ for some } x,y \\ &\iff (a,x) \in \Phi,\ (x,y) \in \Psi, \text{ and } (y,b) \in \theta \text{ for some } x,y \\ &\iff (a,b) \in (\theta \circ \Psi) \circ \Phi, \end{aligned}$$

from which the desired equation follows.

Equation (2.2). We have

$$\begin{aligned}(a,b)\in(\Psi\circ\Phi)^{-1} &\iff (b,a)\in\Psi\circ\Phi\\ &\iff (b,x)\in\Phi \text{ and } (x,a)\in\Psi \text{ for some } x\\ &\iff (a,x)\in\Psi^{-1} \text{ and } (x,b)\in\Phi^{-1} \text{ for some } x\\ &\iff (a,b)\in\Phi^{-1}\circ\Psi^{-1}.\end{aligned}$$

Equation (2.3). We have

$$(a,b)\in(\Phi^{-1})^{-1} \iff (b,a)\in\Phi^{-1} \iff (a,b)\in\Phi.$$

Equations (2.4). We have, for $\Phi\subseteq A\times B$,

$$(a,b)\in\Phi\circ\Delta_A \iff (a,x)\in\Delta_A \text{ and } (x,b)\in\Phi \text{ for some } x \iff (a,b)\in\Phi.$$

Hence, $\Phi\circ\Delta_A=\Phi$, as claimed. Similarly,

$$(a,b)\in\Delta_B\circ\Phi \iff (a,x)\in\Phi \text{ and } (x,b)\in\Delta_B \text{ for some } x \iff (a,b)\in\Phi.$$

2.4. For $a\in A$, we have

$$\begin{aligned}(a,a)\in\Phi^{-1}\circ\Phi \iff (a,b)\in\Phi \text{ and } (b,a)\in\Phi^{-1} \text{ for some}\\ b\in B \iff (a,b)\in\Phi \text{ for some } b\in B.\end{aligned}$$

Hence, the correspondence $\Phi\subseteq A\times B$ satisfies Condition (2.5) if, and only if, for each element $a\in A$ there exists at least one element $b\in B$ such that $(a,b)\in\Phi$.

Similarly, for $x,y\in B$, we have

$$\begin{aligned}(x,y)\in\Phi\circ\Phi^{-1} \iff (x,a)\in\Phi^{-1} \text{ and } (a,y)\in\Phi \text{ for some}\\ a\in A \iff (a,x)\in\Phi \text{ and } (a,y)\in\Phi \text{ for some } a\in A.\end{aligned}$$

Hence, Φ satisfies (2.6) if, and only if, $(a,x)\in\Phi$ and $(a,y)\in\Phi$ for $a\in A$ and $x,y\in B$ implies $x=y$.

Combining these two observations, we see that the correspondence $\Phi\subseteq A\times B$ satisfies both (2.5) and (2.6) if, and only if, for each element $a\in A$ there exists precisely one element $b\in B$ such that $(a,b)\in\Phi$; thus, if, and only if, Φ is a function from A to B.

2.5. If $|A|=m$, then $|A^n|=|A|^n=m^n$. Hence, there are $|\mathscr{B}(A^n)|=2^{m^n}$ n-ary relations on the set A.

2.6. If $B\subseteq A$ is any subset of the given set A, then its characteristic function χ_B is defined, and is a member of 2^A; thus, the map $\Phi:\mathscr{B}(A)\to 2^A$ is well-defined. Let $B_1,B_2\subseteq A$, and suppose that $\Phi(B_1)=\Phi(B_2)$; that is, $\chi_{B_1}=\chi_{B_2}$. Then

$$B_1=\{x\in A:\chi_{B_1}(x)=1\}=\{x\in A:\chi_{B_2}(x)=1\}=B_2.$$

Hence, the map Φ is injective. Finally, let $f:A\to 2$ be arbitrary. Then setting

$$B:=\{x\in A: f(x)=1\}$$

defines a subset B of A and, for $x\in A$, we have

$$\Phi(B)(x)=\chi_B=\left\{\begin{matrix}1, & x\in B\\ 0, & x\notin B\end{matrix}\right\}=f(x).$$

It follows that $\Phi(B) = f$, so that Φ is surjective, and thus a bijection, as claimed.

2.7. For $x, y \in A$, we have

$$\begin{aligned} x \equiv y \pmod{\mathfrak{q}_{P_\mathfrak{q}}} &\iff x \text{ and } y \text{ lie in the same class of the partition } P_\mathfrak{q} \\ &\iff x^\mathfrak{q} = y^\mathfrak{q} \\ &\iff x \equiv y \pmod{\mathfrak{q}}. \end{aligned}$$

Thus, we have $\mathfrak{q}_{P_\mathfrak{q}} = \mathfrak{q}$, as claimed. Also, for $x \in A$,

$$\begin{aligned} [x]_{P_{\mathfrak{q}_P}} &= \{y \in A : x \equiv y \pmod{\mathfrak{q}_P}\} \\ &= \{y \in A : [x]_P = [y]_P\} \\ &= [x]_P. \end{aligned}$$

Here, $[x]_P$ denotes the class of the partition P of the set A containing the element x. Hence, $P_{\mathfrak{q}_P} = P$, as desired. It follows that the maps φ and ψ are indeed bijections inverse to each other.

2.8. For each $x \in X$, set $f(x) := f_i(x)$, where $i \in I$ is any index for which $x \in X_i$. In this way we obtain a well-defined mapping $f : X \to Y$. Indeed, if also $x \in X_j$, then

$$f_i(x) = f_i|_{X_i \cap X_j}(x) = f_j|_{X_i \cap X_j}(x) = f_j(x)$$

by our hypothesis. Evidently, the map f thus defined is an extension of each f_i. Moreover, any map $f' : X \to Y$ extending f_i for each $i \in I$ must coincide with f, since necessarily

$$f(x) = f_i(x) = f'(x), \quad x \in X_i.$$

2.9. We have

$$3 \times 3 = \{(0,0),(0,1),(0,2),(1,0),(1,1),(1,2),(2,0),(2,1),(2,2)\}.$$

Since equivalence relations are reflexive, any equivalence relation $\mathfrak{q} \subseteq 3 \times 3$ must contain the elements $(0,0)$, $(1,1)$, and $(2,2)$. Indeed,

$$\mathfrak{q}_1 = \{(0,0),(1,1),(2,2)\} = \Delta_3$$

is the smallest equivalence relation on the set 3. If $\mathfrak{q} \supset \Delta_3$ then, since $\mathfrak{q}$ is symmetric, we have to add at least two elements; and the 5-element relations resulting in this way are automatically transitive. Thus, we obtain the equivalence relations

$$\begin{aligned} \mathfrak{q}_2 &= \{(0,0),(1,1),(2,2),(0,1),(1,0)\}, \\ \mathfrak{q}_3 &= \{(0,0),(1,1),(2,2),(0,2),(2,0)\}, \\ \mathfrak{q}_4 &= \{(0,0),(1,1),(2,2),(1,2),(2,1)\}. \end{aligned}$$

If $|\mathfrak{q}| > 5$, then $\mathfrak{q} = \mathfrak{q}_5 = 3 \times 3$ by transitivity of $\mathfrak{q}$. Hence, there are precisely five distinct equivalence relations on the set 3. Their corresponding partitions of 3 are, in order:

$$\begin{aligned}\mathfrak{p}_1 &= \{\{0\},\{1\},\{2\}\},\\ \mathfrak{p}_2 &= \{\{0,1\},\{2\}\},\\ \mathfrak{p}_3 &= \{\{0,2\},\{1\}\},\\ \mathfrak{p}_4 &= \{\{0\},\{1,2\}\},\\ \mathfrak{p}_5 &= \{\{0,1,2\}\}.\end{aligned}$$

2.10. Set $\hat{\mathfrak{q}} := \bigcap \mathscr{E}$. Then, by definition, $\hat{\mathfrak{q}} \subseteq A \times A$ is a binary relation on A. If $x \in A$ is any element, then $(x,x) \in \mathfrak{q}$ for each $\mathfrak{q} \in \mathscr{E}$ by reflexivity of $\mathfrak{q}$, thus $(x,x) \in \hat{\mathfrak{q}}$, so that $\hat{\mathfrak{q}}$ is reflexive. Next, suppose that $(x,y) \in \hat{\mathfrak{q}}$. Then $(x,y) \in \mathfrak{q}$ for all $\mathfrak{q} \in \mathscr{E}$, so that $(y,x) \in \mathfrak{q}$ for all $\mathfrak{q} \in \mathscr{E}$ by symmetry of $\mathfrak{q}$. It follows that $(y,x) \in \hat{\mathfrak{q}}$, hence $\hat{\mathfrak{q}}$ is symmetric. Finally, suppose that $(x,y),(y,z) \in \hat{\mathfrak{q}}$. Then we have $(x,y),(y,z) \in \mathfrak{q}$ for each $\mathfrak{q} \in \mathscr{E}$. By transitivity of each such $\mathfrak{q}$, $(x,z) \in \mathfrak{q}$ holds for all $\mathfrak{q} \in \mathscr{E}$, so that $(x,z) \in \hat{\mathfrak{q}}$, showing that $\hat{\mathfrak{q}}$ is transitive as well, thus an equivalence relation, as claimed.

2.11. Let $(a,b) \in A \times B$ be an arbitrary element. Then, since both $\equiv_A$ and $\equiv_B$ are reflexive relations on A and B, respectively, we have $a \equiv_A a$ and $b \equiv_B b$, thus $(a,b) \equiv (a,b)$ by definition of the relation $\equiv$. Hence, $\equiv$ is itself reflexive. Next, suppose that $(a_1,b_1) \equiv (a_2,b_2)$ for some $a_1,a_2 \in A$ and $b_1,b_2 \in B$. By definition of $\equiv$, this means that $a_1 \equiv_A a_2$ and $b_1 \equiv_B b_2$. Since $\equiv_A$ and $\equiv_B$ are both symmetric, we get $a_2 \equiv_A a_1$ and $b_2 \equiv_B b_1$, implying $(a_2,b_2) \equiv (a_1,b_1)$ by definition of $\equiv$. Hence, the binary relation $\equiv$ on $A \times B$ is itself symmetric. Finally, suppose that $(a_1,b_1) \equiv (a_2,b_2)$ and that $(a_2,b_2) \equiv (a_3,b_3)$ for some $a_1,a_2,a_3 \in A$ and $b_1,b_2,b_3 \in B$. By definition of $\equiv$, this means that $a_1 \equiv_A a_2$ as well as $a_2 \equiv_A a_3$, and that $b_1 \equiv_B b_2$ and $b_2 \equiv_B b_3$. Transitivity of $\equiv_A$ and $\equiv_B$ now yields $a_1 \equiv_A a_3$ and $b_1 \equiv_B b_3$, respectively. By definition of $\equiv$, this in turn implies $(a_1,b_1) \equiv (a_3,b_3)$, and thus transitivity of $\equiv$. Hence, $\equiv$ is an equivalence relation on the set $A \times B$ as claimed.

3.1. If $x \in X$ and $y \in Y$, then $x \leq x$ and $y \leq' y$, since $\leq$ and $\leq'$ are reflexive, thus $(x,y) \preceq (x,y)$, so that the relation $\preceq$ on $X \times Y$ is reflexive. Next, suppose that $(x_1,y_1) \preceq (x_2,y_2)$ and $(x_2,y_2) \preceq (x_1,y_1)$. Then, by definition of the relation $\preceq$, we have $x_1 \leq x_2$, $x_2 \leq x_1$, $y_1 \leq' y_2$, and $y_2 \leq' y_1$. Since $\leq$ and $\leq'$ are both antisymmetric, we infer that $x_1 = x_2$ and $y_1 = y_2$, thus $(x_1,y_1) = (x_2,y_2)$, so that $\preceq$ is antisymmetric as well. Finally, suppose that $(x_1,y_1) \preceq (x_2,y_2)$ and $(x_2,y_2) \preceq (x_3,y_3)$. Then, by definition of the relation $\preceq$, we have $x_1 \leq x_2$, $x_2 \leq x_3$, $y_1 \leq' y_2$, and $y_2 \leq' y_3$. Since $\leq$ and $\leq'$ are both transitive, we conclude that $x_1 \leq x_3$ and $y_1 \leq' y_3$, so that $(x_1,y_1) \preceq (x_3,y_3)$. Hence, $\preceq$ is transitive, thus an order relation, as claimed.

3.2. Let $\mathscr{O} \subseteq \mathscr{B}(A^2)$ be a non-empty collection of order relations R on the set A, and set $\leq = \bigcap_{R \in \mathscr{O}} R$. If $x \in A$, then $(x,x) \in R$ for each $R \in \mathscr{O}$, thus $x \leq x$ since $\mathscr{O} \neq \emptyset$, so that the relation $\leq$ on A is reflexive. Next, suppose that $x \leq y$ and $y \leq x$ for some $x,y \in A$. Then xRy and YRx hold for all $R \in \mathscr{O}$, thus $x = y$, since $\mathscr{O} \neq \emptyset$. This shows that $\leq$ is antisymmetric. Finally, suppose that $x \leq y$ and $y \leq z$ for some $x,y,z \in A$. Then xRy and yRz hold for all $R \in \mathscr{O}$, implying xRz for all $R \in \mathscr{O}$, thus $x \leq z$, which shows that $\leq$ is transitive, hence an order relation, as claimed.

3.3. This is shown by induction on the size of the finite subset of A considered. Since any singleton subset of A is obviously bounded above, the induction starts. Suppose that every n-element subset of A has an upper bound, where $n \geq 1$, and consider a subset $B \subseteq A$ of size $n+1$, say, $B = \{b_1, b_2, \ldots, b_n, b_{n+1}\}$. By the induction hypothesis, the subset $B' = \{b_1, b_2, \ldots, b_n\}$ has an upper bound $b \in A$. Moreover, since we assume that any 2-element subset of A has an upper bound in A, so has the subset $B'' := \{b, b_{n+1}\}$. Let b' be this upper bound of B'' in A. It follows that b' is an upper bound for the set B in A; hence, any finite subset of A has an upper bound in A, so $(A, \leq)$ is directed upwards, as claimed.

3.4. Suppose that a_1 and a_2 are both greatest elements of the ordered set $(A, \leq)$. Then $a_1 \leq a_2$, since a_2 is a greatest element of A, and $a_2 \leq a_1$, since a_1 is also a greatest element. Hence, $a_1 = a_2$ by antisymmetry of the relation $\leq$.

3.5. Let a be a maximal element of A, and let $a' \in A$ be an arbitrary element. Since A is directed (upwards), the subset $\{a, a'\}$ of A has an upper bound a'' in A. If $a'' \neq a$, then $a < a''$, contradicting our hypothesis that a is a maximal element in A. Hence, $a'' = a$, so that $a' \leq a$, showing that a is the greatest element of A.

3.6. Suppose that $(A, \leq)$ is well-ordered, and let $a, b \in A$ with $a \neq b$. If $a \not< b$ and $b \not< a$, then both a and b are minimal elements of the subset $\{a, b\}$ of A, contradicting our hypothesis. Hence, one of these relations must hold, implying that A is totally ordered.

3.7. Assume for a contradiction that $A \setminus B \neq \emptyset$, and let a_0 be a minimal element of $A \setminus B$ (such an element exists since A is assumed to satisfy the minimum condition). Thus, every element $a' \in A$ with $a' < a_0$ is such that $a' \in B$. By our hypothesis on the subset B, this implies $a_0 \in B$, so $a_0 \notin A \setminus B$, a contradiction. It follows that $A \setminus B = \emptyset$, or $B = A$, as desired.

3.8. This is proved by induction on the size of the subset B of the lattice $(A, \leq)$ in question. A one-element subset $B = \{b_1\} \subseteq A$ satisfies $\inf B = b_1 = \sup B$ by definition of infimum and supremum. Suppose inductively that every n-element subset of A has an infimum and a supremum, where $n \geq 1$, and consider a subset $B = \{b_1, b_2, \ldots, b_n, b_{n+1}\} \subseteq A$ of size $n+1$. By the induction hypothesis, there exist elements $b', b'' \in A$ with $b' = \inf B'$ and $b'' = \sup B'$, where $B' := \{b_1, b_2, \ldots, b_n\}$. Then $b' \wedge b_{n+1}$ exists since $(A, \leq)$ is a lattice, and is the infimum of the subset B; similarly, $b'' \vee b_{n+1}$ exists and is the supremum of B. We focus on the first of these statements. By definition of the infimum, we have in particular $b' \leq b_1, b_2, \ldots, b_n$, so $b' \wedge b_{n+1}$ is a lower bound for the subset B. Moreover, if $b_0 \in A$ is any lower bound for B then, in particular, $b_0 \leq b_1, b_2, \ldots, b_n$, so $b_0 \leq b'$ by definition of $\inf B'$, and we also have $b_0 \leq b_{n+1}$. Thus, b_0 is a lower bound for the subset $\{b', b_{n+1}\}$ of A, implying $b_0 \leq b' \wedge b_{n+1}$, so that indeed $b' \wedge b_{n+1} = \inf B$, as claimed. The argument for $b'' \vee b_{n+1}$ is similar, and is left to the reader.

3.9. There is no telling what example the reader may come up with here. However, one possible solution is as follows. Let $A = (\mathbb{N}^*, |)$, $B = (\mathbb{N}^*, \leq)$, and $f = 1_{\mathbb{N}^*}$. Here, $\mathbb{N}^* := \mathbb{N} \setminus \{0\}$, while $m \mid n$ means that m *divides* n; that is, there exists some $k \in \mathbb{N}^*$ such that $m \cdot k = n$. Clearly, if in this setting $m \mid n$, then we have $m \leq n$. Thus, the map $f = 1_{\mathbb{N}^*} : (\mathbb{N}^*, |) \to (\mathbb{N}^*, \leq)$ is a bijective order morphism, and, on a set-theoretic level, $f^{-1} = f = 1_{\mathbb{N}^*}$. However, f^{-1} considered as a map f^{-1}: $(\mathbb{N}^*, \leq) \to (\mathbb{N}^*, |)$ is not order-preserving; for instance, we have $2 \leq 3$, while 2 does not divide 3.

3.10. Again, we let $A = (\mathbb{N}^*, |)$, $B = (\mathbb{N}^*, \leq)$, and $f = 1_{\mathbb{N}^*} : A \to B$, with notation as in the solution of Exercise 7. Then A and B are lattices; indeed, $m \wedge n = \gcd\{m,n\}$ and $m \vee n = \operatorname{lcm}\{m,n\}$ for $m,n \in A$, while $m \wedge n = \min\{m,n\}$ and $m \vee n = \max\{m,n\}$ for $m,n \in B$. Moreover, as observed before, $f : A \to B$ is an order morphism. However, f is not a lattice homomorphism since, for instance, $\gcd\{10,12\} = 2$, while $\min\{10,12\} = 10$.

3.11. Let A and B be lattices, and let $f : A \to B$ be an order isomorphism. We have to show that both f and f^{-1} are lattice homomorphisms. By symmetry, it suffices to consider the map f. Let $a_1, a_2 \in A$ be arbitrary. Then $a_1 \wedge a_2 \leq a_1, a_2$ by definition of the infimum, thus $f(a_1 \wedge a_2) \leq f(a_1), f(a_2)$, since f is order-preserving. Next, let $b \leq f(a_1), f(a_2)$ be any lower bound, and let $b = f(a)$ (such a exists since f is assumed to be surjective). Then $a = f^{-1}(b) \leq a_1, a_2$, since f^{-1} is order-preserving; thus, $a \leq a_1 \wedge a_2$ by definition of the infimum. Since f is order-preserving, it follows that $b = f(a) \leq f(a_1 \wedge a_2)$, so that $f(a_1 \wedge a_2) = f(a_1) \wedge f(a_2)$, as desired. The argument for the supremum is similar.

3.12. Set

$$\mathscr{I} := \{x \in A : x \leq f(x) \text{ for all } f \in F\} \text{ and } a_0 := \sup \mathscr{I}.$$

For $x \in \mathscr{I}$, we have $x \leq f(x)$ for all $f \in F$, as well as $x \leq a_0$. Hence, since $f \in F$ is order-preserving, $f(x) \leq f(a_0)$, thus $x \leq f(a_0)$ for all $x \in \mathscr{I}$ and all $f \in F$. Therefore, each point $f(a_0)$ is an upper bound for the set $\mathscr{I}$, implying

$$a_0 \leq f(a_0), \quad f \in F, \tag{14.19}$$

by definition of a_0, and thus $a_0 \in \mathscr{I}$. Applying $g \in F$ to (14.19), we find that $g(a_0) \leq g(f(a_0)) = f(g(a_0))$ for all $f, g \in F$, so that $g(a_0) \in \mathscr{I}$ for all $g \in F$. Hence, $f(a_0) \leq a_0$ for all $f \in F$ which, together with (14.19) shows that a_0 is a common fixed point for all $f \in F$, so $a_0 \in \mathscr{F}$ and $\mathscr{F} \neq \emptyset$. Moreover, we see that

$$\sup \mathscr{F} = \sup \mathscr{I} \in \mathscr{F}. \tag{14.20}$$

The only part of (14.20) requiring a further argument is the equality $\sup \mathscr{F} = \sup \mathscr{I}$: since $\mathscr{F} \subseteq \mathscr{I}$, we have $\sup \mathscr{F} \leq \sup \mathscr{I}$. On the other hand, we have just seen that $a_0 = \sup \mathscr{I} \in \mathscr{F}$, thus $\sup \mathscr{F} \geq a_0$, implying the desired equation.

Next, consider the dual lattice $A' = (A, \leq')$, where $a \leq' b :\Longleftrightarrow b \leq a$. Like A, the dual A' is a complete lattice, and the maps $f \in F$, considered as maps on A', are again order-preserving. Thus, applying to A' the result (14.20) established for A, we find that, translated back into $(A, \leq)$,

$$\inf \mathscr{F} = \inf\{x \in A : f(x) \leq x \text{ for all } f \in F\} \in \mathscr{F}. \tag{14.21}$$

Now let $X \subseteq \mathscr{F}$ be an arbitrary subset, and consider the interval $B := [\sup X, \sup A]$ of A. Clearly, B inherits the structure of a complete lattice from A. Moreover, for $x \in X$, we have $x \leq \sup X$, thus $x = f(x) \leq f(\sup X)$ for all $f \in F$. Consequently, $z \in B$ implies

$$\sup X \leq f(\sup X) \leq f(z), \quad f \in F;$$

in particular, $f(z) \in B$ for all $f \in F$. Hence, we may restrict the domain of $f \in F$ to B to obtain commuting order-preserving maps $f' = f|_B : B \to B$ for $f \in F$. Applying (14.21) to the complete lattice B and the maps f', we conclude that the infimum u of the common fixed points of all the maps f' is itself a fixed point of every map f'. Clearly, u is a fixed point of every map $f \in F$, so $u \in \mathscr{F}$. By construction, u is an upper bound of X, and if $u' \in \mathscr{F}$ is any upper bound of X, then $u' \in B$ and $u \leq u'$. Hence, u is the least common fixed point of the maps $f \in F$, which is an upper bound of the set X; that is, u is the supremum of X in $\mathscr{F}$. Since X was an arbitrary subset of $\mathscr{F}$, we conclude from Proposition 3.2 that $(\mathscr{F}, \leq)$ is a complete lattice.

3.13. Let $\mathscr{X} \subseteq \mathscr{S}^-(A)$ be a family of left segments of the ordered set A. We claim that $X_0 := \bigcap \mathscr{X}$ and $Y_0 := \bigcup \mathscr{X}$ are again left segments of A. Let $x \in X_0$ and let $y \in A$ be such that $y \leq x$. Then $x \in X$ for all $X \in \mathscr{X}$, so $y \in X$ for all $X \in \mathscr{X}$, hence $y \in X_0$. Thus, X_0 is a left segment of A, as desired, and a similar argument shows that Y_0 is also a left segment of A. It follows that $\inf \mathscr{X} = X_0$ and $\sup \mathscr{X} = Y_0$. Since $\mathscr{X} \subseteq \mathscr{S}^-(A)$ was an arbitrary subset, we conclude that $\mathscr{S}^-(A)$ is a complete lattice under inclusion, as claimed.

3.14. Suppose that $g(\sup X)$ covers $f(\sup X)$, and let

$$Y := \{y \in A : y \leq \sup X \text{ and } g(y) < g(\sup X)\}.$$

Then $\sup Y \leq \sup X$, and we have $g(y) \leq f(\sup X)$ for all $y \in Y$ since $g(y) \not> f(\sup X)$ and B is totally ordered. If $\sup Y = \sup X$, then $\sup Y \neq 0$; in particular, $Y \neq \emptyset$. Since g is lower semi-continuous, we find that $g(\sup X) = g(\sup Y) \leq f(\sup X)$, which contradicts the case assumption. Hence, $\sup Y < \sup X$. As in Case (i) of the proof, let

$$Z = \{z \in X : \sup Y < z\}.$$

If Z were empty, then we would have $x \leq \sup Y$ for all $x \in X$, since A is totally ordered, thus $\sup X \leq \sup Y$ contradicting our last finding that $\sup Y < \sup X$. Therefore, $Z \neq \emptyset$. Also, by definition of Z, we have $\sup Z = \sup X$. Since f is upper semi-continuous, it follows from

$$g(\sup X) > f(\sup X) = f(\sup Z)$$

that $f(z) < g(\sup X)$ for some $z = z_0 \in Z$, since otherwise we would have $f(\sup X) \geq g(\sup X)$, contradicting our case assumption. It now follows from (3.9) and the definition of the set Z that $\sup Y < z_0$ and $g(z_0) < g(\sup X)$; in particular, $z_0 \in Y$. However, the assertions $\sup Y < z_0$ and $z_0 \in Y$ clearly contradict each other. This contradiction shows that (3.10) cannot hold in case (ii) either.

4.1. Let $\mathfrak{P}$ be a property of finite type for subsets of a set A; that is, $\mathfrak{P} \subseteq \mathscr{B}(A)$, consider $\mathfrak{P}$ as partially ordered by inclusion, and let $\mathscr{C}$ be a chain in $\mathfrak{P}$. Let $U = \bigcup_{C \in \mathscr{C}} C$, and let T be a finite subset of U. Then $T \subseteq C$ for some $C \in \mathscr{C}$, since T is finite (so must lie in a union of finitely many $C \in \mathscr{C}$), and $\mathscr{C}$ is a chain by hypothesis. Since $C \in \mathfrak{P}$, we get $T \in \mathfrak{P}$ as $\mathfrak{P}$ is assumed to be of finite type. Since T was an arbitrary finite subset of U, and $\mathfrak{P}$ is of finite type, $U \in \mathfrak{P}$; that is, $(\mathfrak{P}, \subseteq)$ satisfies the hypothesis of Assertion (c) in Theorem 4.1. It follows now from (c) that the ordered set $(\mathfrak{P}, \subseteq)$ has a maximal element; that is, A contains a subset enjoying Property $\mathfrak{P}$, which is maximal among all such subsets of A, whence (d).

4.2. Reflexivity of $\preceq$ is clear. Next, suppose that $(S_1,\leq_1)\preceq(S_2,\leq_2)$ and $(S_2,\leq_2)\preceq(S_1,\leq_1)$. By definition of the relation $\preceq$, this implies $S_1=S_2=:S$ and $x\leq_1 y$ holds if, and only if, $x\leq_2 y$ holds by Condition (ii), so that the order relations $\leq_1$ and $\leq_2$ on S coincide. Hence, $(S_1,\leq_1)=(S_2,\leq_2)$; that is, $\preceq$ is anti-symmetric. Finally, suppose that $(S_1,\leq_1)\preceq(S_2,\leq_2)$ and $(S_2,\leq_2)\preceq(S_3,\leq_3)$. We want to show that $(S_1,\leq_1)\preceq(S_3,\leq_3)$. First, Condition (i) gives $S_1\subseteq S_2$ and $S_2\subseteq S_3$, so that $S_1\subseteq S_3$, as desired. Second, let $x,y\in S_1$ be arbitrary. If $x\leq_1 y$ then we have $x,y\in S_2$ and $x\leq_2 y$ by Condition (ii), hence $x\leq_3 y$ by another application of Condition (ii). Conversely, suppose that $x,y\in S_1$ and $x\leq_3 y$. Then $x,y\in S_2$, so $x\leq_2 y$, implying $x\leq_1 y$, again by repeated application of Condition (ii). Hence, for $x,y\in S_1$ we have $x\leq_1 y$ if, and only if, $x\leq_3 y$, as required. Third, Let $x\in S_1$ and $y\in S_3\setminus S_1$. We want to show that, in this situation, $x\leq_3 y$. If $y\in S_2$, then $y\in S_2\setminus S_1$, thus $x\leq_2 y$ by Condition (iii). Hence, in this case, $x,y\in S_2$ and $x\leq_2 y$, implying $x\leq_3 y$, as desired. It remains to deal with the case where $x\in S_1$ and $y\in S_3\setminus S_2$. Then $x\in S_2$, and $x\leq_3 y$ follows again. All in all, we have shown that $(S_1,\leq_1)\preceq(S_3,\leq_3)$, and $\preceq$ is transitive as well, hence an order relation on $\mathscr{W}$, as claimed.

4.3. As is well known, a polynomial ring $K[x]$ in one variable is Euclidean, with the complexity measure given by the degree function. Applying the Euclidean algorithm, we obtain polynomials $g(x),r(x)\in K[x]$, such that

$$f(x)=(x-\xi)g(x)+r(x),$$

and we have $r(x)=0$ or $\deg(r)<1$ (so that $r(x)$ is a non-zero element of the field K). Setting $x=\xi$, we find that $r(\xi)=0$, which forces $r(x)=0$ as a polynomial. Thus, we have the factorisation $f(x)=(x-\xi)g(x)$, as required.

4.4. Suppose that $\langle f\rangle\leq\mathfrak{J}\leq K[x]$. Since $K[x]$ is a principal ideal domain, there exists a polynomial $g=g(x)\in K[x]$ such that $\mathfrak{J}=\langle g\rangle$, and the above inclusion implies that $f(x)=g(x)h(x)$ for some $h=h(x)\in K[x]$. Since f is irreducible, this equation in turn implies $g\in K$ or $h\in K$; that is, $\mathfrak{J}=K[x]$, or $g=h^{-1}f$, and so $\mathfrak{J}=\langle g\rangle\leq\langle f\rangle$, thus $\mathfrak{J}=\langle f\rangle$. Since $\deg(f)\geq 1$, the ideal $\langle f\rangle$ is proper, thus a maximal ideal in $K[x]$, as claimed.

4.5. It is clear that $\bar{\Lambda}=\Lambda/\mathfrak{J}$ is a commutative ring with identity element $\bar{1}$, so we only need to prove existence of multiplicative inverses. Let $\bar{a}\in\bar{\Lambda}$ be such that $\bar{a}\neq\bar{0}$. Then $\mathfrak{J}$ together with the element a generate the ring Λ; thus every element $\lambda\in\Lambda$ may be written in the form

$$\lambda=x+ra,\quad(x\in\mathfrak{J},\,r\in\Lambda).$$

In particular, there exist elements $r\in\Lambda$ and $x\in\mathfrak{J}$, such that $1=x+ra$, implying $\bar{r}\bar{a}=\bar{1}$. Hence, $\bar{a}$ has a multiplicative inverse, and $\bar{\Lambda}=\Lambda/\mathfrak{J}$ is a field, as required.

4.6. For instance, let $S=4=\{0,1,2,3\}$ and $\rho=\{(0,1),(1,2),(2,3),(3,0)\}$. Then ρ is certainly an anti-symmetric relation on S. Suppose that ρ' is an order relation on S such that $\rho\subseteq\rho'$. Then, by transitivity of ρ', $(0,2)\in\rho'$, and thus $(0,3)\in\rho'$. Since $(3,0)\in\rho\subseteq\rho'$, ρ' is not anti-symmetric, contradicting our assumption.

4.7. Let

$$\mathfrak{R}=\left\{\rho':\rho'\text{ is an order relation on }S\text{ and }\rho\subseteq\rho'\right\}.$$

Since $\rho \in \mathfrak{R}$, we have $\mathfrak{R} \neq \emptyset$. Consider $\mathfrak{R}$ as being ordered by inclusion. We claim that $(\mathfrak{R}, \subseteq)$ is inductively ordered. Indeed, let $\mathscr{C}$ be a non-empty chain in $(\mathfrak{R}, \subseteq)$, and set $\rho_0 := \bigcup_{\rho' \in \mathscr{C}} \rho'$. Then $\rho_0 \subseteq S \times S$ is a binary relation on S, $\rho \subseteq \rho_0$ as $\mathscr{C} \neq \emptyset$, and we have $\Delta_S \subseteq \rho_0$, since each $\rho' \in \mathscr{C}$ is reflexive. Hence, ρ_0 is a reflexive relation on S containing ρ. If $x\rho_0 y$ and $y\rho_0 x$, then $x\rho' y$ and $y\rho' x$ for each $\rho' \in \mathscr{C}$, hence $x = y$ since $\mathscr{C} \neq \emptyset$. Thus, ρ_0 is anti-symmetric. Next, suppose that $x\rho_0 y$ and $y\rho_0 z$. Then $x\rho' y$ and $y\rho' z$ hold for all $\rho' \in \mathscr{C}$; consequently, we have $x\rho' z$ for all $\rho' \in \mathscr{C}$, hence $x\rho_0 z$, and ρ_0 is transitive. It follows that ρ_0 is an order relation on S containing ρ, so that $\rho_0 \in \mathfrak{R}$, and $\rho' \subseteq \rho_0$ for all $\rho' \in \mathscr{C}$. Hence, $(\mathfrak{R}, \subseteq)$ is inductively ordered, as claimed.

By Zorn's lemma, $(\mathfrak{R}, \subseteq)$ has a maximal element $\bar{\rho}$. We claim that $\bar{\rho}$ is total, which would finish the proof, since $\rho \subseteq \bar{\rho}$ by definition of $\mathfrak{R}$.

Suppose for a contradiction that $s_0 \in S$ is not comparable to all $s \in S$ under $\bar{\rho}$, say $(s_0, s_1), (s_1, s_0) \notin \bar{\rho}$. Let

$$\bar{\rho}' := \bar{\rho} \cup \left\{(x,y) \in S^2 : (x, s_0) \in \bar{\rho} \text{ and } (s_1, y) \in \bar{\rho}\right\}.$$

Then $\bar{\rho}'$ is a reflexive relation on S, and $\rho \subseteq \bar{\rho} \subset \bar{\rho}'$ by construction. We claim that $\bar{\rho}'$ is anti-symmetric and transitive, thus an order relation on S properly containing $\bar{\rho}$, contradicting maximality of $\bar{\rho}$, which would finish the proof, showing that $\bar{\rho}$ is total as claimed.

Suppose that $(x,y), (y,x) \in \bar{\rho}'$. We want to show that, necessarily, $x = y$. There are four cases to consider:

(i) $(x,y), (y,x) \in \bar{\rho}$. Then $x = y$, since $\bar{\rho}$ is anti-symmetric.

(ii) $(x, s_0), (s_1, y), (y, x) \in \bar{\rho}$. Then $(s_1, s_0) \in \bar{\rho}$ by transitivity of $\bar{\rho}$, a contradiction. Thus, this situation does not occur.

(iii) $(x,y), (y, s_0), (s_1, x) \in \bar{\rho}$. Again, we conclude that $(s_1, s_0) \in \bar{\rho}$, again contradicting our hypothesis, so that this case does not occur.

(iv) $(x, s_0), (s_1, y), (y, s_0), (s_1, x) \in \bar{\rho}$. Again, this implies $(s_1, s_0) \in \bar{\rho}$ by transitivity of $\bar{\rho}$, so that this case does not occur.

It follows that $x = y$, implying that $\bar{\rho}'$ is anti-symmetric as claimed.

Next, suppose that $(x,y), (y,z) \in \bar{\rho}'$. We want to show that $(x,z) \in \bar{\rho}'$. Again, there are four cases to consider.

(i) $(x,y), (y,z) \in \bar{\rho}$. Then $(x,z) \in \bar{\rho} \subseteq \bar{\rho}'$, since $\bar{\rho}$ is transitive.

(ii) $(x,y) \in \bar{\rho}$ and $(y, s_0), (s_1, z) \in \bar{\rho}$. Then $(x, s_0) \in \bar{\rho}$ and $(s_1, z) \in \bar{\rho}$, forcing $(x,z) \in \bar{\rho}'$.

(iii) $(y,z) \in \bar{\rho}$ and $(x, s_0), (s_1, y) \in \bar{\rho}$, implying $(x, s_0) \in \bar{\rho}$ and $(s_1, z) \in \bar{\rho}$ by transitivity of $\bar{\rho}$, thus again forcing $(x,z) \in \bar{\rho}'$.

(iv) $(x,s_0),(s_1,y) \in \bar{\rho}$ and $(y,s_0),(s_1,z) \in \bar{\rho}$. Again this implies $(s_1,s_0) \in \bar{\rho}$ by transitivity of $\bar{\rho}$, a further contradiction.

It follows that $(x,z) \in \bar{\rho}'$, so that $\bar{\rho}'$ is transitive, thus an order relation properly containing $\bar{\rho}$, contradicting maximality of $\bar{\rho}$. Hence, $\bar{\rho}$ is a total order relation on S extending the given order relation ρ, as claimed.

4.8. Given a set $A \subseteq V$ of linearly independent vectors, consider the set $\mathscr{S}$ of all linear independent subsets $B \subseteq V$ such that $A \subseteq B$. Since $A \in \mathscr{S}$, the set $\mathscr{S}$ is not empty, and $\mathscr{S}$ becomes an ordered set under inclusion. If $\{B_i\}_{i \in I}$ is any non-empty chain in $(\mathscr{S}, \subseteq)$, then one checks immediately that $B := \bigcup_i B_i \in \mathscr{S}$, so that $\{B_i\}$ has an upper bound in $\mathscr{S}$. By Zorn's Lemma, $(\mathscr{S}, \subseteq)$ contains a maximal element B_0, and the last part in the proof of Proposition 4.12 applies verbatim to show that B_0 is a basis of V. The fact that $B_0 \supseteq A$ holds by construction.

4.9. We first consider the case where X is a tree, and show by induction on the number of vertices that $m = n-1$ in that case. Indeed, for $n = 1$, we must have $m = 0$ to avoid circuits, so the result holds. Next, suppose that X is a finite tree with $n \geq 2$ vertices, and that our claim holds for trees with $n-1$ vertices. Starting from a vertex v_0 of X, and moving away from v_0 along a reduced path, we must end at a terminal vertex v_1 of X, with $S_{v_1} = \{e_1, \overline{e_1}\}$, say. Cutting off the geometric edge $\{e_1, \overline{e_1}\}$, and deleting the vertex v_1, we obtain a subgraph $X' \leq X$ with $n-1$ vertices and $2m-2$ edges. By Part (c) of Lemma 4.13, X' is again a tree thus, by the induction hypothesis, $n-2 = (2m-2)/2$, implying $n-1 = m$, as desired.

Passing to the general case of a non-empty finite connected graph X with n vertices and $2m$ edges, we choose a spanning tree $T \leq X$ of X according to Proposition 4.14. Then $|V(T)| = |V(X)| = n$ and

$$2m = |E(X)| \geq |E(T)| = 2(n-1),$$

with equality if, and only if, $X = T$.

4.10. (i) Using the defining properties of a graph morphism, we find for $e \in E$ that

$$\varphi(t(e)) = \varphi(o(\bar{e})) = o'(\psi(\bar{e})) = o'(\overline{\psi(e)}') = t'(\psi(e)),$$

as claimed.

(ii) If $\Phi = (\varphi, \psi) : X \to X'$ is a graph isomorphism, then the maps $\varphi : V \to V'$ and $\psi : E \to E'$ are necessarily bijective. Conversely, suppose that φ and ψ are both bijections, and set $\Phi' := (\varphi^{-1}, \psi^{-1})$. We claim that Φ' is a graph morphism from X' to X. To see this, we compute, for $e' \in E'$, that

$$\varphi^{-1}(o'(e')) = \varphi^{-1}(o'(\psi(e))) = \varphi^{-1}(\varphi(o(e))) = o(e) = o(\psi^{-1}(e')).$$

Similarly, we find that

$$\psi^{-1}(\overline{e'}') = \psi^{-1}(\overline{\psi(e)}') = \psi^{-1}(\psi(\bar{e})) = \bar{e} = \overline{\psi^{-1}(e')}, \quad e' \in E',$$

whence our claim. Since

$$\Phi' \circ \Phi = (\varphi^{-1} \circ \varphi, \psi^{-1} \circ \psi) = (1_V, 1_E)$$

and

$$\Phi \circ \Phi' = (\varphi \circ \varphi^{-1}, \psi \circ \psi^{-1}) = (1_{V'}, 1_{E'})$$

by definition of the inverse map, Φ is a graph isomorphism with inverse morphism Φ'.

5.1. Trivially, A is a left segment of itself, thus the order isomorphism $A \cong A$ given by the identity map 1_A shows that $o(A) \leq o(A)$ for each ordered set A in the universe U. Hence, $\leq$ is reflexive.

Next, suppose that $o(A) \leq o(B)$ and $o(B) \leq o(C)$. By definition, this means that there exists a left segment B_0 of B and a left segment C_0 of C, such that $A \cong B_0$ and $B \cong C_0$, with corresponding order isomorphisms $\varphi : A \to B_0$ and $\psi : B \to C_0$. Let $C_0' := \psi(B_0) \subseteq C_0$. Then $\chi = \psi \circ \phi$ provides an order isomorphism from A onto C_0'. It remains to show that C_0' is a left segment of C, in which case we may conclude that $o(A) \leq o(C)$, as claimed. To see this, let $y \leq x \in C_0'$, where $y \in C$. We want to show that $y \in C_0'$. Since $C_0' \subseteq C_0$, and C_0 is a left segment of C, we have $y \in C_0$. Hence, $y = \psi(b)$ for some $b \in B$, while $x = \psi(\varphi(a))$ for some $a \in A$. As ψ is an order isomorphism, the fact that $\psi(b) \leq \psi(\varphi(a))$ implies that $b \leq \varphi(a) \in B_0$. Since B_0 is a left segment of B, we conclude that $b \in B_0$, thus $y = \psi(b) \in C_0'$, as desired.

5.2. Let $a, b \in A$, and suppose that $a \leq b$. Then

$$\varphi(a) = J_a = \{x \in A : x < a\} \subseteq \{x \in A : x < b\} = J_b,$$

since $x < a \leq b$ implies $x < b$. Hence, φ is an order-preserving map. Moreover, if $a < b$, then $J_a \subset J_b$, since $a \notin J_a$ by definition, while $a \in J_b$. Thus, if $J_a = J_b$, then $a < b$ cannot hold, nor can $a > b$ be true. Since A is totally ordered, it follows that $a = b$, so that φ is injective, as claimed.

5.3. The elements of F are subsets of $A \times B$, so F is naturally ordered by inclusion. Let $(f_i)_{i \in I}$ be a chain in F, where $f_i : A_i \to B_i$ is an order isomorphism between the left segment A_i of A and the left segment B_i of B. Set

$$A' := \bigcup_{i \in I} A_i, \quad B' := \bigcup_{i \in I} B_i, \quad \text{and} \quad f := \bigcup_{i \in I} f_i.$$

We claim that (i) A' is a left segment of A, that (ii) B' is a left segment of B, and that (iii) f is an order isomorphism of A' onto B'.

Indeed, if $x \in A'$ and $y \in A$ is such that $y \leq x$, then $x \in A_i$ for some $i \in I$, so $y \in A_i \subseteq A'$, since A_i is a left segment of A. Hence, A' itself is a left segment of A, as claimed. This proves (i), and the proof of (ii) is completely analogous.

(iii) Let $x \in A'$. Then $x \in A_i$ for at least one $i \in I$. If $x \in A_i$ and $x \in A_j$ for $i, j \in I$ and $i \neq j$ then we have $f_i \subseteq f_j$, or $f_j \subseteq f_i$, since (f_i) is a chain in F. Thus, $f : A' \to B'$ is a well-defined function. If $x, y \in A'$ and $x \leq y$, then we have $f(x) \leq f(y)$, since (f_i) is a chain, and the f_i themselves are order-preserving. Hence, f is order preserving. A similar argument shows that

$$f' := \bigcup_{i \in I} f_i^{-1}$$

is a well-defined order-preserving map from B' onto A'. Clearly, $f' = f^{-1}$, so that $f : A' \to B'$ is an order isomorphism, and thus $f \in F$. Since, by construction, $f_i \subseteq f$ for all $i \in I$, we conclude that the set $(F, \subseteq)$ is inductively ordered, as claimed.

5.4. Let $x \in A_0'$, $y \in A$, and suppose that $y \leq x$. If $x \in A_0$, then $y \in A_0 \subseteq A_0'$, since A_0 is a left segment of A. If, on the other hand, $x = a$, then either $y = a \in A_0'$, or $y < a$, in which case $y \in A_0 \subseteq A_0'$, since

$$y \in A \setminus A_0 \iff y \geq a$$

by definition of the element a. Hence, A_0' is a left segment of A, and the corresponding statement concerning $B_0' \subseteq B$ follows in a similar manner.

To see that $f_0' : A_0' \to B_0'$ is order-preserving, let $x, y \in A_0'$, and suppose that $x \leq y$. If $x, y \in A_0$, then we have

$$f_0'(x) = f_0(x) \leq f_0(y) = f_0'(y),$$

since f_0 is order-preserving. If $x = a = y$, then $f_0'(x) = b = f_0'(y)$, again as desired. Finally, noting that $y = a$ and $x \in A_0$ is impossible (as it would imply $a \in A_0$), let us suppose that $x = a$ and $y \in A_0$, so that $y < a$. If we had $b \leq f_0'(y) = f_0(y) \in B_0$, then it would follow that $b \in B_0$, as B_0 is a left segment of B, a contradiction. Since B is totally ordered, we must have $f_0'(y) < b = f_0'(a) = f_0'(x)$, and the proof that f_0' is order-preserving is complete.

5.5. By our hypothesis, there exist bijective maps $\varphi : A \to A'$ and $\psi : B \to B'$. Define a map $\chi : A \times B \to A' \times B'$ by sending an ordered pair (a,b) to $\chi(a,b) := (\varphi(a), \psi(b))$. If $\chi(a_1,b_1) = \chi(a_2,b_2)$, then $\varphi(a_1) = \varphi(a_2)$ and $\psi(b_1) = \psi(b_2)$, implying $a_1 = a_2$ and $b_1 = b_2$ (and hence $(a_1,b_1) = (a_2,b_2)$), since φ and ψ are injective. Hence, χ itself is injective. Now, let $(a,b') \in A' \times B'$ be an arbitrary element. As the maps φ and ψ are surjective, there exist elements $a \in A$ and $b \in B$, such that $\varphi(a) = a'$ and $\psi(b) = b'$, thus

$$\chi(a,b) = (\varphi(a), \psi(b)) = (a', b'),$$

showing that χ is surjective as well, hence a bijection. It follows that $|A \times B| = |A' \times B'|$, as desired.

5.6. By our hypothesis, we have bijective maps $\varphi : A \to A'$ and $\psi : B \to B'$. Define a map $\chi : A \cup B \to A' \cup B'$ via

$$\chi(x) = \begin{Bmatrix} \varphi(x), & x \in A \\ \psi(x), & x \in B \end{Bmatrix} \qquad (x \in A \cup B).$$

Note that χ is well-defined, since A and B are disjoint. If $\chi(x_1) = \chi(x_2)$, then we have $x_1, x_2 \in A$ or $x_1, x_2 \in B$, since $A' \cap B' = \emptyset$. To fix ideas, suppose that $x_1, x_2 \in A$. Then

$$\varphi(x_1) = \chi(x_1) = \chi(x_2) = \varphi(x_2),$$

implying $x_1 = x_2$, since φ is injective. A similar argument using injectivity of the map ψ works in the case where $x_1, x_2 \in B$. Now let $y \in A' \cup B'$ be an arbitrary element. If $y \in A'$, then there exists some element $a \in A$ such that $y = \varphi(a) = \chi(a)$, and an analogous arguments involving ψ serves if $y \in B'$. Hence, χ is a bijection, implying that $|A \cup B| = |A' \cup B'|$, as claimed.

5.7. (i) Let A, A', B, B' be sets such that $|A| = |A'|$ and $|B| = |B'|$. Then there exist bijections $\varphi : A \to A'$ and $\psi : B \to B'$. Define a map $\chi : B^A \to (B')^{A'}$ via

$$\chi(f) := \psi \circ f \circ \varphi^{-1}, \quad (f \in B^A).$$

If $\chi(f_1) = \chi(f_2)$ for maps $f_1, f_2 \in B^A$, then

$$\psi \circ f_1 \circ \varphi^{-1} = \psi \circ f_2 \circ \varphi^{-1}.$$

Multiplying this last equation from left by ψ^{-1}, and from right by φ yields $f_1 = f_2$; thus, χ is injective. Next, let $f' \in (B')^{A'}$ be an arbitrary element. Then

$$f := \psi^{-1} \circ f' \circ \varphi \in B^A,$$

and we have $\chi(f) = f'$, showing that χ is surjective, thus, a bijection.

(ii) We give the proof of Equation (5.4); the other two identities are established in a similar fashion.

Choose sets A, B, C, such that $|A| = \alpha$, $|B| = \beta$, and $|C| = \gamma$. Given $f \in A^{B\times C}$, define a map $\chi_f : C \to A^B$ via

$$\chi_f(c)(b) := f(b,c), \quad (b \in B, c \in C).$$

The map $\chi : A^{B\times C} \to (A^B)^C$ given by $f \mapsto \chi_f$ is easily shown to be bijective, thus establishing the desired identity $\alpha^{\beta\gamma} = (\alpha^\beta)^\gamma$ for cardinals.

Indeed, if $\chi(f_1) = \chi(f_2)$ for some $f_1, f_2 \in A^{B\times C}$, then

$$\begin{aligned} f_1(b,c) = \chi_{f_1}(c)(b) = \chi(f_1)(c)(b) = \chi(f_2)(c)(b) &= \chi_{f_2}(c)(b) \\ &= f_2(b,c), \quad (b \in B, c \in C); \end{aligned}$$

hence, $f_1 = f_2$, so that χ is injective.

Finally, let $g : C \to A^B$ be arbitrary. Define a map $f : B \times C \to A$ via

$$f(b,c) := g(c)(b), \quad (b \in B, c \in C).$$

Then, for $b \in B$ and $c \in C$,

$$\chi(f)(c)(b) = \chi_f(c)(b) = f(b,c) = g(c)(b),$$

showing that $\chi(f) = g$, as desired.

5.8. Let $(A, \varphi_A) \in \mathscr{A}$ be an arbitrary element. Then, trivially, $A \subseteq A$ and $\varphi_A|_A = \varphi_A$, so that $(A, \varphi_A) \preceq (A, \varphi_A)$. Hence, $\preceq$ is reflexive.

Next, suppose that $(A, \varphi_A), (B, \varphi'_B) \in \mathscr{A}$ satisfy $(A, \varphi_A) \preceq (B, \varphi'_B)$ as well as $(B, \varphi'_B) \preceq (A, \varphi_A)$. By definition of $\preceq$, this implies $A \subseteq B$, $B \subseteq A$, and $\varphi'_B|_A = \varphi_A$. It follows that $A = B$ and $\varphi_A = \varphi'_B$, so that $(A, \varphi_A) = (B, \varphi'_B)$, showing that $\preceq$ is anti-symmetric.

Finally, suppose that $(A, \varphi_A), (B, \varphi'_B), (C, \varphi''_C) \in \mathscr{A}$ satisfy $(A, \varphi_A) \preceq (B, \varphi'_B)$ and $(B, \varphi'_B) \preceq (C, \varphi''_C)$. By definition of the relation $\preceq$, this implies $A \subseteq B$, $B \subseteq C$, $\varphi'_B|_A = \varphi_A$, and $\varphi''_C|_B = \varphi'_B$. It follows that $A \subseteq C$, and that $\varphi''_C|_A = \varphi'_B|_A = \varphi_A$, so that $(A, \varphi_A) \preceq (C, \varphi''_C)$. This shows that $\preceq$ is transitive as well, thus an order relation, as claimed.

5.9. Let $\tilde{A} := \bigcup_{i\in I} A_i$, and let $x, y \in \tilde{A}$, say, $x \in A_i$ and $y \in A_j$ for some $i, j \in I$. Since $\mathscr{C}$ is a chain in $\mathscr{A}$, the elements (A_i, φ_{A_i}) and (A_j, φ_{A_j}) are comparable; to fix ideas, suppose that $(A_i, \varphi_{A_i}) \preceq (A_j, \varphi_{A_j})$. Then, in particular, $A_i \subseteq A_j$, so that $x, y \in A_j$, and we set

$$\varphi(x,y) := \varphi_{A_j}(x,y) \in A_j \subseteq \tilde{A}.$$

If $x, y \in A_k$ as well, for some $(A_k, \varphi_{A_k}) \in \mathscr{C}$, then we have $(A_j, \varphi_{A_j}) \preceq (A_k, \varphi_{A_k})$ or $(A_k, \varphi_{A_k}) \preceq (A_j, \varphi_{A_j})$. It follows that $A_j \subseteq A_k$ or $A_k \subseteq A_j$ and, correspondingly, that $\varphi_{A_k}|_{A_j} = \varphi_{A_j}$ or $\varphi_{A_j}|_{A_k} = \varphi_{A_k}$. In both cases, we find that

$$\varphi_{A_j}(x,y) = \varphi_{A_k}(x,y),$$

so that the map $\tilde{\varphi} : \tilde{A} \times \tilde{A} \to \tilde{A}$ is well defined. Surjectivity of $\tilde{\varphi}$ is clear, since each φ_{A_i} is surjective, and $\tilde{\varphi}(x,y) = \varphi_{A_i}(x,y)$ for $x, y \in A_i$. Finally, suppose that $\tilde{\varphi}(x_1,y_1) = \tilde{\varphi}(x_2,y_2)$ for some $x_1, x_2, y_1, y_2 \in \tilde{A}$. Then there exists some index $i \in I$ such that $x_1, x_2, y_1, y_2 \in A_i$, from which we conclude that $\tilde{\varphi}(x_1,y_1) = \varphi_{A_i}(x_1,y_1)$ and $\tilde{\varphi}(x_2,y_2) = \varphi_{A_i}(x_2,y_2)$, implying $\varphi_{A_i}(x_1,y_1) = \varphi_{A_i}(x_2,y_2)$, and thus $(x_1,y_1) = (x_2,y_2)$, since φ_{A_i} is injective. Hence, $\tilde{\varphi}$ is a bijection, as claimed.

5.10. Let $B_1 = \{x_\rho\}_{\rho \in R}$ and $B_2 = \{y_\sigma\}_{\sigma \in S}$ be two bases of the vector space V over the field K of cardinalities $|R| = \beta_1$ and $|S| = \beta_2$, respectively. We assume that the vector space V is not finitely generated, so that both β_1 and β_2 are transfinite cardinals. Expressing the members of one basis in terms of the other, we have equations

$$x_\rho = \sum_{\sigma \in S} a_{\rho,\sigma} y_\sigma, \quad \rho \in R, \tag{14.22}$$

and

$$y_\sigma = \sum_{\rho \in R} b_{\sigma,\rho} x_\rho, \quad \sigma \in S, \tag{14.23}$$

with coefficients $a_{\rho,\sigma}, b_{\sigma,\rho} \in K$, where, for each given ρ or σ only finitely many of the coefficients $a_{\rho,\sigma}$, respectively $b_{\sigma,\rho}$, are non-zero. In the equations (14.22), each y_σ has at least one coefficient $a_{\rho,\sigma} \neq 0$; that is, y_σ occurs in the decomposition of at least one of the vectors x_ρ. Indeed, if all a_{ρ,σ_0} were zero for some $\sigma = \sigma_0$, then we would have

$$y_{\sigma_0} = \sum_{\rho \in R} b_{\sigma_0,\rho} x_\rho = \sum_{\rho} b_{\sigma_0,\rho} \Big(\sum_{\sigma} a_{\rho,\sigma} y_\sigma \Big) = \sum_{\sigma \neq \sigma_0} \Big(\sum_{\rho} a_{\rho,\sigma} b_{\sigma_0,\rho} \Big) y_\sigma,$$

which expresses y_{σ_0} as a linear combination of finitely many basis vectors y_σ with $\sigma \neq \sigma_0$, contradicting the linear independence of the vectors y_σ. Using Equations (14.22), assign to each vector $x_\rho \in B_1$ the set S_ρ of those finitely many vectors y_σ for which $a_{\rho,\sigma} \neq 0$; that is, those finitely many y_σ occurring in the decomposition of the vector x_ρ. By what was shown above, we have

$$B_2 = \bigcup_{\rho \in R} S_\rho.$$

Hence, we have

$$\beta_2 = |B_2| = \Big| \bigcup_{\rho \in R} S_\rho \Big| \leq |R| \cdot \aleph_0 = \beta_1 \cdot \aleph_0 = \max\{\beta_1, \aleph_0\} = \beta_1,$$

since β_1 is a transfinite cardinal; cf. Part b(i) of Proposition 5.5. Here, $\aleph_0$ denotes the cardinality of the set of natural numbers. We have thus shown that $\beta_2 \leq \beta_1$, and an analogous argument yields the inequality $\beta_1 \leq \beta_2$, whence the desired result; that is, $|B_1| = |B_2|$, establishing invariance of the dimension of the vector space V.

7.1. Symmetry of ρ_σ is clear from the definition. Also, if $s_1 = s_2$, then $\sigma^m(s_1) = \sigma^m(s_2)$ for all $m \geq 0$; in particular, we have $s_1\,\rho_\sigma s_1$, so that ρ_σ is reflexive. Suppose that $s_1, s_2, s_3 \in S$ are such that $s_1\,\rho_\sigma s_2$ and $s_2\,\rho_\sigma s_3$. By definition of ρ_σ, there exist non-negative integers k, ℓ, m, and n such that

$$\sigma^k(s_1) = \sigma^\ell(s_2) \quad \text{and} \quad \sigma^m(s_2) = \sigma^n(s_3). \tag{14.24}$$

Now there are three cases to consider: (i) $\ell = m$, (ii) $\ell < m$, and (iii) $\ell > m$. If $\ell = m$, then we infer from (14.24) that $\sigma^k(s_1) = \sigma^m(s_2) = \sigma^n(s_3)$, implying $s_1\,\rho_\sigma s_3$, as desired. Next, suppose that $\ell < m$. Then we may apply the map $\sigma^{m-\ell}$ to the first equation in (14.24), and find that

$$\sigma^{k+m-\ell}(s_1) = \sigma^m(s_2) = \sigma^n(s_3),$$

again showing that $s_1\,\rho_\sigma s_3$. If, on the other hand, $\ell > m$, then we apply the map $f^{\ell-m}$ to the second equation in (14.24) to obtain

$$\sigma^k(s_1) = \sigma^\ell(s_2) = \sigma^{n+\ell-m}(s_3),$$

implying again that $s_1\,\rho_\sigma s_3$. Hence, ρ_σ is an equivalence relation on S as claimed, thus corresponds to a partition π_σ of S in the way discussed in Chapter § 2.

7.2. Let $s_0 \in P$ be an arbitrary but fixed element, and let $s \in P$ be chosen arbitrarily. Then $s\,\rho_\sigma s_0$, hence there exist non-negative integers m and n such that $\sigma^m(s) = \sigma^n(s_0)$. Applying σ to this equation, we find that

$$\sigma^{m+1}(s) = \sigma^m(\sigma(s)) = \sigma^{n+1}(s_0),$$

implying $\sigma(s)\,\rho_\sigma s_0$; that is, $\sigma(s) \in P$. Since s was chosen arbitrarily in P, it follows that $\sigma(P) \subseteq P$. Similarly, making use of the fact that $\sigma \circ \sigma^{-1} = 1_S$, we get

$$\sigma^m(s) = (\sigma^m \circ \sigma \circ \sigma^{-1})(s) = \sigma^{m+1}(\sigma^{-1}(s)) = \sigma^n(s_0),$$

which proves that $s'\,\rho_\sigma s_0$ for $s' \in \sigma^{-1}(s)$. Since $s \in P$ was chosen arbitrarily, we conclude that $\sigma^{-1}(P) \subseteq P$, as claimed.

7.3. By Corollary 7.30, an element φ of the centraliser $C_G(\sigma)$ is determined once we have defined a length-preserving bijection $\Phi : \pi_\sigma \to \pi_\sigma$ on the set π_σ of cycles of σ, as well as an arbitrary point s_P in each cycle $P \in \pi_\sigma$. As there are λ_ν cycles of length ν in the disjoint cycle decomposition of σ, there are $\prod_{\nu=1}^n \lambda_\nu!$ ways of choosing Φ, and, given a cycle P of length ν, there are ν ways of choosing a point in P. It follows that

$$|C_G(\sigma)| = \prod_{\nu=1}^{n} \left[\nu^{\lambda_\nu}\,\lambda_\nu!\right].$$

8.1. First, $\preceq$ is clearly reflexive. Second, suppose that $(H, +_H), (H', +_{H'}) \in \mathscr{H}$ are such that $(H, +_H) \preceq (H', +_{H'})$ and $(H', +_{H'}) \preceq (H, +_H)$. Then, in particular, we have $H \subseteq H'$ and $H' \subseteq H$, thus $H = H'$. Moreover, we have

$$h_1 +_{H'} h_2 = h_1 +_H h_2, \quad (h_1, .h_2 \in H),$$

since $+_{H'}$ extends $+_H$ by hypothesis. Hence, $+_{H'} = +_H$, and so $(H, +_H) = (H', +_{H'})$, showing that $\preceq$ is antisymmetric. Third, let $(H, +_H)$, $(H', +_{H'})$, $(H'', +_{H''}) \in \mathscr{H}$ be such that $(H, +_H) \preceq (H', +_{H'})$ and $(H', +_{H'}) \preceq (H'', +_{H''})$. Then, in particular, we have $H \subseteq H'$ and $H' \subseteq H''$, thus $H \subseteq H''$. Also, for $h_1, h_2 \in H$,

$$h_1 +_{H''} h_2 = h_1 +_{H'} h_2 = h_1 +_H h_2,$$

since $+_{H''}$ extends $+_{H'}$ and $+_{H'}$ extends $+_H$ by hypothesis. It follows that $+_{H''}$ extends $+_H$, and that $(H,+_H) \preceq (H'',+_{H''})$, showing that $\preceq$ is transitive, thus an order relation, as claimed.

8.2. Let $\mathscr{C} = \{(H_\nu,+_{H_\nu})\}_{\nu\in N}$ be a non-empty chain in $(\mathscr{H},\preceq)$, and set $\tilde{H} := \bigcup_{\nu\in N} H_\nu$. Let $h_1,h_2 \in \tilde{H}$, say $h_1 \in H_\mu$ and $h_2 \in H_\nu$ for indices $\mu,\nu \in N$. Since $\mathscr{C}$ is a chain, the elements $(H_\mu,+_{H_\mu})$ and $(H_\nu,+_{H_\nu})$ are comparable via $\preceq$; to fix ideas, say $(H_\mu,+_{H_\mu}) \preceq (H_\nu,+_{H_\nu})$, so that, in particular, $H_\mu \subseteq H_\nu$. Then $h_1,h_2 \in H_\nu$, and it makes sense to define $h_1 +_{\tilde{H}} h_2$ as in (8.21). If also $h_1,h_2 \in H_\kappa$ for some $\kappa \in N$, then $(H_\nu,+_{H_\nu})$ and $(H_\kappa,+_{H_\kappa})$ are comparable as members of the chain $\mathscr{C}$, say $(H_\kappa,+_{H_\kappa}) \preceq (H_\nu,+_{H_\nu})$. Then

$$h_1 +_{H_\kappa} h_2 = h_1 +_{H_\nu} h_2 = h_1 +_{\tilde{H}} h_2,$$

since $+_{H_\nu}$ extends $+_{H_\kappa}$ by definition of $\preceq$. Thus, the sum $h_1 +_{\tilde{H}} h_2$ is well defined. Moreover, since each $(H_\nu,+_{H_\nu}) \in \mathscr{C}$ is an abelian 2-group, we have $h_1 +_{\tilde{H}} h_2 = h_2 +_{\tilde{H}} h_1$, so that $(\tilde{H},+_{\tilde{H}})$ is abelian, and we also have $2x := x +_{\tilde{H}} x = 0$ for each $x \in \tilde{H}$. Hence, $(\tilde{H},+_{\tilde{H}}) \in \mathscr{H}$, as required, and $(\tilde{H},+_{\tilde{H}})$ is obviously an upper bound for $\mathscr{C}$.

8.3. (i) Since $(R^*,+)$ is the direct sum of the abelian groups $(R,+)$ and $(F,+)$, $(R^*,+)$ is again an abelian group; that is, the ring axioms for addition all hold. We need to check that multiplication in R^* is associative, distributive over addition, and that every element of R^* is an idempotent. We have

$$\begin{aligned}[(r,\bar{x})\cdot(s,\bar{y})]\cdot(t,\bar{z}) &= (rs+\bar{x}s+\bar{y}r,\, \bar{x}\bar{y})\cdot(t,\bar{z})\\ &= ((rs+\bar{x}s+\bar{y}r)t+\bar{x}\bar{y}t+\bar{z}(rs+\bar{x}s+\bar{y}r),\, (\bar{x}\bar{y})\bar{z})\\ &= (rst+\bar{x}st+\bar{y}rt+\bar{x}\bar{y}t+\bar{z}rs+\bar{x}\bar{z}s+\bar{y}\bar{z}r,\, (\bar{x}\bar{y})\bar{z}),\end{aligned}$$

while

$$\begin{aligned}(r,\bar{x})\cdot[(s,\bar{y})\cdot(t,\bar{z})] &= (r,\bar{x})\cdot(st+\bar{y}t+\bar{z}s,\, \bar{y}\bar{z})\\ &= (r(st+\bar{y}t+\bar{z}s)+\bar{x}(st+\bar{y}t+\bar{z}s)+\bar{y}\bar{z}r,\, \bar{x}(\bar{y}\bar{z}))\\ &= (rst+\bar{y}rt+\bar{z}rs+\bar{x}st+\bar{x}\bar{y}t+\bar{x}\bar{z}s+\bar{y}\bar{z}r,\, \bar{x}(\bar{y}\bar{z}))\\ &= [(r,\bar{x})\cdot(s,\bar{y})]\cdot(t,\bar{z}),\end{aligned}$$

as required. Next, we need to check the distributive law. We have

$$\begin{aligned}(r,\bar{x})\cdot[(s,\bar{y})+(t,\bar{z})] &= (r,\bar{x})\cdot(s+t,\bar{y}+\bar{z})\\ &= (r(s+t)+\bar{x}(s+t)+(\bar{y}+\bar{z})r,\, \bar{x}(\bar{y}+\bar{z}))\\ &= (rs+rt+\bar{x}s+\bar{x}t+\bar{y}r+\bar{z}r,\, \bar{x}\bar{y}+\bar{x}\bar{z}),\end{aligned}$$

while

$$\begin{aligned}(r,\bar{x})(s,\bar{y}) + (r,\bar{x})(t,\bar{z}) &= (rs+\bar{x}s+\bar{y}r,\, \bar{x}\bar{y}) + (rt+\bar{x}t+\bar{z}r,\, \bar{x}\bar{z})\\ &= (rs+rt+\bar{x}s+\bar{x}t+\bar{y}r+\bar{z}r,\, \bar{x}\bar{y}+\bar{x}\bar{z})\\ &= (r,\bar{x})\cdot[(s,\bar{y})+(t,\bar{z})],\end{aligned}$$

again as desired. It remains to check that each element of R^* is an idempotent. Indeed, for $r \in R$ and $\bar{x} \in F$, we have

$$(r,\bar{x})^2 = (r,\bar{x})\cdot(r,\bar{x}) = (r^2+2\bar{x}r,\bar{x}^2) = (r,\bar{x}),$$

since R is a Boolean ring by hypothesis, thus satisfies $r^2 = r$ and $2r = 0$ for all $r \in R$ (the latter by Lemma 8.2). It follows that R^* is a Boolean ring, as claimed.

(ii) For $r \in R$ and $\overline{x} \in F$, we have

$$(r,\overline{x}) \cdot (0,\overline{1}) = (r \cdot 0 + \overline{x} \cdot 0 + \overline{1} \cdot r, \overline{x} \cdot \overline{1}) = (r,\overline{x}),$$

which shows that $(0,\overline{1})$ is indeed an identity element for R^*.

(iii) We have to check that the map $\varphi : R \to R^*$ given by $\varphi(r) = (r,\overline{0})$ respects addition and multiplication, and is injective. We compute that, for $r,s \in R$,

$$\varphi(r+s) = (r+s,\overline{0}) = (r,\overline{0}) + (s,\overline{0}) = \varphi(r) + \varphi(s),$$

and that

$$\varphi(rs) = (rs,\overline{0}) = (r,\overline{0}) \cdot (s,\overline{0}) = \varphi(r) \cdot \varphi(s).$$

The fact that φ is injective is clear.

8.4. For elements $a,b,c \in G$, we have

$$(a \circ b) \circ c = a \circ c = a = a \circ b = a \circ (b \circ c),$$

whence our claim.

9.1. (i) Since $0 \in P \cup \{0\}$, relation $\leq$ is reflexive. Next, if $a \leq b$ and $b \leq a$, then $b - a \in P \cup \{0\}$ and $a - b = -(b-a) \in P \cup \{0\}$, implying $b - a = 0$ or $a = b$, since $P \cap (-P) = \emptyset$ by definition of a positive cone. Hence, $\leq$ is anti-symmetric. Further, suppose that $a \leq b$ and $b \leq c$, so that $b - a \in P \cup \{0\}$ and $c - b \in P \cup \{0\}$. As P (and thus $P \cup \{0\}$) is closed under addition, we find that

$$c - a = (c-b) + (b-a) \in P \cup \{0\},$$

which is equivalent to the assertion that $a \leq c$, whence transitivity of $\leq$. It follows that $\leq$ is an order relation on G. If $a,b \in G$ are arbitrary elements, then $b - a \in P \cup \{0\}$, or $b - a \in (-P)$ by definition of a positive cone. In the first case, we have $a \leq b$ while, in the second case, $a - b \in P \cup \{0\}$, thus $b \leq a$. Consequently, the relation $\leq$ is a total order on G. Finally, Equation (9.25) clearly holds, and if $a,b,c \in G$ with $a \leq b$, then

$$(b+c) - (a+c) = b - a \in P \cup \{0\},$$

so that $a + c \leq b + c$, as required.

(ii) We have to show that

$$P := \{a \in G : a > 0\}$$

is closed under addition, and that G is the disjoint union of the sets P, $-P$, and $\{0\}$. For $a,b \in P$, we have $a > 0$ and $b > 0$, thus by (9.24), $a + b > b > 0$, so that $a + b \in P$, as required. Also, as $\leq$ is a total order by hypothesis, we have $a \geq 0$ or $a \leq 0$ for $a \in G$. Hence, for an arbitrary element $a \in G$, one of the statements: $a > 0$, $a = 0$, or $a < 0$ holds true. In the first case, $a \in P$, while in the last case $-a \in P$, thus $a \in (-P)$, and so

$$G = P \cup (-P) \cup \{0\}.$$

It is clear by definition of P that $P \cap \{0\} = \emptyset = (-P) \cap \{0\}$. Suppose that $a \in P \cap (-P)$. Then, in particular, $0 \leq a \leq 0$, implying $a = 0$ by anti-symmetry of the relation $\leq$, again contradicting the definition of P.

9.2. Let $\{H'_i\}_{i \in I}$ be a non-empty chain of quasi-cones in G containing the given quasi-cone H, and set $H' := \bigcup_{i \in I} H'_i$. Then $H \subseteq H' \subseteq G$ by construction, and $0 \notin H'$, as $0 \notin H'_i$ for all $i \in I$. Let $a, b \in H'$ be arbitrary elements, say $a \in H'_i$ and $b \in H'_j$ for $i, j \in I$, where (without loss of generality) $H'_i \subseteq H'_j$. Then $a, b \in H'_j$, so $a + b \in H'_j \subseteq H'$, since H'_j is a quasi-cone in G, so H' is closed under addition, as required. It follows that $H' \in \mathscr{Q}$, and H' is an upper bound for the chain $\{H'_i\}$ with respect to inclusion by construction, as desired. It follows that the ordered set $(\mathscr{Q}, \subseteq)$ is inductively ordered, as claimed.

9.3. If K is not formally real, then $-1 \in K$ can be represented as a sum of squares, say,

$$a_1^2 + a_2^2 + \cdots + a_n^2 = -1.$$

Adding 1 to both sides then yields

$$a_1^2 + \cdots + a_n^2 + 1^2 = 0,$$

which violates Condition (9.31). Conversely, suppose that

$$a_1^2 + a_2^2 + \cdots + a_n^2 = 0, \tag{14.25}$$

and that not all summands in (14.25) are equal to zero; without loss of generality, say, $a_1^2 \neq 0$. Then the inverse a_1^{-2} of a_1^2 exists in K, and multiplying both sides of (14.25) by a_1^{-2}, we find that

$$1 + (a_2 a_1^{-1})^2 + \cdots + (a_n a_1^{-1})^2 = 0,$$

so that -1 can be expressed as a sum of squares, showing that K is not formally real.

9.4. If $s_1 = a_1^2 + \cdots + a_r^2$ and $s_2 = b_1^2 + \cdots + b_s^2$ with $r, s \in \mathbb{N} - \{0\}$ and $a_i, b_j \in K - \{0\}$, then

$$s_1 + s_2 = a_1^2 + a_2^2 + \cdots + a_r^2 + b_1^2 + b_2^2 + \cdots + b_s^2$$

is a sum of $r + s$ squares a_i^2, b_j^2, where $r + s \in \mathbb{N} - \{0\}$ and $a_i, b_j \in K - \{0\}$, so $s_1 + s_2 \in S$, as desired. Also,

$$s_1 s_2 = \left(\sum_i a_i^2\right)\left(\sum_j b_j^2\right) = \sum_{i,j} (a_i b_j)^2$$

is a sum of rs squares, where $rs \in \mathbb{N} - \{0\}$ and $a_i b_j \in K - \{0\}$, since fields have no zero divisors. It follows that $s_1 s_2 \in S$, and the proof is complete.

9.5. Let $\mathscr{C} = \{A_i\}_{i \in I}$ be a non-empty chain in $\mathscr{Q}$, and let $\tilde{A} := \bigcup_i A_i$. Then $S \subseteq \tilde{A} \subseteq K - \{0\}$, as each A_i satisfies this. Moreover, if $a, b \in \tilde{A}$, to fix ideas say $a \in A_i$, $b \in A_j$ for $i, j \in I$, and (without loss of generality) $A_i \subseteq A_j$, then $a, b \in A_j$, so $a + b, ab \in A_j \subseteq \tilde{A}$, since each A_i is closed under the field operations of K. Hence, $\tilde{A} \in \mathscr{Q}$, and $\tilde{A}$ is an upper bound for $\mathscr{C}$ under inclusion, as desired.

9.6. Let K be a formally real field. It suffices to rule out the possibility that K has positive characteristic. However, if $\mathrm{char}(K) = p > 0$, then we have

$$-1 \equiv p-1 \equiv \underbrace{1^2 + \cdots + 1^2}_{\text{p}-1\text{ summands}} \mod p,$$

so that, in the prime field of K (and thus in K), -1 is a sum of squares, contradicting our hypothesis that K be formally real. Hence, $\mathrm{char}(K) = 0$, as was to be shown.

9.7. If $\mathbb{Q}$ is dense in K and $a \in K$ is such that $a \leq 0$, then $1 > a$. If, on the other hand, $a > 0$ in K, then $0 < (2a)^{-1} < a^{-1}$; thus, by our density assumption, there exists some $\alpha \in \mathbb{Q}$ with $(2a)^{-1} \leq \alpha \leq a^{-1}$. Setting $\alpha = \frac{m}{n}$ with $m, n \in \mathbb{N}$, we get $a \leq \frac{n}{m} \leq n$, and so $a < n+1 \in \mathbb{N}$. This shows that, assuming the prime field $\mathbb{Q}$ of K to be dense in K, the ordered field K is archimedean.

Conversely, suppose that the ordered field K is archimedean, and let $a, b \in K$ be elements such that $a < b$. If $a = 0$ or $b = 0$, or if $a < 0 < b$, we may take $\alpha = 0 \in \mathbb{Q}$. Thus, there are only two non-trivial cases to consider.

(I) $0 < a < b$. Here we have $(b-a)^{-1} < n$ for some $n \in \mathbb{N}$, implying $0 < \frac{1}{n} < b - a$. Also, with this value of n, we have $na < m$ for some $m \in \mathbb{N}$. Choosing the least such m, we have $\frac{m-1}{n} \leq a$. Hence,

$$\frac{m}{n} \leq a + \frac{1}{n} < a + (b-a) = b.$$

It follows that $a < \frac{m}{n} < b$, as desired.

(II) $a < b < 0$. Then we have $0 < -b < -a$, and the argument of Case (I) provides us with some rational number α such that $-b < \alpha < -a$. Thus, we have $a < -\alpha < b$, again as desired. Consequently, K being archimedean implies that the prime field $\mathbb{Q}$ of K is dense in K, finishing the argument.

9.8. (i) In a compatible ordering of the ring of integers $\mathbb{Z}$ we must have $0 < 1$, since $1 < 0$ implies $1 = 1 \cdot 1 > 0 \cdot 1 = 0$, a contradiction. Thus, we have $n > 0$ and $-n < 0$ for all positive integers n, which shows that the ring $\mathbb{Z}$ has exactly one ordering compatible with addition and multiplication, given by the positive cone $P = \mathbb{N}$.

(ii) Suppose that the quotient field K carries an ordering extending the given ordering on R. If $a = \frac{p}{q}$ is an arbitrary element of K, where $p, q \in R$ and $q \neq 0$, then we have

$$a > 0 \iff pq > 0. \tag{14.26}$$

Hence, the ordering on K is uniquely determined by the given ordering on R. Conversely, defining an order relation on K via (14.26) yields an ordering on K which extends the given ordering on R.

(iii) Since $\mathbb{Q}$ is formally real, it possesses an ordering $\leq$ compatible with the field arithmetic by Proposition 9.2 and, by Part (i), this relation $\leq$ induces the natural ordering on the ring of integers $\mathbb{Z}$. By Part (ii), it follows that the ordering on $\mathbb{Q}$ is uniquely determined, as claimed.

9.9. Let n be any natural number in the field K. Then the polynomial $t-n$ is positive in $K[t]$, since its leading coefficient is $1>0$. Thus, we have $t>n$ in $K(t)$ for each $n\in\mathbb{N}$, which shows that $K(t)$ is not archimedean.

10.1. The ideals $\langle p\rangle = p\mathbb{Z}$ in $\mathbb{Z}$ satisfy $\mathbb{Z}/\langle p\rangle \cong K_p$ and $\bigcap_{p\,\text{prime}}\langle p\rangle = 0$. Applying Proposition 10.1, the result follows.

10.2. By definition, a simple ring is non-zero, and its only non-zero ideal is the full ring itself; in particular, the intersection of the non-zero ideals is non-zero, so that a simple ring is sub-directly irreducible by Proposition 10.2.

Now let K be a skew field, and let R be a non-zero ideal in K. Choose a non-zero element a in R. Multiplication by the element $a^{-1}\in K$ then shows that $1\in R$, and hence $R=K$. In particular, K is a simple ring, thus sub-directly irreducible by the first part of the exercise.

10.3. Suppose that the intersection of all non-zero ideals in the ring S is non-zero, and let

$$S\le \bigoplus_{\nu\in N}{}^{*}R_\nu \xrightarrow{\pi_\nu} R_\nu$$

be an arbitrary subdirect decomposition of S, where $A_\nu=\ker(\pi_\nu|_S)$, $S/A_\nu\cong R_\nu$, and $\bigcap_{\nu\in N}A_\nu=0$. By our hypothesis, there exists some index $\nu_0\in N$ such that $A_{\nu_0}=0$, implying that $\pi_{\nu_0}|_S : S\to R_\nu$ is an isomorphism. Hence, the above representation of S as a subdirect sum of the rings R_ν is trivial. Since our representation of S as a subdirect sum was chosen arbitrarily, it follows that S is sub-directly irreducible.

Conversely, let S be a sub-directly irreducible ring, let $\mathscr{A}=\{A_\nu\}_{\nu\in N}$ be the collection of all non-zero ideals in S, and set $R_\nu := S/A_\nu$. Let $R:=\bigoplus_{\nu\in N}{}^{*}R_\nu$, and define a map $\psi : S\to R$ by

$$(\psi(s))_\nu := s+A_\nu, \quad (s\in S,\ \nu\in N).$$

Clearly, ψ is a ring homomorphism, and, for each $\nu\in N$, we have $\pi_\nu(\psi(S))=R_\nu$. Also,

$$\ker(\psi)=\bigcap_{\nu\in N}A_\nu.$$

If the right-hand intersection is zero, then we have defined a non-trivial representation of S as a subdirect sum of rings, contradicting our hypothesis. Hence, the intersection of all non-zero ideals in S is non-zero, as claimed.

10.4. Let $\mathscr{C}=\{J_\alpha\}_{\alpha\in A}$ be a non-empty chain in the ordered set $(\mathscr{Q}_a,\subseteq)$; that is, a non-empty collection of ideals J_α in R such that $a\notin J_\alpha$, and such that any two ideals J_α, J_β for $\alpha,\beta\in A$ are comparable under inclusion. Set $J:=\bigcup_{\alpha\in A}J_\alpha$. Then J is certainly a non-empty subset of R, and $a\notin J$. Next, let $a,b\in J$ be arbitrary elements, to fix ideas say $a\in J_\alpha$ and $b\in J_\beta$, and suppose without loss of generality that $J_\alpha\subseteq J_\beta$. Then $a,b\in J_\beta$, so $a-b\in J_\beta\subseteq J$, since J_β is a subring of R. Thus, J is closed under subtraction in R. Moreover, since each J_α is an ideal in R, $a\in J_\alpha$ and $b\in R$ implies $ab, ba\in J_\alpha\subseteq J$. Hence, J is again an ideal in R not containing the element a, so $J\in\mathscr{Q}_a$, and is obviously an upper bound for $\mathscr{C}$, as desired.

10.5. Let $\mathfrak{A} = (A;F)$ be an algebra, and let $(\mathfrak{B}_\mu)_{\mu \in M}$ be a family of subalgebras of $\mathfrak{A}$, such that $\mathfrak{B} = (B_\mu;F)$, and such that

$$\bigcap_{\mu \in M} B_\mu \neq \emptyset.$$

Moreover, let $b_0, b_1, \ldots, b_{n_i-1} \in B$. Then $b_0, b_1, \ldots, b_{n_i-1} \in B_\mu$ for each $\mu \in M$, thus

$$f_i(b_0, b_1, \ldots, b_{n_i-1}) \in B_\mu, \quad (i \in I,\, \mu \in M),$$

so that

$$f_i(b_0, b_1, \ldots, b_{n_i-1}) \in B, \quad (i \in I),$$

which says that $\mathfrak{B} = (B;F)$ is a subalgebra of $\mathfrak{A}$.

Now, let $H \subseteq A$ be a non-empty subset, and let

$$\mathscr{B} = \{\mathfrak{B}_\mu\}_{\mu \in M}$$

be the family of all subalgebras $\mathfrak{B}_\mu = (B_\mu;F) \leq \mathfrak{A}$ such that $H \subseteq B_\mu$. Then $\mathscr{B}$ is non-empty, since $\mathfrak{A} \in \mathscr{B}$; and we have

$$\bigcap_{\mu \in M} B_\mu \supseteq H \neq \emptyset.$$

Hence, by the previous argument,

$$(\langle H \rangle;F) := \Big(\bigcap_{\mu \in M} B_\mu;F\Big)$$

is a subalgebra of $\mathfrak{A}$. By definition, this is the smallest subalgebra of $\mathfrak{A}$ containing the set H.

10.6. Let $\mathfrak{A} = (A;F)$ and $F = (f_i)_{i \in I}$, and let θ be a congruence relation on $\mathfrak{A}$. As the operations of the quotient algebra $\mathfrak{A}/\theta$ are defined in terms of representatives, we need to show that these operations are well-defined; that is, that they are independent of the particular representatives chosen. So, given an index $i \in I$, let $a_0, b_0, a_1, b_1, \ldots, a_{n_i-1}, b_{n_i-1} \in A$ be such that $a_\nu \equiv b_\nu(\theta)$ for all ν such that $0 \leq \nu < n_i$. Then, by the substitution property (10.37) of θ, we have

$$f_i(a_0, a_1, \ldots, a_{n_i-1}) \equiv f_i(b_0, b_1, \ldots, b_{n_i-1})\,(\theta), \quad (i \in I),$$

and therefore

$$\big[f_i(a_0, a_1, \ldots, a_{n_i-1})\big]_\theta = \big[f_i(b_0, b_1, \ldots, b_{n_i-1})\big]_\theta, \quad (i \in I),$$

whence our claim.

10.7. Let $i \in I$, and let $b_0, b_1, \ldots, b_{n_i-1}$ be elements of $\varphi(A)$. Then there exist elements $a_0, a_1, \ldots, a_{n_i-1} \in A$ such that

$$\varphi(a_\nu) = b_\nu, \quad (0 \leq \nu < n_i).$$

Since

$$\begin{aligned} g_i(b_0, b_1, \ldots, b_{n_i-1}) &= g_i(\varphi(a_0), \varphi(a_1), \ldots, \varphi(a_{n_i-1})) \\ &= \varphi(f_i(a_0, a_1, \ldots, a_{n_i-1}) \in \varphi(A), \quad (i \in I) \end{aligned}$$

by definition of a homomorphism, $\varphi(A)$ is indeed closed under the operations g_i of $\mathfrak{B}$, as claimed.

10.8. Let $i \in I$, and let $a_0, a_1, \ldots, a_{n_i-1}$ be elements in A. Then we have

$$\varphi(f_i(a_0, a_1, \ldots, a_{n_i-1})) = g_i(\varphi(a_0), \varphi(a_1), \ldots, \varphi(a_{n_i-1}))$$

since $\varphi : \mathfrak{A} \to \mathfrak{B}$ is a homomorphism. Moreover, for $i \in I$ and elements $b_0, b_1, \ldots, b_{n_i-1} \in B$, we have

$$\psi(g_i(b_0, b_1, \ldots, b_{n_i-1})) = h_i(\psi(b_0), \psi(b_1), \ldots, \psi(b_{n_i-1})),$$

as $\psi : \mathfrak{B} \to \mathfrak{C}$ is a homomorphism too. Setting $b_\nu = \varphi(a_\nu)$ for $0 \leq \nu < n_i$, we find that

$$\begin{aligned}(\psi \circ \varphi)(f_i(a_0, a_1, \ldots, a_{n_i-1})) &= \psi(\varphi(f_i(a_0, a_1, \ldots, a_{n_i-1}))) \\ &= \psi(g_i(\varphi(a_0), \varphi(a_1), \ldots, \varphi(a_{n_i-1}))) \\ &= h_i(\psi(\varphi(a_0)), \psi(\varphi(a_1)), \ldots, \psi(\varphi(a_{n_i-1}))) \\ &= h_i((\psi \circ \varphi)(a_0), (\psi \circ \varphi)(a_1), \ldots, (\psi \circ \varphi)(a_{n_i-1})),\end{aligned}$$

which shows that the composition of maps $\psi \circ \varphi : A \to C$ induces a homomorphism $\psi \circ \varphi : \mathfrak{A} \to \mathfrak{C}$, as desired.

10.9. Let $\nu \in N$ and $i \in I$, and let $p_0, p_1, \ldots, p_{n_i-1}$ be elements of P. Then we have

$$\begin{aligned}\pi_\nu(f_i(p_0, p_1, \ldots, p_{n_i-1})) &= f_i(p_0, p_1, \ldots, p_{n_i-1})(\nu) \\ &= f_i^{(\nu)}(p_0(\nu), p_1(\nu), \ldots, p_{n_i-1}(\nu)) \\ &= f_i^{(\nu)}(\pi_\nu(p_0), \pi_\nu(p_1), \ldots, \pi_\nu(p_{n_i-1})),\end{aligned}$$

which shows that, for each $\nu \in N$, the map $\pi_\nu : P \to A_\nu$ respects all the operations f_i of the direct product algebra $\mathfrak{P} = \prod_\nu \mathfrak{A}_\nu$, thus is a homomorphism. Surjectivity of π_ν is clear, so that $\pi_\nu : \mathfrak{P} \to \mathfrak{A}_\nu$ is indeed an epimorphism, as claimed.

10.10. We have mappings $\varphi : A \to B$ and $\psi : B \to C$; thus, the composition $\psi \circ \varphi : A \to C$ is a well-defined map of sets. Moreover, by hypothesis, φ and ψ are algebra homomorphisms. Hence, for $i \in I$, $(a_0, a_1, \ldots, a_{n_i-1}) \in A^{n_i}$, and $(b_0, b_1, \ldots, b_{n_i-1}) \in B^{n_i}$, we have

$$\varphi(f_i(a_0, a_1, \ldots, a_{n_i-1})) = g_i(\varphi(a_0), \varphi(a_1), \ldots, \varphi(a_{n_i-1})), \tag{14.27}$$

as well as

$$\psi(g_i(b_0, b_1, \ldots, b_{n_i-1})) = h_i(\psi(b_0), \psi(b_1), \ldots, \psi(b_{n_i-1})). \tag{14.28}$$

From (14.27) and (14.28) we get

$$\begin{aligned}(\psi \circ \varphi)(f_i(a_0, a_1, \ldots, a_{n_i-1})) &= \psi(\varphi(f_i(a_0, a_1, \ldots, a_{n_i-1}))) \\ &= \psi(g_i(\varphi(a_0), \varphi(a_1), \ldots, \varphi(a_{n_i-1}))) \\ &= h_i(\psi(\varphi(a_0)), \psi(\varphi(a_1)), \ldots, \psi(\varphi(a_{n_i-1}))) \\ &= h_i((\psi \circ \varphi)(a_0), (\psi \circ \varphi)(a_1), \ldots, (\psi \circ \varphi)(a_{n_i-1})),\end{aligned}$$

which implies that the map $\psi \circ \varphi$ is indeed a homomorphism from the algebra $\mathfrak{A}$ to the algebra $\mathfrak{C}$, as claimed.

10.11. For $\mu \in N$, the map $\pi_\mu : \prod_{\nu \in N} A_\nu \to A_\mu$ is given by $\pi_\mu(p) = p(\mu)$. As $p \in P$ is running through all choice functions, π_μ is clearly surjective. Thus, it suffices to prove that π_μ is a homomorphism of (universal) algebras, in order to be able to conclude that each π_μ is an epimorphism.

Let $i \in I$, determining the operation f_i on $\mathfrak{P}$, let $p_0, p_1, \dots, p_{n_i-1} \in \prod_{\nu\in N} \mathfrak{A}_\nu$ be choice functions, and let $\mu \in N$ be a fixed index. Then we have

$$\begin{aligned}\pi_\mu(f_i(p_0,p_1,\dots,p_{n_i-1})) &= f_i(p_0,p_1,\dots,p_{n_i-1})(\mu)\\ &= f_i^{(\mu)}(p_0(\mu),p_1(\mu),\dots,p_{n_i-1}(\mu))\\ &= f_i^{(\mu)}(\pi_\mu(p_0),\pi_\mu(p_1),\dots,\pi_\mu(p_{n_i-1})),\end{aligned}$$

whence the result. Here, the first and last equalities come from the definition of the map π_μ, while the second equality uses the definition of operations on a direct product; cf. (10.40).

10.12. Clearly, the condition stated (that translations of the algebra $\mathfrak{A} = (A;F)$ have the substitution property) is necessary for θ to be a congruence relation. Conversely, suppose that every translation of $\mathfrak{A}$ has the substitution property, and let $x_1, x_2, \dots, x_{n_i}, y_1, y_2, \dots, y_{n_i} \in A$ be such that $x_k \equiv y_k\ (\theta)$ for $k = 1, 2, \dots, n_i$. Then, for each index $i \in I$, we have successive equivalences modulo θ

$$f_i(x_1,x_2,\dots,x_{n_i}) \equiv f_i(y_1,x_2,\dots,x_{n_i}) \equiv f_i(y_1,y_2,x_3,\dots,x_{n_i}) \equiv \cdots \equiv f_i(y_1,y_2,\dots,y_{n_i}).$$

By transitivity of θ, it follows that

$$f_i(x_1,x_2,\dots,x_{n_i}) \equiv f_i(y_1,y_2,\dots,y_{n_i})\ (\theta), \quad (i \in I),$$

which shows that θ is indeed a congruence relation, as claimed.

10.13. Given a (non-empty) family $\mathscr{F} = (\theta_\beta)_{\beta\in B}$ of equivalence relations on A (considered as subsets of A^2), we need to exhibit $\inf_{\beta\in B} \theta_\beta$ and $\sup_{\beta\in B} \theta_\beta$. Let

$$\theta := \bigcap_{\beta\in B} \theta_\beta \quad \text{and} \quad \hat{\theta} := \Big\langle \bigcup_{\beta\in B} \theta_\beta \Big\rangle,$$

where $\langle\,\rangle$ denotes transitive closure. In detail, this means that

$$x \equiv y\ (\theta) \iff x \equiv y\ (\theta_\beta) \text{ for all } \beta \in B, \tag{14.29}$$

and

$$\begin{aligned}x \equiv y\ (\hat{\theta}) \iff &\text{ for some chain } x = z_0, z_1, \dots, z_m = y \text{ in } A \text{ and associated}\\ &\text{indices } \beta(j) \in B, \text{ we have } z_{j-1} \equiv z_j\ (\theta_{\beta(j)}) \text{ for } j = 1,2,\dots,m. \end{aligned}\tag{14.30}$$

We have to show that θ and $\hat{\theta}$ are equivalence relations on A, such that $\inf_\beta \theta_\beta = \theta$ and $\sup_{beta} \theta_\beta = \hat{\theta}$, respectively.

We first consider the relation θ. For $x \in A$, we have $x \equiv x\ (\theta_\beta)$ for all β, as each θ_β is reflexive by hypothesis; thus, $x \equiv x\ (\theta)$ by (14.29). Next, suppose that $x \equiv y\ (\theta)$. By definition of θ, we deduce that $x \equiv y\ (\theta_\beta)$ for all β, so that $y \equiv x\ (\theta_\beta)$ by symmetry of the θ_β. It follows that $y \equiv x\ (\theta)$, whence symmetry of θ. Third, suppose that $x \equiv y\ (\theta)$ and $y \equiv z\ (\theta)$. By definition of θ, we get $x \equiv y\ (\theta_\beta)$ and $y \equiv z\ (\theta_\beta)$ for all $\beta \in B$; thus, $x \equiv z\ (\theta_\beta)$ for all β by transitivity of the θ_β, which in turn yields $x \equiv z\ (\theta)$ by (14.29); that is, transitivity of θ. Hence, θ is an equivalence relation on A, as required, and we have $\theta \subseteq \theta_\beta$ $(\beta \in B)$ by definition of θ, so that θ is a lower bound for $\mathscr{F}$. Finally, if θ' is any lower bound of $\mathscr{F}$, that is, $\theta' \subseteq \theta_\beta$ $(\beta \in B)$, then

$$\theta' \subseteq \bigcap_{\beta \in B} \theta_\beta = \theta,$$

which shows that $\theta = \inf_{\beta \in B} \theta_\beta$, as desired.

We now turn to the relation $\hat{\theta}$. Since θ_β is reflexive for any $\beta \in B$, the trivial chain $x = z_0, z_1 = x$ shows that $x \equiv x\ (\hat{\theta})$, whence reflexivity of $\hat{\theta}$. Next, if $x \equiv y\ (\hat{\theta})$, then we have a chain $x = z_0, z_1, \ldots, z_m = y$ in A, and corresponding indices $\beta(j)$, such that $z_{j-1} \equiv z_j\ (\theta_{\beta(j)})$ for $j = 1, 2, \ldots, m$. Reversing this chain, and applying symmetry of the $\theta_{\beta(j)}$, we obtain a corresponding chain

$$y = z_m, z_{m-1}, \ldots, z_0 = x$$

in A, such that $z_j \equiv z_{j-1}\ (\theta_{\beta(j)})$ which, by (14.30), yields $y \equiv x\ (\hat{\theta})$; that is, symmetry of $\hat{\theta}$. Third, if $x \equiv y\ (\hat{\theta})$ and $y \equiv z\ (\hat{\theta})$, then we have chains

$$x = z_0, z_1, \ldots, z_m = y \quad \text{and} \quad y = z'_0, z'_1, \ldots, z'_n = z,$$

such that $z_{j-1} \equiv z_j\ (\theta_{\beta(j)})$ for $1 \le j \le m$, and $z'_{k-1} \equiv z'_k\ (\theta_{\gamma(k)})$ for $1 \le k \le n$. Concatenating these two chains, we obtain the chain

$$x = z_0, z_1, \ldots, z_m, z'_0, z'_1, \ldots, z'_n = z.$$

Since $z_m \equiv z'_0\ (\theta_\beta)$ for any $\beta \in B$, it follows that $x \equiv z\ (\hat{\theta})$, whence transitivity of $\hat{\theta}$. We have thus shown that $\hat{\theta}$ is an equivalence relation.

In order to complete the proof, it remains to see that $\hat{\theta} = \sup_{\beta \in B} \theta_\beta$. For $\beta \in B$ and $x, y \in A$ such that $x \equiv y\ (\theta_\beta)$, the trivial chain $x = z_0, z_1 = y$ shows that $x \equiv y\ (\hat{\theta})$, which implies

$$\theta_\beta \subseteq \hat{\theta}, \quad (\beta \in B);$$

so that $\hat{\theta}$ is an upper bound for the family of equivalence relations $\mathscr{F}$. Finally, suppose that $\hat{\theta}'$ is any equivalence relation on A such that $\theta_\beta \subseteq \hat{\theta}'$ for all $\beta \in B$, and let x, y be such that $x \equiv y\ (\hat{\theta})$. Then we have a chain $x = z_0, z_1, \ldots, z_m = y$ such that $z_{j-1} \equiv z_j\ (\theta_{\beta(j)})$ for $1 \le j \le m$, and our hypothesis on $\hat{\theta}'$ implies that $z_{j-1} \equiv z_j\ (\hat{\theta}')$ for $j = 1, 2, \ldots, m$. Transitivity of $\hat{\theta}'$ thus yields that $x \equiv y\ (\hat{\theta}')$, finishing the proof that $(\mathfrak{E}(A), \subseteq)$ is a complete lattice.

10.14. We first check that $\hat{\theta}/\theta$ is well defined; that is, assuming that $a \equiv a'\ (\theta)$ and $b \equiv b'\ (\theta)$, then $a \equiv b\ (\hat{\theta})$ if, and only if, $a' \equiv b'\ (\hat{\theta})$. Indeed, suppose that $a \equiv b\ (\hat{\theta})$. Then, by symmetry of θ plus the fact that $\theta \le \hat{\theta}$, we have

$$a' \equiv a\ (\hat{\theta}) \wedge a \equiv b\ (\hat{\theta}) \wedge b \equiv b'\ (\hat{\theta}),$$

and transitivity of $\hat{\theta}$ gives $a' \equiv b'\ (\hat{\theta})$, as desired. Conversely, if $a' \equiv b'\ (\hat{\theta})$, then we have (by the same kind of reasoning)

$$a \equiv a'\ (\hat{\theta}) \wedge a' \equiv b'\ (\hat{\theta}) \wedge b' \equiv b\ (\hat{\theta}),$$

which implies $a \equiv b\ (\hat{\theta})$, again by transitivity of $\hat{\theta}$. Hence $\hat{\theta}$ is well defined.

Next, we need to convince ourselves that $\hat{\theta}/\theta$ is an equivalence relation; this, however, is clear by definition of $\hat{\theta}/\theta$, since $\hat{\theta}$ itself is an equivalence relation.

It remains to show that $\hat{\theta}/\theta$ has the substitution property. Suppose to this end that

$$[a_\nu]_\theta \equiv [b_\nu]_\theta \ (\hat{\theta}/\theta), \quad (i \in I, 0 \le \nu < n_i).$$

By definition of $\hat{\theta}/\theta$, this means that

$$a_\nu \equiv b_\nu \ (\hat{\theta}), \quad (i \in I, 0 \le \nu < n_i),$$

which, in turn, yields

$$f_i(a_0, a_1, \ldots, a_{n_i-1}) \equiv f_i(b_0, b_1, \ldots, b_{n_i-1}) \ (\hat{\theta}),$$

since $\hat{\theta}$ is a congruence relation by hypothesis. Consequently, by definition of the relation $\hat{\theta}/\theta$ plus that of the algebra $\mathfrak{A}/\theta$, we have

$$\begin{aligned} f_i([a_0]_\theta, [a_1]_\theta, \ldots, [a_{n_i-1}]_\theta) = [f_i(a_0, a_1, \ldots, a_{n_i-1})]_\theta &\equiv [f_i(b_0, b_1, \ldots, b_{n_i-1})]_\theta \\ &= f_i([b_0]_\theta, [b_1]_\theta, \ldots, [b_{n_i-1}]_\theta) \ (\hat{\theta}/\theta), \end{aligned}$$

which completes the proof of the lemma.

10.15. Let $\theta \in \Theta(\mathfrak{A})$ be any congruence relation on the algebra $\mathfrak{A} = (A; F)$, let $\mathfrak{B} = (B; F)$ be a subalgebra of $\mathfrak{A}$, and set

$$\theta' := \theta \cap B^2.$$

We first want to show that θ' is an equivalence relation on the set B. Let $x \in B$. Then we have $x \in A$, so $x \equiv x \ (\theta)$ since θ is reflexive; thus,

$$(x, x) \in \theta \cap B^2 = \theta',$$

showing that $x \equiv x \ (\theta')$, so that θ' is itself reflexive. Next, suppose that $x \equiv y \ (\theta')$. Then $(x, y) \in B^2$ and $x \equiv y \ (\theta)$. Since θ is symmetric, we have $y \equiv x \ (\theta)$, hence

$$(y, x) \in \theta \cap B^2 = \theta',$$

which gives $y \equiv x \ (\theta')$, proving that θ' is symmetric. Third, assume that $x \equiv y \ (\theta')$ and $y \equiv z \ (\theta')$. Then $x, y, z \in B$, as well as $x \equiv y \ (\theta)$ and $y \equiv z \ (\theta)$. Transitivity of θ yields $x \equiv z \ (\theta)$, so that

$$(x, z) \in \theta \cap B^2 = \theta',$$

which shows that $x \equiv z \ (\theta')$; whence transitivity of θ'.

Finally, let $f_i \in F$ be an n_i-ary operation on $\mathfrak{B}$, and let $x_\nu \equiv y_\nu \ (\theta')$ for $0 \le \nu < n_i$. Then we have

$$x_0, x_1, \ldots, x_{n_i-1}, y_0, y_1, \ldots, y_{n_i-1} \in B,$$
$$x := f_i(x_0, x_1, \ldots, x_{n_i-1}),\ y := f_i(y_0, y_1, \ldots, y_{n_i-1}) \in B$$

(as $\mathfrak{B}$ is a subalgebra of $\mathfrak{A}$), and $x \equiv y \ (\theta)$, as θ is a congruence relation on $\mathfrak{A}$. Consequently,

$$(x, y) \in \theta \cap B^2 = \theta',$$

proving that θ' respects the operations on $\mathfrak{B}$, whence our claim.

10.16. Let $(\mathscr{C}_{a,b}, \subseteq)$ be the poset of all congruence relations θ on $\mathfrak{A}$ such that $a \not\equiv b \ (\theta)$. As 0, the equality relation, is in $\mathscr{C}_{a,b}$, we have $\mathscr{C}_{a,b} \neq \emptyset$. We want to apply Zorn's lemma to prove existence of a maximal element in $\mathscr{C}_{a,b}$. Let $C \leq \mathscr{C}_{a,b}$ be any non-empty chain, and set

$$\hat{\theta} := \bigcup_{\theta \in C} \theta.$$

It suffices to show that $\hat{\theta}$ is a congruence relation on $\mathfrak{A}$. First, let $x \in A$. Then $x \equiv x\ (\theta)$ for each $\theta \in C \neq \emptyset$, thus $x \equiv x\ (\hat{\theta})$, so that $\hat{\theta}$ is reflexive. Next, suppose that $x \equiv y\ (\hat{\theta})$. Then $x \equiv y\ (\theta)$ for some $\theta \in C$, hence $y \equiv x\ (\theta)$ for such θ and, consequently, $y \equiv x\ (\hat{\theta})$ by definition of $\hat{\theta}$, whence symmetry of $\hat{\theta}$. Third, suppose that $x \equiv y\ (\hat{\theta})$ and $y \equiv z\ (\hat{\theta})$. Then there exist congruence relations $\theta_1, \theta_2 \in C$ such that $x \equiv y\ (\theta_1)$ and $y \equiv z\ (\theta_2)$. As C is a chain in $\mathscr{C}_{a,b}$, we have $\theta_1 \subseteq \theta_2$ or $\theta_2 \subseteq \theta_1$, implying $x \equiv z\ (\theta_2)$, respectively $x \equiv z\ (\theta_1)$. In each case, we get $x \equiv z\ (\hat{\theta})$, whence transitivity of $\hat{\theta}$. So far, we have shown that $\hat{\theta}$ is an equivalence relation on the set A.

It remains to see that $\hat{\theta}$ respects the operations $f_i \in F$ of $\mathfrak{A}$. By Lemma 10.6, it suffices to consider translations g_γ. If $x \equiv y\ (\hat{\theta})$, then we have $x \equiv y\ (\theta)$ for some $\theta \in C$. As θ is a congruence relation, we get $g_\gamma(x) \equiv g_\gamma(y)\ (\theta)$, which in turn implies $g_\gamma(x) \equiv g_\gamma(y)\ (\hat{\theta})$, finishing the proof.

11.1. (LD1): if $X = \{x_1, x_2, \ldots, x_m\}$ and $x_\mu \in X$, then we have

$$x_\mu = 0 \cdot x_1 + \cdots + 0 \cdot x_{\mu-1} + 1 \cdot x_\mu + 0 \cdot x_{\mu+1} + \cdot + 0 \cdot x_m,$$

which shows that x_μ is linearly dependent on X.

(LD2): as y is linearly dependent on $X = \{x_1, x_2, \ldots, x_m\}$ by hypothesis, we have a relation

$$y = \beta_1 x_1 + \beta_2 x_2 + \cdots + \beta_m x_m, \quad (\beta_1, \beta_2, \ldots, \beta_m \in K).$$

Moreover, since y is not linearly dependent on $X \setminus \{x_m\}$, we must have $\beta_m \neq 0$. Thus,

$$x_m = \beta_m^{-1} y - \beta_m^{-1} \beta_1 x_1 - \cdots - \beta_m^{-1} \beta_{m-1} x_{m-1},$$

which shows that x_m is linearly dependent on $\{x_1, \ldots, x_{m-1}, y\}$, as claimed.

(LD3): by hypothesis, we have relations $z = \sum_{\nu=1}^{n} \alpha_\nu y_\nu$ and $y_\nu = \sum_{\mu=1}^{m} \beta_{\mu,\nu} x_\mu$ for $\nu \in [n]$. Hence,

$$z = \sum_\nu \alpha_\nu \sum_\mu \beta_{\mu,\nu} x_\mu = \sum_\mu \Big(\sum_\nu \alpha_\nu \beta_{\mu,\nu} \Big) x_\mu,$$

which shows that z is linearly dependent on $\{x_1, \ldots, x_m\}$, as desired.

11.2. Linear equivalence is symmetric by definition, and is reflexive by Property (LD1). Suppose that, for $X, Y, Z \in \mathscr{B}_{\text{fin}}(V)$, $X = \{x_1, \ldots, x_r\}$ is linearly equivalent with $Y = \{y_1, \ldots, y_s\}$, while Y is linearly equivalent with $Z = \{z_1, \ldots, z_t\}$. This means that each x_ρ is linearly dependent on Y, and each y_σ is linearly dependent on X; also, each element y_σ is linearly dependent on Z, while every z_τ is linearly dependent on Y. It follows by (LD3) that each x_ρ is linearly dependent on Z, and that each z_τ is linearly dependent on X. Hence, X and Z are linearly equivalent, whence transitivity of linear equivalence.

11.3. (i) As z is linearly dependent on the set Y by hypothesis, we have a linear relation

$$z = \alpha_1 y_1 + \alpha_2 y_2 + \cdots + \alpha_m y_m \tag{14.31}$$

with scalars $\alpha_1, \ldots, \alpha_m \in K$. However, as $Y \subseteq Z$, Equation (14.31) may also be interpreted as a dependence relation over Z; thus, z is linearly dependent on Z.

(ii) Since the set $\{x_1, \ldots, x_{n-1}, x_n\}$ is not linearly independent, one of its vectors must be linearly dependent on the others. If this is true of x_n, then we are done. If instead this is true of, say, x_{n-1}, then x_{n-1} is linearly dependent on $\{x_1, \ldots, x_{n-2}, x_n\}$, but not on $\{x_1, \ldots, x_{n-2}\}$. By (LD2), it follows that x_n is linearly dependent on $\{x_1, \ldots, x_{n-1}\}$, as claimed.

(iii) Choose $X' \subseteq X$ such that it is linearly independent and of maximal cardinality among linear independent subsets. Then, if $x \in X'$, the element x is linearly dependent on X' by (LD1) while, for $x \in X \setminus X'$, the element x is linearly dependent on X' by (ii) plus maximality of X'.

(iv) We argue by induction on $s = |Y|$. For $s = 0$, there is nothing to show (Y is empty and nothing is exchanged). Suppose that our claim holds for $|Y| = s-1$ with some positive integer s, and that $\{y_1, \ldots, y_{s-1}\}$ may be exchanged against $\{x_{i_1}, \ldots, x_{i_{s-1}}\}$. This exchange produces a set

$$\{y_1, \ldots, y_{s-1}, x_k, x_\ell, \ldots\}, \tag{14.32}$$

which is linearly equivalent to $\{x_1, \ldots, x_r\}$. Now y_s is linearly dependent on the system of vectors $\{x_1, \ldots, x_r\}$ by hypothesis, thus is also linearly dependent on the equivalent system (14.32). Consequently, there must exist a subset of (14.32) of minimal cardinality, on which y_s is still linearly dependent. This smallest subset of (14.32) cannot consist solely of vectors y_j, since $\{y_1, \ldots, y_s\}$ is linearly independent by hypothesis. Thus, this minimal subset $\{y_j, \ldots, x_k\}$ must contain at least one vector x_k, which we name x_{i_s}. By (LD2), $x_k = x_{i_s}$ is linearly dependent on the system resulting from $\{y_j, \ldots, x_k\}$ by exchanging x_k for y_s. By (i) above, x_k is also linearly dependent on the larger system $\{y_1, \ldots, y_{s-1}, y_s, x_\ell, \ldots\}$ resulting from (14.32) by replacing x_k with y_s. This last system is linearly equivalent with (14.32), thus also with $\{x_1, \ldots, x_r\}$, completing the induction step.

11.4 Condition (i) is immediate by definition of $\mathscr{D}$. If $\mathscr{D}(x,A)$ and $\overline{\mathscr{D}}(x, A \setminus \{a\})$, then $x \in A$ and $x \notin A \setminus \{a\}$, so that $x = a$ and, consequently, $A \setminus \{a\} \cup \{x\} = A$. It follows that $\mathscr{D}(a, A \setminus \{a\} \cup \{x\})$ holds by hypothesis, whence Condition (ii). If we have $\mathscr{D}(x,A)$ and $\mathscr{D}(A,B)$, then $x \in A$ and $A \subseteq B$, thus $x \in B$, implying $\mathscr{D}(x,B)$, whence Condition (iii). Also, if $\mathscr{D}(x,A)$, then $x \in A$, hence $x \in A' = \{x\} \subseteq A$, so that $\mathscr{D}(x,A')$ with $A' = \{x\}$, showing that Condition (iv) holds as well. Consequently, $\mathscr{D}$ is a dependence relation on S, thus $(S, \mathscr{D})$ is a $\mathscr{D}$-set, as claimed. Finally, by definition of $\mathscr{D}$, we have

$$\overline{\mathscr{D}}(a, A \setminus \{a\}), \quad (A \subseteq S, a \in A),$$

which shows that each subset A of S is $\mathscr{D}$-independent, as desired.

11.5. Let S be a $\mathscr{D}$-set, let $A \subseteq S$ be $\mathscr{D}$-independent, and suppose that the subset $A' \subseteq A$ is $\mathscr{D}$-dependent. Then there exists some finite subset $A'' \subseteq A'$ such that A'' is

$\mathscr{D}$-dependent. Since $A'' \subseteq A$, it follows that A itself is $\mathscr{D}$-dependent, a contradiction. Hence, subsets of a $\mathscr{D}$-independent set are themselves $\mathscr{D}$-independent, as claimed.

11.6. Let $x \in V$ be a vector, and let $A \subseteq V$ be a set of vectors. If $x \in A$ then, a fortiori, $x \in \langle A \rangle$, thus $\mathscr{D}(x,A)$ holds by definition of $\mathscr{D}$, whence Condition (i).

If $\mathscr{D}(x,A)$ and $\overline{\mathscr{D}}(x,A \setminus \{a\})$ for some $a \in A$, then we have $x \in \langle A \rangle$, but $x \notin \langle A \setminus \{a\} \rangle$. It follows that the vector x can be written as

$$x = \alpha_1 a_1 + \alpha_2 a_2 + \cdots + \alpha_r a_r + \alpha a$$

with coefficients $\alpha_1, \ldots, \alpha_r, \alpha \in K$ and $\alpha \neq 0$, and with $a_1, \ldots, a_r \neq a$. Hence,

$$a = \alpha^{-1}\left(x - \sum_{\rho=1}^{r} \alpha_\rho a_\rho\right) \in \langle (A \setminus \{a\}) \cup \{x\} \rangle,$$

implying $\mathscr{D}(a, (A \setminus \{a\}) \cup \{x\})$, as desired, whence Condition (ii).

Next, suppose that $\mathscr{D}(x,A)$ and $\mathscr{D}(A,B)$ hold. Then $x \in \langle A \rangle$ and $\langle A \rangle \leq \langle B \rangle$, thus $x \in \langle B \rangle$, so that $\mathscr{D}(x,B)$, showing that Condition (iii) holds as well.

Finally, suppose that we have $\mathscr{D}(x,A)$ for some vector $x \in V$ and some subset A of V. The $x \in \langle A \rangle$, so x may be written in the form

$$x = \alpha_1 a_1 + \alpha_2 a_2 + \cdots + \alpha_r a_r$$

with suitable constants $\alpha_1, \ldots, \alpha_r \in K$ and vectors $a_1, \ldots, a_r \in A$. We conclude that $x \in \langle A' \rangle$, where $A' := \{a_1, \ldots, a_r\} \subseteq A$, a finite subset of A. This shows that $\mathscr{D}(x,A')$ with some suitable finite subset A' of A, establishing Condition (iv), thus proving our claim that $(V, \mathscr{D})$ is a $\mathscr{D}$-set.

11.7. The proof is by induction on n. For $n = 1$, there is nothing to show. Suppose that $r(A) = r(A \cup \{x_1\}) = \cdots = r(A \cup \{x_{n-1}\})$ implies $r(A) = r(A \cup \{x_1, \ldots, x_{n-1}\})$ for some integer $n \geq 2$, each finite subset $A \subseteq S$, and arbitrary elements $x_1, \ldots, x_{n-1} \in S$, and suppose that $r(A) = r(A \cup \{x_1\}) = \cdots = r(A \cup \{x_n\})$. By the induction hypothesis, we have

$$r(A) = r(A \cup \{x_1, \ldots, x_{n-2}\} \cup \{x_{n-1}\}) = r(A \cup \{x_1, \ldots, x_{n-2}\} \cup \{x_n\}).$$

Applying Axiom (R3) with A replaced by $A \cup \{x_1, \ldots, x_{n-2}\}$, $x = x_{n-1}$, and $y = x_n$, we get

$$r(A) = r(A \cup \{x_1, \ldots, x_{n-2}\} \cup \{x_{n-1}, x_n\}) = r(A \cup \{x_1, \ldots, x_n\}),$$

as desired. Our claim follows now by induction.

11.8. Suppose that $A \subseteq S$ is r-independent; that is, we have $r(A') = |A'|$ for each finite subset A' of A. Suppose further that A is $\mathscr{D}$-dependent, so that there exists some element $a \in A$ with $\mathscr{D}(a, A \setminus \{a\})$. Hence, by definition of $\mathscr{D}$, there exists a finite subset $A'' \subseteq A \setminus \{a\}$ such that $r(A'') = r(A'' \cup \{a\})$. However, both A'' and $A'' \cup \{a\}$ are finite subsets of the r-independent set A, thus we obtain the equation

$$|A''| = |A'' \cup \{a\}| = |A''| + 1,$$

a contradiction. Consequently, each r-independent subset A of S is also $\mathscr{D}$-independent.

Conversely, suppose that $A \subseteq S$ is $\mathscr{D}$-independent, and that there exists some finite subset A' of A such that $r(A') < |A'|$. In view of Axiom (R1), there exists a subset $A'' \subseteq A'$ and an element $a \in A'' \setminus A'$ such that $r(A'') = |A''|$ and $r(A'' \cup \{a\}) = r(A'')$. By definition of the correspondence $\mathscr{D}$, it follows that $\mathscr{D}(a, A \setminus \{a\})$, implying that A is not $\mathscr{D}$-independent, contradicting our hypothesis. Hence, A is r-independent, as desired.

11.9. If $A = \emptyset$, then the only $\mathscr{D}$-independent subset of A is $\emptyset$, thus $R(\emptyset) = |\emptyset| = 0$, whence (R1). Next, suppose that A is a finite subset of S, and let A' be a maximal $\mathscr{D}$-independent subset of A', so that $r(A) = |A'|$. Enlarging A by adding an element x, A' is still a $\mathscr{D}$-independent subset of the larger set $A \cup \{x\}$; in this situation, either A' is a maximal $\mathscr{D}$-independent set of $A \cup \{x\}$, or $A' \cup \{x\}$ is, thus $r(A \cup \{x\}) = r(A) + \delta$ with $\delta \in \{0,1\}$, proving (R2). Finally, suppose that

$$r(A) = r(A \cup \{x\}) = r(A \cup \{y\}), \quad x, y \notin A,$$

and let A' be a maximal $\mathscr{D}$-independent subset of A. Then A' is also a $\mathscr{D}$-independent subset of the sets $A \cup \{x\}$ and $A \cup \{y\}$ and, by our hypothesis, neither $A' \cup \{x\}$ nor $A' \cup \{y\}$ are $\mathscr{D}$-independent. Consequently, A' is also a maximal $\mathscr{D}$-independent subset of $A \cup \{x, y\}$, so that

$$r(A \cup \{x, y\}) = |A'| = r(A),$$

as desired. The case where one of the elements x, y is in A, while the other is not, is handled in a similar way, while the case where $x, y \in A$ is trivial.

11.10. Let $H \subseteq L$, and let $h \in H$ be an element. Then h satisfies the polynomial equation $x - h = 0$ with coefficients $1, -h \in K[H]$. Hence, $\mathscr{D}$ satisfies Axiom (D1).

Next, let $H \subseteq L$ be a subset, let $h \in H$, and let $a \in L$. If $\mathscr{D}(a, H)$ and $\overline{\mathscr{D}}(a, H \setminus \{h\})$, then a is algebraically dependent (over K) on H, but not on $H \setminus \{h\}$. Consider the field $K' := K[H \setminus \{h\}] \leq L$. Then a is algebraically dependent on $\{h\}$ over K'; thus, there exists a polynomial relation of the form

$$\alpha_0(h) + \alpha_1(h)a + \cdots + \alpha_m(h)a^m = 0, \tag{14.33}$$

where $\alpha_0(h), \ldots, \alpha_m(h)$ are polynomials in h with coefficients in K', not all zero. Rewriting (14.33) in terms of powers of h, we obtain a relation

$$\beta_0(a) + \beta_1(a)h + \cdots + \beta_n(a)h^n = 0, \tag{14.34}$$

where the coefficients $\beta_0(a), \ldots, \beta_n(a)$ are polynomials in a over K'; that is, $\beta_j(a) \in K'[a] = K[(H \setminus \{h\}) \cup \{a\}]$. If all polynomials $\beta_0(a), \ldots, \beta_n(a)$ were identically zero in a, the the left-hand side of (14.33) would also be identically zero in a, so that $\alpha_0(h), \ldots, \alpha_m(h) = 0$, contradicting our hypothesis. It follows that not all coefficients in (14.34) vanish, showing that h is algebraically dependent over K on the set $(H \setminus \{h\}) \cup \{a\}$, as desired, whence (D2).

If $a \in L$ is algebraically dependent over K on $H_1 \subseteq L$, then a is algebraic over the subfield $K[H_1]$ of L, hence also over the field $K[H_1 \cup H_2]$, where $H_2 \subseteq L$ is another subset of L. If, moreover, each element of H_1 is algebraically dependent over K on H_2, then the field $K[H_1 \cup H_2]$ is algebraic over $K[H_2]$, so that a is also algebraic over $K[H_2]$. Hence, a is algebraically dependent over K on H_2, showing that Axiom (D3) holds.

Finally, suppose that $a \in L$ is algebraically dependent over K on the set $H \subseteq L$. Then there exists a polynomial relation

$$\alpha_0(H) + \alpha_1(H)a + \cdots + \alpha_m(H)a^m = 0$$

with coefficients $\alpha_0(H), \ldots, \alpha_m(H) \in K[H]$. Let $H' = \{h_1, \ldots, h_r\}$ be the set of those elements from H occurring in at least one of the polynomials $\alpha_i(H)$ over K. Then $\alpha_0(H), \ldots, \alpha_m(H) \in K[H']$, so that a is algebraically dependent over K on the finite subset H' of H, whence (D4).

11.11. If $A \subseteq S$ and $x \in A$ then, a fortiori, $x \in \langle\langle A \rangle\rangle_G$, so that $\mathscr{D}(x,A)$ holds, whence Axiom (D1). Also, if $\mathscr{D}(x,A)$ and $\mathscr{D}(A,B)$, then $x \in \langle\langle A \rangle\rangle_G$ and $\langle\langle A \rangle\rangle_G \leq \langle\langle B \rangle\rangle_G$, thus $x \in \langle\langle B \rangle\rangle_G$, implying $\mathscr{D}(x,B)$. Hence, Axiom (D3) holds as well. Next, suppose that $\mathscr{D}(x,A)$. Then

$$x = a_1^{g_1} a_2^{g_2} \cdots a_r^{g_r},$$

where $a_1, \ldots, a_r \in A$, $g_1, \ldots, g_r \in G$, and $a^g := g^{-1}ag$. Setting $A' := \{a_1, a_2, \ldots, a_r\}$, we obtain a finite subset of A such that $x \in \langle\langle A' \rangle\rangle_G$, implying $\mathscr{D}(x,A')$, whence Axiom (D4). Finally, to prove (D2), suppose that $\mathscr{D}(x,A)$ and $\overline{\mathscr{D}}(x, A \setminus \{a\})$ for $x \in S$, $A \subseteq S$, and some element $a \in A$. Then there exist elements $a_1, \ldots, a_s \in A$ and $g_1, \ldots, g_s, g \in G$, such that

$$x = a_1^{g_1} \cdots a_r^{g_r} \cdot a^g \cdot a_{r+1}^{g_{r+1}} \cdots a_s^{g_s} = a_1^{g_1} \cdots a_r^{g_r} a_{r+1}^{g_{r+1}} \cdots a_s^{g_s} \cdot \hat{a} = \hat{b} \cdot \hat{a},$$

where $\hat{a} \in \langle\langle a \rangle\rangle_G \setminus \{1\}$ and $\hat{b} \in \langle\langle A \setminus \{a\} \rangle\rangle_G$. Since a is a distinguished element of G, $\langle\langle a \rangle\rangle_G$ is a minimal normal subgroup of G, so that $\langle\langle \hat{a} \rangle\rangle_G = \langle\langle a \rangle\rangle_G$; in particular, $a \in \langle\langle \hat{a} \rangle\rangle_G$. It follows that $a \in \langle\langle (A \setminus \{a\}) \cup \{x\} \rangle\rangle_G$, so that $\mathscr{D}(a, (A \setminus \{a\}) \cup \{x\})$, as desired.

11.12. We assume the result stated in the previous exercise, and adopt the same notation. For each $\mu \in M$ and $\nu \in N$ choose non-trivial elements $g_\mu \in G_\mu$ and $h_\nu \in H_\nu$. By simplicity of the factors, we have $\langle\langle g_\mu \rangle\rangle_G = G_\mu$ and $\langle\langle H_\nu \rangle\rangle_G = H_\nu$, and the factors G_μ, H_ν are minimal normal subgroups of G. Hence, the sets $\hat{M} = \{g_\mu\}_{\mu \in M}$ and $\hat{N} = \{h_\nu\}_{\nu \in N}$ consist of distinguished elements of G. Moreover, we have

$$g_{\mu_0} \notin \langle\langle \hat{M} \setminus \{g_{\mu_0}\} \rangle\rangle_G = \prod_{\mu \in M \setminus \{\mu_0\}} G_\mu, \quad \mu_0 \in M,$$

with a similar statement holding for $\hat{N}$. Hence, $\hat{M}$ and $\hat{N}$ are $\mathscr{D}$-independent sets of distinguished elements of G. Since also maximality of $\hat{M}$ and $\hat{N}$ is clear, Corollary 11.3 yields

$$|M| = |\hat{M}| = |\hat{N}| = |N|,$$

as desired.

12.1. If L is a Λ-submodule of M then, in particular, L is closed under addition, scalar multiplication, and under the operation $\mathbf{x} \mapsto -\mathbf{x}$ of forming the additive inverse of an element $\mathbf{x} \in L$. Hence, Conditions (S1) and (S2) hold.

Conversely, suppose that L is a non-empty subset of the Λ-module M, satisfying (S1) and (S2). Since L is non-empty, say $\mathbf{x_0} \in L$, we have $\mathbf{0} = 0 \cdot \mathbf{x_0} \in L$. Also, for $\mathbf{x} \in L$, we have $-\mathbf{x} = (-1) \cdot \mathbf{x} \in L$, so that L is closed under forming additive inverses. It follows now from (S1) that L is closed under addition as well, hence forms an abelian group under the restriction to L of addition in M. Finally, Axioms

(M1)–(M4) hold for arbitrary scalars and elements of L, since they hold for arbitrary scalars and all elements of M. Consequently, L is a Λ-submodule of M, as claimed.

12.2. Since $I \neq \emptyset$ by hypothesis, we have $\mathbf{0} \in \bigcap \mathscr{S}$, in particular, $\bigcap \mathscr{S} \neq \emptyset$. If $\mathbf{x}, \mathbf{y} \in \bigcap \mathscr{S}$, then $\mathbf{x}, \mathbf{y} \in L_i$ for each index i, thus $\mathbf{x} - \mathbf{y} \in L_i$ for all i, so that $\mathbf{x} - \mathbf{y} \in \bigcap \mathscr{S}$. Hence, $\bigcap \mathscr{S}$ satisfies Condition (S1). Also, if $\mathbf{x} \in \bigcap \mathscr{S}$ and $\lambda \in \Lambda$, then $\mathbf{x} \in L_i$ for all i, thus $\lambda \mathbf{x} \in L_i$ for all i, so $\lambda \mathbf{x} \in \bigcap \mathscr{S}$, whence (S2). It follows that $\bigcap \mathscr{S}$ is a Λ-submodule of M, as claimed.

12.3. (a) By definition, we have

$$\langle \emptyset \rangle = \bigcap_{L \leq M} L = \{\mathbf{0}\}.$$

(b) Denote the right-hand side of (12.53) by $\tilde{S}$. If $S = \emptyset$, then $\tilde{S}$ only contains the empty sum, so $\tilde{S} = \{\mathbf{0}\} = \langle S \rangle$, in accordance with (a). Thus, we may suppose that $S \neq \emptyset$. Since Λ contains an identity element, and M is assumed to be unitary we have $S \subseteq \tilde{S}$. Moreover, it is clear that $\tilde{S}$ satisfies Conditions (S1) and (S2), thus is a Λ-submodule of M containing the set S, so that $\tilde{S} \geq \langle S \rangle$ by definition of $\langle S \rangle$. On the other hand, each element of $\tilde{S}$ is contained in $\langle S \rangle$, since $\langle S \rangle$ is a Λ-submodule containing S. Hence, $\langle S \rangle = \tilde{S}$, as desired.

12.4. We have $0 \in \mathscr{A}(M)$, thus, in particular, $\mathscr{A}(M) \neq \emptyset$. Next, let $\lambda, \mu \in \mathscr{A}(M)$. Then, for an arbitrary vector $\mathbf{x} \in M$, we have

$$(\lambda - \mu)\mathbf{x} = \lambda \mathbf{x} - \mu \mathbf{x} = \mathbf{0} - \mathbf{0} = \mathbf{0},$$

so that $\lambda - \mu \in \mathscr{A}(M)$. Next, let $\lambda \in \mathscr{A}(M)$, and let $\mu \in \Lambda$ be an arbitrary scalar. Then, for each vector $\mathbf{x} \in M$,

$$(\mu\lambda)\mathbf{x} = \mu(\lambda \mathbf{x}) = \mu \cdot \mathbf{0} = \mathbf{0},$$

thus, $\mu\lambda \in \mathscr{A}(M)$, showing that $\mathscr{A}(M)$ is a left ideal of Λ. Also,

$$(\lambda\mu)\mathbf{x} = \lambda(\mu \mathbf{x}) = \mathbf{0},$$

so that $\mathscr{A}(M)$ is a two-sided ideal of Λ, as claimed.

12.5. Since

$$\varphi(\mathbf{0}) = \varphi(\mathbf{0} + \mathbf{0}) = \varphi(\mathbf{0}) + \varphi(\mathbf{0}),$$

we have $\varphi(\mathbf{0}) = \mathbf{0}$, which shows that $\mathbf{0} \in \ker(\varphi)$ and $\mathbf{0} \in \mathrm{image}(\varphi)$; in particular,

$$\ker(\varphi), \mathrm{image}(\varphi) \neq \emptyset.$$

Next, let $\mathbf{x}, \mathbf{y} \in \ker(\varphi)$. Then $\varphi(\mathbf{x}) = \mathbf{0} = \varphi(\mathbf{y})$, thus by linearity of φ,

$$\varphi(\mathbf{x} - \mathbf{y}) = \varphi(\mathbf{x}) - \varphi(\mathbf{y}) = \mathbf{0} - \mathbf{0} = \mathbf{0},$$

so that $\mathbf{x} - \mathbf{y} \in \ker(\varphi)$. Similarly, for $\mathbf{x} \in \ker(\varphi)$ and $\lambda \in \Lambda$, we have

$$\varphi(\lambda \mathbf{x}) = \lambda \varphi(\mathbf{x}) = \lambda \cdot \mathbf{0} = \mathbf{0},$$

showing that $\lambda \mathbf{x} \in \ker(\varphi)$. It follows that $\ker(\varphi)$ is a non-empty subset of the Λ-module M satisfying Conditions (S1) and (S2), implying that $\ker(\varphi)$ is a Λ-submodule of M, as claimed.

Now let $\mathbf{x}', \mathbf{y}' \in \text{image}(\varphi)$. Then there exist vectors $\mathbf{x}, \mathbf{y} \in M$ such that $\varphi(\mathbf{x}) = \mathbf{x}'$ and $\varphi(\mathbf{y}) = \mathbf{y}'$. Hence,

$$\varphi(\mathbf{x}-\mathbf{y}) = \varphi(\mathbf{x}) - \varphi(\mathbf{y}) = \mathbf{x}' - \mathbf{y}' \in \text{image}(\varphi),$$

showing that image(φ) satisfies Condition (S1). Also, for $\mathbf{x}' \in \text{image}(\varphi)$ and $\lambda \in \Lambda$, we have $\varphi(\mathbf{x}) = \mathbf{x}'$ for some suitable vector $\mathbf{x} \in M$, thus

$$\varphi(\lambda \mathbf{x}) = \lambda \varphi(\mathbf{x}) = \lambda \mathbf{x}' \in \text{image}(\varphi),$$

which proves that image(φ) also satisfies Condition (S2). Hence, image(φ) is a Λ-submodule of N, as desired.

12.6. If φ is injective, then $\ker(\varphi) = \{\mathbf{0}\}$, since $\varphi(\mathbf{0}) = \mathbf{0}$. Conversely, suppose that $\ker(\varphi) = \{\mathbf{0}\}$, and assume that $\varphi(\mathbf{x}_1) = \varphi(\mathbf{x}_2)$ with $\mathbf{x}_1, \mathbf{x}_2 \in M$. By linearity of φ,

$$\varphi(\mathbf{x}_1 - \mathbf{x}_2) = \varphi(\mathbf{x}_1) - \varphi(\mathbf{x}_2) = \varphi(\mathbf{x}_1) - \varphi(\mathbf{x}_1) = \mathbf{0},$$

implying that $\mathbf{x}_1 - \mathbf{x}_2 \in \ker(\varphi) = \{\mathbf{0}\}$; thus, $\mathbf{x}_1 = \mathbf{x}_2$, so that φ is injective, as claimed.

12.7. For $\mathbf{x}, \mathbf{y} \in M$ and $\lambda \in \Lambda$, set

$$(\mathbf{x}+L) + (\mathbf{y}+L) := (\mathbf{x}+\mathbf{y}) + L,$$
$$\lambda(\mathbf{x}+L) := (\lambda \mathbf{x}) + L.$$

To see that these operations are well defined, let $\mathbf{x}' \sim \mathbf{x}$ and $\mathbf{y}' \sim \mathbf{y}$. Then there exist vectors $\mathbf{z}_1, \mathbf{z}_2 \in L$, such that $\mathbf{x}' = \mathbf{x} + \mathbf{z}_1$ and $\mathbf{y}' = \mathbf{y} + \mathbf{z}_2$, thus

$$(\mathbf{x}'+\mathbf{y}') + L = (\mathbf{x}+\mathbf{z}_1+\mathbf{y}+\mathbf{z}_2) + L = (\mathbf{x}+\mathbf{y}) + L,$$

showing that addition on M/L is well-defined. Also, for $\mathbf{x}, \mathbf{x}' \in M$, $\mathbf{x} \sim \mathbf{x}'$, and $\lambda \in \Lambda$, we have

$$(\lambda \mathbf{x}') + L = \lambda(\mathbf{x}+\mathbf{z}) + L = (\lambda \mathbf{x}) + (\lambda \mathbf{z}) + L = (\lambda \mathbf{x}) + L$$

for some suitable $\mathbf{z} \in L$, showing that scalar multiplication on M/L is well-defined. It is now obvious that, with these operations, M/L is again a Λ-module, and that the surjective map $\pi : M \to M/L$ given by $\pi(\mathbf{x}) = \mathbf{x} + L$ is Λ-linear.

12.8. (i) Let $\psi : N \to L$ and $\chi : N \to L$ be Λ-linear maps such that $\psi \circ \varphi = \chi \circ \varphi$, and suppose that φ is surjective. Furthermore, let $\mathbf{x}' \in N$ be arbitrary. Then there exists some vector $\mathbf{x} \in M$ with $\varphi(\mathbf{x}) = \mathbf{x}'$, thus

$$\psi(\mathbf{x}') = \psi(\varphi(\mathbf{x})) = \chi(\varphi(\mathbf{x})) = \chi(\mathbf{x}'),$$

so that $\psi = \chi$. Hence, φ is a Λ-epimorphism.

Conversely, suppose that φ is a Λ-epimorphism, and assume, for a contradiction, that φ is not surjective. Choose a vector $\mathbf{x}' \in N - \text{image}(\varphi)$, let $\psi : N \to N/\text{image}(\varphi) =: L$ be the canonical projection associated with the Λ-submodule image(φ) of N, and let $\chi : N \to L$ be the zero map. Then $\psi(\mathbf{x}') \neq \mathbf{0} = \chi(\mathbf{x}')$, so that $\psi \neq \chi$, while

$$\psi(\varphi(\mathbf{x})) = \mathbf{0} = \chi(\varphi(\mathbf{x})), \quad \mathbf{x} \in M,$$

contradicting our hypothesis that φ is a Λ-epimorphism. Hence, φ is surjective, as claimed.

(ii) Let $\psi, \chi : L \to M$ be Λ-linear maps such that $\varphi \circ \psi = \varphi \circ \chi$, and suppose that φ is injective. Then, for each $\mathbf{x} \in L$, the fact that $\varphi(\psi(\mathbf{x})) = \varphi(\chi(\mathbf{x}))$ implies $\psi(\mathbf{x}) = \chi(\mathbf{x})$ by injectivity of φ, so that $\psi = \chi$, as desired.

Conversely, suppose that φ is a Λ-monomorphism, let $\psi : \ker(\varphi) \to M$ be the inclusion map, and let $\chi : \ker(\varphi) \to M$ be the zero map. Then $\varphi \circ \psi$ and $\varphi \circ \chi$ both equal the zero map $\ker(\varphi) \to N$. Since φ is a Λ-monomorphism by hypothesis, we conclude that $\psi = \chi$, implying $\ker(\varphi) = \{\mathbf{0}\}$. By Exercise 12.6, it follows that φ is injective, as claimed.

(iii) If φ is a Λ-isomorphism, then the map φ' is its set-theoretic inverse, thus φ is a bijection.

Conversely, suppose that φ is bijective, and let $\varphi' : N \to M$ be its set-theoretic inverse. Then $\varphi' \circ \varphi = 1_M$ and $\varphi \circ \varphi' = 1_N$, so it only remains to show that φ' is Λ-linear. Given vectors $\mathbf{x}', \mathbf{y}' \in N$, let $\mathbf{x}, \mathbf{y} \in M$ be such that $\varphi(\mathbf{x}) = \mathbf{x}'$ and $\varphi(\mathbf{y}) = \mathbf{y}'$ ($\mathbf{x}$ and $\mathbf{y}$ exist since φ is surjective by hypothesis). Then, by linearity of φ,

$$\varphi'(\mathbf{x}' + \mathbf{y}') = \varphi'(\varphi(\mathbf{x}) + \varphi(\mathbf{y})) = \varphi'(\varphi(\mathbf{x} + \mathbf{y})) = \mathbf{x} + \mathbf{y} = \varphi'(\mathbf{x}') + \varphi'(\mathbf{y}'),$$

as desired. Also, for $\lambda \in \Lambda$,

$$\varphi'(\lambda \mathbf{x}') = \varphi'(\lambda \varphi(\mathbf{x})) = \varphi'(\varphi(\lambda \mathbf{x})) = \lambda \mathbf{x} = \lambda \varphi'(\mathbf{x}').$$

Hence, φ' is Λ-linear, implying that φ is a Λ-isomorphism.

12.9. Define a map $\bar{\varphi} : M/\ker(\varphi) \to \operatorname{image}(\varphi)$ sending $\mathbf{x} + \ker(\varphi)$ to $\varphi(\mathbf{x}) \in \operatorname{image}(\varphi)$. If $\mathbf{x}' = \mathbf{x} + \mathbf{y}$ for some $\mathbf{y} \in \ker(\varphi)$, then

$$\varphi(\mathbf{x}') = \varphi(\mathbf{x} + \mathbf{y}) = \varphi(\mathbf{x}) + \varphi(\mathbf{y}) = \varphi(\mathbf{x}),$$

showing that $\bar{\varphi}$ is well-defined. Clearly, $\bar{\varphi}$ is also surjective and Λ-linear. It remains to see that $\bar{\varphi}$ is injective. Let $\mathbf{x} \in M$ be such that $\bar{\varphi}(\mathbf{x} + \ker(\varphi)) = \mathbf{0}$. Then, by definition of $\bar{\varphi}$, we have $\varphi(\mathbf{x}) = \mathbf{0}$, thus $\mathbf{x} \in \ker(\varphi)$, and $\mathbf{x} + \ker(\varphi) = \mathbf{0}$. Hence, $\ker(\bar{\varphi}) = \{\mathbf{0}\}$, so $\bar{\varphi}$ is injective, thus an isomorphism.

12.10. Suppose that M is semi-simple, let

$$M = \bigoplus_{\nu \in N} M_\nu \tag{14.35}$$

be a decomposition of M as a (discrete) direct sum of simple Λ-modules M_ν, and let L be an arbitrary submodule of M. We claim that the family of submodules

$$\mathscr{S} := \{K : K \leq M,\ K \cap L = \{\mathbf{0}\}\} \tag{14.36}$$

is inductively ordered by inclusion.

Indeed, let $\mathscr{C} = \{K_i\}_{i \in I}$ be a non-empty chain of the ordered set $(\mathscr{S}, \subseteq)$, and let $K' := \bigcup_{i \in I} K_i$. Since $K_i \cap L = \{\mathbf{0}\}$ for all $i \in I$, and I is non-empty, we have $K' \cap L = \{\mathbf{0}\}$; in particular, $K' \neq \emptyset$. Let $x, y \in K'$. Then $x \in K_i$, $y \in K_j$, and (without loss of generality) $K_i \subseteq K_j$, since $\mathscr{C}$ is a chain. Thus $x - y \in K_j \subseteq K'$. Also, for $\lambda \in \Lambda$ and $x \in K'$, we have $x \in K_i$ for some index i, so that $\lambda x \in K_i \subseteq K'$. This shows that K' is a submodule of M, as desired. Finally, by construction, $K' \supseteq K_i$ for all $i \in I$, so that K' is an upper bound for the chain $\mathscr{C}$. Hence, $\mathscr{S}$ is inductively ordered by inclusion, as claimed.

Now let K_0 be a maximal element of $\mathscr{S}$ (which exists by Zorn's lemma). We claim that

$$M = L \oplus K_0. \tag{14.37}$$

Clearly, $L \oplus K_0 \leq M$. Suppose that there exists an index $\nu_0 \in N$ such that $M_{\nu_0} \not\leq L \oplus K_0$. Since M_{ν_0} is simple, we have

$$M_{\nu_0} \cap (L \oplus K_0) = \{\mathbf{0}\},$$

implying

$$L \cap (K_0 + M_{\nu_0}) = \{\mathbf{0}\}, \tag{14.38}$$

since $L \cap K_0 = \{\mathbf{0}\}$ by choice of K_0. However, since $M_{\nu_0} \not\leq K_0$ by hypothesis, $K_0 < K_0 + M_{\nu_0}$, so that Equation (14.38) contradicts maximality of the submodule K_0. This contradiction shows that $M_\nu \leq L \oplus K_0$ for all $\nu \in N$, implying $L \oplus K_0 = M$ as desired. Hence, every submodule of M is a direct summand, as desired.

12.11. Let $\mathscr{C} = \{L_i\}_{i \in I}$ be a non-empty chain in $(\mathscr{R}, \subseteq)$, and let $L := \bigcup_{i \in I} L_i$. We need to show that L is a semi-simple submodule of M. Since $\mathscr{C}$ is non-empty, we have $\mathbf{0} \in L$; in particular, L is non-empty. Let $\mathbf{x}, \mathbf{y} \in L$ and $\lambda \in \Lambda$. Then $\mathbf{x} \in L_i$, $\mathbf{y} \in L_j$ for some $i, j \in I$, and (without loss of generality) $L_i \leq L_j$, since $\mathscr{C}$ is a chain. Thus, $\mathbf{x} - \mathbf{y} \in L_j \subseteq L$. Also, $\lambda \mathbf{x} \in L_i \subseteq L$. Hence, L is a submodule of M. Moreover, by definition, L is the sum of the semi-simple submodules L_i, thus the sum of simple modules. By Lemma 12.3, L is semi-simple, thus $L \in \mathscr{R}$, and L is an upper bound for the chain $\mathscr{C}$ by construction.

12.12. Let $\mathbf{U} \subseteq \tilde{L}_i$, and let $\mathbf{x} \in \mathbf{U}$. Then, a fortiori, $\mathbf{x} \in \langle \mathbf{U} \rangle$, so that $\mathscr{D}(\mathbf{x}, \mathbf{U})$ by definition of $\mathscr{D}$, whence (D1).

Next, let $\mathbf{U} \subseteq \tilde{L}_i$, $\mathbf{x} \in \tilde{L}_i$, and suppose that $\mathscr{D}(\mathbf{x}, \mathbf{U})$ and $\overline{\mathscr{D}}(\mathbf{x}, \mathbf{U} \setminus \{\mathbf{y}\})$ for some $\mathbf{y} \in \mathbf{U}$. Then we have $\mathbf{x} \in \langle \mathbf{U} \rangle$, so that $\mathbf{x}$ has a representation of the form

$$\mathbf{x} = \lambda_1 \mathbf{y_1} + \cdots + \lambda_r \mathbf{y_r} + \lambda \mathbf{y}$$

with distinct vectors $\mathbf{y_1}, \ldots, \mathbf{y_r}, \mathbf{y} \in \mathbf{U}$ and $\lambda_1, \ldots, \lambda_r, \lambda \in \Lambda$. Moreover, the second hypothesis implies that $\lambda \mathbf{y} \neq \mathbf{0}$. It follows that $\lambda \mathbf{y} \in \langle (\mathbf{U} \setminus \{\mathbf{y}\}) \cup \{\mathbf{x}\} \rangle$, implying that $\mathbf{y} \in \langle (\mathbf{U} \setminus \{\mathbf{y}\}) \cup \{\mathbf{x}\} \rangle$, since $\langle \mathbf{y} \rangle$ is simple. Hence, $\mathscr{D}(\mathbf{y}, (\mathbf{U} \setminus \{\mathbf{y}\}) \cup \{\mathbf{x}\})$, as required, so that Axiom (D2) holds as well.

Now, let $\mathbf{U}, \mathbf{V} \subseteq \tilde{L}_i$, $\mathbf{x} \in \tilde{L}_i$, and suppose that $\mathscr{D}(\mathbf{x}, \mathbf{U})$ and $\mathscr{D}(\mathbf{U}, \mathbf{V})$. Then we have $\mathbf{x} \in \langle \mathbf{U} \rangle$ and $\langle \mathbf{U} \rangle \leq \langle \mathbf{V} \rangle$, so that $\mathbf{x} \in \langle \mathbf{V} \rangle$, showing that $\mathscr{D}(\mathbf{x}, \mathbf{V})$, whence (D3).

Finally, let $\mathbf{U} \subseteq \tilde{L}_i$, $\mathbf{x} \in \tilde{L}_i$, and suppose that $\mathscr{D}(\mathbf{x}, \mathbf{U})$. Then we have $\mathbf{x} \in \langle \mathbf{U} \rangle$, thus $\mathbf{x}$ has a representation of the form

$$\mathbf{x} = \lambda_1 \mathbf{y_1} + \cdots + \lambda_r \mathbf{y_r}, \quad (\mathbf{y}_\rho \in \mathbf{U}, \lambda_\rho \in \Lambda).$$

Hence, setting $\mathbf{U}_0 := \{\mathbf{y_1}, \ldots, \mathbf{y_r}\}$, we have $\mathbf{x} \in \langle \mathbf{U}_0 \rangle$, so that $\mathscr{D}(\mathbf{x}, \mathbf{U}_0)$, where $\mathbf{U}_0$ is a finite subset of $\mathbf{U}$, as required. This shows that Axiom (D4) holds as well, thus finishing the proof that $\mathscr{D}$ is a dependence relation on $\tilde{L}_i$.

12.13. We have $0 \in \Lambda'$ as $0 \cdot \mathbf{x} = \mathbf{0} \in L'$, and $\Lambda' > \{0\}$, since $\langle \mathbf{x} \rangle \cap L' \neq \{\mathbf{0}\}$ by hypothesis, so that $\lambda \mathbf{x} \in L'$ for some $\lambda \neq 0$.

Next, let $\lambda', \mu' \in \Lambda'$. Then, by definition of Λ', we have $\lambda'\mathbf{x}, \mu'\mathbf{x} \in L'$, so that

$$\lambda'\mathbf{x} - \mu'\mathbf{x} = (\lambda' - \mu')\mathbf{x} \in L',$$

showing that $\lambda' - \mu' \in \Lambda'$. Also, for arbitrary elements $\lambda \in \Lambda$ and $\lambda' \in \Lambda'$, we have $\lambda'\mathbf{x} \in L'$, thus

$$\lambda(\lambda'\mathbf{x}) = (\lambda\lambda')\mathbf{x} \in L',$$

showing that $\lambda\lambda' \in \Lambda'$, as desired. Hence, Λ' is a non-zero left ideal of Λ, as claimed.

12.14. Recall that the map $\tilde{\varphi} : \langle L', \mathbf{x}\rangle \to M$ defined in the proof of Lemma 12.8 is given by

$$\tilde{\varphi}(\mathbf{x}' + \lambda\mathbf{x}) = \varphi(\mathbf{x}') + \lambda\mathbf{x}_0, \quad (\mathbf{x}' \in L', \lambda \in \Lambda)$$

for some suitable vector $\mathbf{x}_0 \in M$. Hence, making use of linearity of the map $\varphi : L' \to M$, we have, for $\mathbf{x}_1', \mathbf{x}_2' \in L'$ and $\lambda_1, \lambda_2 \in \Lambda$,

$$\begin{aligned}\tilde{\varphi}((\mathbf{x}_1' + \lambda_1\mathbf{x}) + (\mathbf{x}_2' + \lambda_2\mathbf{x})) &= \tilde{\varphi}((\mathbf{x}_1' + \mathbf{x}_2') + (\lambda_1 + \lambda_2)\mathbf{x}) \\ &= \varphi(\mathbf{x}_1' + \mathbf{x}_2') + (\lambda_1 + \lambda_2)\mathbf{x}_0 \\ &= (\varphi(\mathbf{x}_1') + \lambda_1\mathbf{x}_0) + (\varphi(\mathbf{x}_2') + \lambda_2\mathbf{x}_0) \\ &= \tilde{\varphi}(\mathbf{x}_1' + \lambda_1\mathbf{x}) + \tilde{\varphi}(\mathbf{x}_2' + \lambda_2\mathbf{x}),\end{aligned}$$

which shows that $\tilde{\varphi}$ is additive. Also, again making use of linearity of the map φ, we have, for $\mathbf{x}_1' \in L'$ and $\lambda, \lambda_1 \in \Lambda$,

$$\begin{aligned}\tilde{\varphi}(\lambda(\mathbf{x}_1' + \lambda_1\mathbf{x})) = \tilde{\varphi}((\lambda\mathbf{x}_1') + (\lambda\lambda_1)\mathbf{x}) &= \varphi(\lambda\mathbf{x}_1') + (\lambda\lambda_1)\mathbf{x}_0 \\ &= \lambda(\varphi(\mathbf{x}_1') + \lambda_1\mathbf{x}_0) = \lambda\tilde{\varphi}(\mathbf{x}_1' + \lambda_1\mathbf{x}),\end{aligned}$$

as required. Hence, the map $\tilde{\varphi}$ is Λ-linear, as claimed.

12.15. Clearly, the relation $\leq$ on the set $\mathscr{E}$ is reflexive. Suppose that $(\tilde{L}_1, \tilde{\varphi}_1) \leq (\tilde{L}_2, \tilde{\varphi}_2)$ and $(\tilde{L}_2, \tilde{\varphi}_2) \leq (\tilde{L}_1, \tilde{\varphi}_1)$. Then $\tilde{L}_1 \subseteq \tilde{L}_2$ and $\tilde{L}_2 \subseteq \tilde{L}_1$, so that $\tilde{L}_1 = \tilde{L}_2$. Also, $\tilde{\varphi}_2 = \tilde{\varphi}_2|_{\tilde{L}_1} = \tilde{\varphi}_1$, thus $(\tilde{L}_1, \tilde{\varphi}_1) = (\tilde{L}_2, \tilde{\varphi}_2)$, showing that $\leq$ is antisymmetric as well. Next, suppose that $(\tilde{L}_1, \tilde{\varphi}_1) \leq (\tilde{L}_2, \tilde{\varphi}_2)$ and $(\tilde{L}_2, \tilde{\varphi}_2) \leq (\tilde{L}_3, \tilde{\varphi}_3)$. Then $\tilde{L}_1 \subseteq \tilde{L}_2$ and $\tilde{L}_2 \subseteq \tilde{L}_3$, thus $\tilde{L}_1 \subseteq \tilde{L}_3$, and

$$\tilde{\varphi}_3|_{\tilde{L}_1} = \tilde{\varphi}_3|_{\tilde{L}_2}|_{\tilde{L}_1} = \tilde{\varphi}_2|_{\tilde{L}_1} = \tilde{\varphi}_1,$$

so that $(\tilde{L}_1, \tilde{\varphi}_1) \leq (\tilde{L}_3, \tilde{\varphi}_3)$. Hence, $\leq$ is an order relation on the set $\mathscr{E}$.

Finally, suppose that $\mathscr{C} = \{(\tilde{L}_i, \tilde{\varphi}_i)\}_{i \in I}$ is a non-empty chain in $(\mathscr{E}, \leq)$, and let $\tilde{L} := \bigcup_{i \in I} \tilde{L}_i$. Since $\mathscr{C}$ is non-empty, we have $L' \subseteq \tilde{L} \subseteq L$. Let $\mathbf{x}, \mathbf{y} \in \tilde{L}$, and let $\lambda \in \Lambda$. Then $\mathbf{x} \in \tilde{L}_i$, $\mathbf{y} \in \tilde{L}_j$ for some $i, j \in I$, and (without loss of generality) $\tilde{L}_i \subseteq \tilde{L}_j$, since $\mathscr{C}$ is a chain. Thus, $\mathbf{x}, \mathbf{y} \in \tilde{L}_j$, so that $\mathbf{x} + \mathbf{y} \in \tilde{L}_j \subseteq \tilde{L}$. Also, we have $\lambda\mathbf{x} \in \tilde{L}_i \subseteq \tilde{L}$, so that $\tilde{L}$ is a Λ-submodule of L containing L'. Next, we define a map $\tilde{\varphi} : \tilde{L} \to M$ by setting

$$\tilde{\varphi}(\mathbf{x}) := \tilde{\varphi}_i(\mathbf{x}), \quad \mathbf{x} \in \tilde{L}_i.$$

By the chain property of $\mathscr{C}$, $\tilde{\varphi}$ is well defined and a Λ-linear map. Also, we have

$$\tilde{\varphi}|_{L'} = \tilde{\varphi}|_{\tilde{L}_i}|_{L'} = \tilde{\varphi}_i|_{L'} = \varphi$$

by construction, so that $(\tilde{L}, \tilde{\varphi}) \in \mathscr{E}$, and $(\tilde{L}, \tilde{\varphi}) \geq (\tilde{L}_i, \tilde{\varphi}_i)$ for all indices $i \in I$, thus is an upper bound for the chain $\mathscr{C}$. Hence, $(\mathscr{E}, \leq)$ is inductively ordered as claimed.

12.16. (a) Let Λ be a ring with identity element, and let X be any set. Define a Λ-module F_X, whose elements are formal sums $\sum_{\rho=1}^{r} \lambda_\rho \mathbf{x}_\rho$ with $r \in \mathbb{N}_0$, $\lambda_\rho \in \Lambda$, and $\mathbf{x}_\rho \in X$, and with addition and scalar multiplication given by

$$\Big(\sum_\rho \lambda_\rho \mathbf{x}_\rho\Big) + \Big(\sum_\rho \lambda'_\rho \mathbf{x}_\rho\Big) := \sum_\rho (\lambda_\rho + \lambda'_\rho)\mathbf{x}_\rho,$$

$$\lambda \cdot \sum_\rho \lambda_\rho \mathbf{x}_\rho := \sum_\rho (\lambda \lambda_\rho)\mathbf{x}_\rho, \quad \lambda \in \Lambda.$$

The axioms of a (unital) Λ-module obviously hold for F_X and, identifying a vector $\mathbf{x} \in X$ with $1 \cdot \mathbf{x} \in F_X$, it is clear that F_X is free with basis X.

(b) Let M be a free Λ-module with basis X, and let $\varphi : X \to N$ be a set-theoretic map of X into the Λ-module N. Setting

$$\hat{\varphi}\Big(\sum_\rho \lambda_\rho \mathbf{x}_\rho\Big) := \sum_\rho \lambda_\rho \varphi(\mathbf{x}_\rho), \quad (\lambda_\rho \in \Lambda, \mathbf{x}_\rho \in X)$$

yields a well-defined map $\hat{\varphi} : M \to N$ extending φ; also, $\hat{\varphi}$ is obviously Λ-linear, as desired.

(c) Let

$$\mathbf{0} \longrightarrow M' \xrightarrow{\alpha} M \xrightarrow{\beta} M'' \longrightarrow \mathbf{0} \tag{14.39}$$

be an exact sequence of Λ-modules and Λ-linear maps, in which the module M'' is free. Let $X = \{\mathbf{x}_i\}_{i \in I}$ be a basis of M''. Since β is surjective by exactness of (14.39), there exists, for each $i \in I$, an element $\mathbf{y}_i \in M$ with $\beta(\mathbf{y}_i) = \mathbf{x}_i$. Moreover, by Part (b), the mapping $\mathbf{x}_i \mapsto \mathbf{y}_i$ extends uniquely to a Λ-linear map $\beta^* : M'' \to M$ and, by definition of β^*, we have

$$\beta\Big(\beta^*\Big(\sum_\rho \lambda_\rho \mathbf{x}_\rho\Big)\Big) = \beta\Big(\sum_\rho \lambda_\rho \mathbf{y}_\rho\Big) = \sum_\rho \lambda_\rho \mathbf{x}_\rho,$$

so that $\beta \circ \beta^* = 1_{M''}$, as desired.

(d) Let Λ be a ring with identity element, let M be a (unital) Λ-module, and let $Y = \{\mathbf{y}_i\}_{i \in I} \subseteq M$ be a generating system for M. Set $X := \{\mathbf{x}_i\}_{i \in I}$ with new symbols $\mathbf{x}_i$, and let F_X be the free Λ-module with basis X. By Part (b), the map $\varphi : X \to Y$ given by $\mathbf{x}_i \mapsto \mathbf{y}_i$ for $i \in I$ extends uniquely to a Λ-linear map $\hat{\varphi} : F_X \to M$, and $\hat{\varphi}$ is surjective, since $Y \subseteq \mathrm{image}(\hat{\varphi})$. Hence, $M \cong F_X/\ker(\hat{\varphi})$; thus, setting $F := F_X$ and $G := \ker(\hat{\varphi})$, we have found a free Λ-module F and a submodule $G \leq F$, such that $F/G \cong M$, as required.

13.1. Since Λ_0 is a left modular ideal of Λ, there exists some element $\lambda_0 \in \Lambda$, such that $\lambda - \lambda\lambda_0 \in \Lambda_0$ for all $\lambda \in \Lambda$. As $\Lambda_1 \geq \Lambda_0$ by hypothesis, we have

$$\lambda - \lambda\lambda_0 \in \Lambda_1, \quad \lambda \in \Lambda,$$

so that λ_0 is a right identity element of Λ modulo Λ_1, implying that Λ_1 is left modular, as claimed, since Λ_1 is assumed to be a left ideal of Λ.

13.2. Suppose that Λ_0 is a proper right modular ideal of the ring Λ, and let λ_0 be a left identity element of Λ modulo Λ_0; that is, $\lambda - \lambda_0\lambda \in \Lambda_0$ for all $\lambda \in \Lambda$. Let $\mathscr{I}$ be the set of all right ideals Λ' of Λ satisfying (i) $\Lambda_0 \leq \Lambda'$ and (ii) $\lambda_0 \notin \Lambda'$, ordered by inclusion. We have $\Lambda_0 \in \mathscr{I}$, since $\lambda_0 \in \Lambda_0$ would imply $\Lambda_0 = \Lambda$, which is not possible, since Λ_0 is assumed to be proper. If $\mathscr{C} = \{\Lambda'_\nu\}_{\nu \in N}$ is a non-empty chain in $\mathscr{I}$, then $\Lambda'' := \bigcup_{\nu \in N} \Lambda'_\nu$ is easily seen to be a right ideal of Λ meeting conditions (i) and (ii); hence, $(\mathscr{I}, \subseteq)$ is inductively ordered. By Zorn's lemma, $\mathscr{I}$ possesses a maximal element Λ^*. We claim that Λ^* is a maximal right ideal of Λ. Indeed, suppose that $\tilde{\Lambda}$ is a right ideal of Λ properly containing Λ^*. Then we must have $\lambda_0 \in \tilde{\Lambda}$, implying $\tilde{\Lambda} = \Lambda$. Hence, Λ^* is a maximal right ideal of Λ containing Λ_0, as desired. Finally, Proposition 4.8 follows from Proposition 13.12, since in a ring with left identity element every right ideal is right modular.

13.3. Let $(\Lambda, \cdot)$ be a multiplicative semigroup, let λ_0 be a left identity element of Λ, and let λ_1 be a right identity element. Then we have

$$\lambda_1 = \lambda_0 \lambda_1 = \lambda_0,$$

so that the element $\lambda' := \lambda_0 = \lambda_1$ is a two-sided identity element of the semigroup Λ. The analogous result holds for rings: if $(\Lambda, +, \cdot)$ is a ring, then $(\Lambda, \cdot)$ is a semigroup, and a left (right) identity element of the ring Λ is a left (right) identity element of the semigroup $(\Lambda, \cdot)$.

13.4. Suppose that $\lambda', \tilde{\lambda}' \in \Lambda'_1$. Then $\lambda' = \lambda - \lambda\lambda_1$ and $\tilde{\lambda}' = \tilde{\lambda} - \tilde{\lambda}\lambda_1$ for some $\lambda, \tilde{\lambda} \in \Lambda$. Hence,

$$\lambda' + \tilde{\lambda}' = (\lambda - \lambda\lambda_1) + (\tilde{\lambda} - \tilde{\lambda}\lambda_1) = (\lambda + \tilde{\lambda}) - (\lambda + \tilde{\lambda})\lambda_1 \in \Lambda'_1,$$

so that Λ'_1 is closed under addition. Next, let $\tilde{\lambda} \in \Lambda$ and let $\mu \in \Lambda'_1$. Then $\mu = \lambda - \lambda\lambda_1$ for some $\lambda \in \Lambda$. It follows that

$$\tilde{\lambda}\mu = \tilde{\lambda}(\lambda - \lambda\lambda_1) = (\tilde{\lambda}\lambda) - (\tilde{\lambda}\lambda)\lambda_1 \in \Lambda'_1;$$

thus, Λ'_1 is a left ideal of Λ. Also, we have

$$\lambda - \lambda\lambda_1 \in \Lambda'_1, \quad \lambda \in \Lambda,$$

so that λ_1 is a right identity element of Λ modulo Λ'_1. Hence, Λ'_1 is a modular left ideal of Λ, as claimed. Finally, suppose that $\lambda_1 \in \Lambda'_1$, and let $\tilde{\lambda} \in \Lambda$ be an arbitrary element. Then $\tilde{\lambda}\lambda_1 \in \Lambda'_1$, since Λ'_1 is a left ideal of Λ, and we have

$$\tilde{\lambda} - \tilde{\lambda}\lambda_1 = \lambda'_1 \in \Lambda'_1$$

by definition of Λ'_1. Hence,

$$\tilde{\lambda} = \lambda'_1 + \tilde{\lambda}\lambda_1 \in \Lambda'_1,$$

so that $\Lambda'_1 = \Lambda$. If, conversely, $\Lambda'_1 = \Lambda$, then obviously $\lambda_1 \in \Lambda'_1$, whence the result.

13.5. Since $0 \cdot \mathbf{x} = \mathbf{0}$, we have $0 \in \mathrm{Ann}_\Lambda(\mathbf{x})$; in particular, $\mathrm{Ann}_\Lambda(\mathbf{x})$ is non-empty. Next, if $\lambda, \mu \in \mathrm{Ann}_\Lambda(\mathbf{x})$, then $\lambda\mathbf{x} = \mathbf{0} = \mu\mathbf{x}$, thus

$$(\lambda - \mu)\mathbf{x} = \lambda\mathbf{x} - \mu\mathbf{x} = \mathbf{0} - \mathbf{0} = \mathbf{0}.$$

Hence, $\lambda - \mu \in \mathrm{Ann}_\Lambda(\mathbf{x})$, and the annihilator of $\mathbf{x}$ is closed under forming differences. Moreover, if $\mu \in \mathrm{Ann}_\Lambda(\mathbf{x})$ and $\lambda \in \Lambda$, then $\mu\mathbf{x} = \mathbf{0}$, and so

$$\mathbf{0} = \lambda \cdot \mathbf{0} = \lambda \cdot \mu\mathbf{x} = (\lambda\mu)\mathbf{x},$$

showing that $\lambda\mu \in \mathrm{Ann}_\Lambda(\mathbf{x})$. Hence, $\mathrm{Ann}_\Lambda(\mathbf{x})$ is a left ideal of Λ as claimed.

13.6. For arbitrary elements $\lambda, \mu, \nu \in \Lambda$, we have

$$\begin{aligned}(\lambda \circ \mu) \circ \nu &= (\lambda \circ \mu) + \nu - (\lambda \circ \mu)\nu \\ &= \lambda + \mu - \lambda\mu + \nu - (\lambda + \mu - \lambda\mu)\nu \\ &= \lambda + \mu + \nu - \lambda\mu - \lambda\nu - \mu\nu + \lambda\mu\nu\end{aligned}$$

and

$$\begin{aligned}\lambda \circ (\mu \circ \nu) &= \lambda + (\mu \circ \nu) - \lambda(\mu \circ \nu) \\ &= \lambda + \mu + \nu - \mu\nu - \lambda(\mu + \nu - \mu\nu) \\ &= \lambda + \mu + \nu - \lambda\mu - \lambda\nu - \mu\nu + \lambda\mu\nu.\end{aligned}$$

Hence, we have

$$(\lambda \circ \mu) \circ \nu = \lambda \circ (\mu \circ \nu), \quad (\lambda, \mu, \nu \in \Lambda),$$

as desired.

13.7. (i) Let $\lambda \in \mathscr{C}(\Lambda_1)$ be a quasi-regular element of Λ_1. Then, by Lemma 13.6, there exists some $\lambda' \in \Lambda_1$ such that $\lambda \circ \lambda' = 0 = \lambda' \circ \lambda$. Thus,

$$\varphi(\lambda) \circ \varphi(\lambda') = 0 = \varphi(\lambda') \circ \varphi(\lambda),$$

so that $\varphi(\lambda)$ is a quasi-regular element of Λ_2. Hence, the restriction $\mathscr{C}(\varphi)$ is a well-defined map from $\mathscr{C}(\Lambda_1)$ to $\mathscr{C}(\Lambda_2)$, and φ respects the circle operation, since it respects addition and multiplication of the rings involved.

(ii) If Λ is a ring, and $1_\Lambda : \Lambda \to \Lambda$ is the identity map on Λ, then, clearly, its restriction $\mathscr{C}(1_\Lambda)$ is the identity map on $\mathscr{C}(\Lambda)$.

(iii) Suppose that Λ_1, Λ_2, and Λ_3 are rings, and that $\varphi_1 : \Lambda_1 \to \Lambda_2$ and $\varphi_2 : \Lambda_2 \to \Lambda_3$ are ring homomorphisms. In this situation, the composition $\varphi_2 \circ \varphi_1 : \Lambda_1 \to \Lambda_3$ is defined, and is again a ring homomorphism; in particular, the induced group homomorphism

$$\mathscr{C}(\varphi_2 \circ \varphi_1) : \mathscr{C}(\Lambda_1) \to \mathscr{C}(\Lambda_3)$$

is well defined. On the other hand, the composition $\mathscr{C}(\varphi_2) \circ \mathscr{C}(\varphi_1)$ of the induced group homomorphisms $\mathscr{C}(\varphi_1) : \mathscr{C}(\Lambda_1) \to \mathscr{C}(\Lambda_2)$ and $\mathscr{C}(\varphi_2) : \mathscr{C}(\Lambda_2) \to \mathscr{C}(\Lambda_3)$ is defined and yields a group homomorphism

$$\mathscr{C}(\varphi_2) \circ \mathscr{C}(\varphi_1) : \mathscr{C}(\Lambda_1) \longrightarrow \mathscr{C}(\Lambda_3).$$

We have, for an arbitrary element $\lambda \in \mathscr{C}(\Lambda_1)$,

$$\begin{aligned}(\mathscr{C}(\varphi_2) \circ \mathscr{C}(\varphi_1))(\lambda) &= \mathscr{C}(\varphi_2)(\mathscr{C}(\varphi_1)(\lambda)) \\ &= \varphi_2|_{\mathscr{C}(\Lambda_2)}(\varphi_1|_{\mathscr{C}(\Lambda_1)}(\lambda)) \\ &= (\varphi_2|_{\mathscr{C}(\Lambda_2)} \circ \varphi_1|_{\mathscr{C}(\Lambda_1)})(\lambda) \\ &= (\varphi_2 \circ \varphi_1)|_{\mathscr{C}(\Lambda_1)}(\lambda) \\ &= \mathscr{C}(\varphi_2 \circ \varphi_1)(\lambda).\end{aligned}$$

It follows that

$$\mathscr{C}(\varphi_2 \circ \varphi_1) = \mathscr{C}(\varphi_2) \circ \mathscr{C}(\varphi_1),$$

as desired.

13.8. Let Λ be a ring with two-sided identity element 1, and let $\mathscr{U}(\Lambda)$ be the group of units of Λ. If $\lambda \in \mathscr{C}(\Lambda)$ then, by Lemma 13.6, there exists some $\lambda' \in \Lambda$, such that $\lambda \circ \lambda' = 0 = \lambda' \circ \lambda$. Thus

$$(1-\lambda)(1-\lambda') = 1-\lambda-\lambda'+\lambda\lambda' = 1-(\lambda\circ\lambda') = 1,$$

and

$$(1-\lambda')(1-\lambda) = 1-\lambda-\lambda'+\lambda'\lambda = 1-(\lambda'\circ\lambda) = 1.$$

Hence, $1-\lambda \in \mathscr{U}(\Lambda)$, and the map

$$\Phi : \mathscr{C}(\Lambda) \longrightarrow \mathscr{U}(\Lambda)$$

given by $\lambda \mapsto 1-\lambda$ is well defined. Also, for $\lambda_1, \lambda_2 \in \mathscr{C}(\Lambda)$,

$$\Phi(\lambda_1\circ\lambda_2) = 1-(\lambda_1\circ\lambda_2) = 1-\lambda_1-\lambda_2+\lambda_1\lambda_2 = (1-\lambda_1)(1-\lambda_2) = \Phi(\lambda_1)\cdot\Phi(\lambda_2),$$

showing that Φ is a group homomorphism.

Conversely, suppose that $\lambda \in \mathscr{U}(\Lambda)$ with inverse λ'. Then

$$\begin{aligned}(1-\lambda)\circ(1-\lambda') &= (1-\lambda)+(1-\lambda')-(1-\lambda)(1-\lambda')\\ &= 2-\lambda-\lambda'-1+\lambda+\lambda'-\lambda\lambda' = 1-\lambda\lambda' = 0,\end{aligned}$$

and

$$\begin{aligned}(1-\lambda')\circ(1-\lambda) &= (1-\lambda')+(1-\lambda)-(1-\lambda')(1-\lambda)\\ &= 2-\lambda'-\lambda-1+\lambda'+\lambda-\lambda'\lambda = 1-\lambda'\lambda = 0.\end{aligned}$$

Hence, $1-\lambda$ is a quasi-regular element of Λ, and we obtain a well-defined map

$$\Psi : \mathscr{U}(\Lambda) \longrightarrow \mathscr{C}(\Lambda), \quad \lambda \longmapsto 1-\lambda.$$

Moreover, for $\lambda_1, \lambda_2 \in \mathscr{U}(\Lambda)$ with inverses λ_1' and λ_2', respectively, we have

$$\Psi(\lambda_1\lambda_2) = 1-\lambda_1\lambda_2 = (1-\lambda_1)\circ(1-\lambda_2) = \Psi(\lambda_1)\circ\Psi(\lambda_2),$$

so that Ψ is again a group homomorphism. Since, obviously, $\Psi\circ\Phi = 1_{\mathscr{C}(\Lambda)}$ and $\Phi\circ\Psi = 1_{\mathscr{U}(\Lambda)}$, the maps Φ and Ψ are group isomorphisms inverse to each other, and our claim (13.9) follows. The particular statement is an immediate consequence of (13.9).

13.9. Let $\lambda = m\zeta \in \mathscr{C}(m\mathbb{Z})$ be non-zero. Then there exists some $\zeta' \in \mathbb{Z}$, such that

$$\lambda\circ\lambda' = (m\zeta)\circ(m\zeta') = 0 = m\zeta+m\zeta'-m^2\zeta\zeta',$$

or, equivalently,

$$\zeta+\zeta' = m\zeta\zeta'. \tag{14.40}$$

In particular, (14.40) implies that $\zeta = 0 \Leftrightarrow \zeta' = 0$, so that we may assume that $\zeta, \zeta' \neq 0$. In this situation, (14.40) first implies that $|\zeta| = |\zeta'|$ (since $\zeta \mid \zeta'$ and $\zeta' \mid \zeta$), and then $\zeta = \zeta'$, as $\zeta, \zeta' \neq 0$. Consequently, (14.40) simplifies to the equivalent condition that $\lambda = m\zeta = 2$, which is only possible for $m = 1, 2$. Hence, for $m > 2$, $\mathscr{C}(m\mathbb{Z})$ is trivial while, for $m = 1, 2$, $\lambda = 2$ is the unique non-zero element of $\mathscr{C}(m\mathbb{Z})$. Summarising, we have found that

$$\mathscr{C}(m\mathbb{Z}) = \begin{Bmatrix} \{0,2\}, & m = 1,2 \\ \{0\}, & m > 2 \end{Bmatrix}, \quad m \geq 1. \tag{14.41}$$

We note that, since $\mathscr{U}(\mathbb{Z}) = \{1, -1\}$, the result of Exercise 5 gives $\mathscr{C}(\mathbb{Z}) = \{0, 2\}$, in accordance with (14.41).

13.10 One checks immediately that

$$[2]_8 \circ [2]_8 = [4]_8 \circ [4]_8 = [6]_8 \circ [6]_8 = [0]_8,$$

so that $\mathfrak{J}(2\mathbb{Z}_8) = 2\mathbb{Z}_8$ and $\mathscr{C}(2\mathbb{Z}_8) = (2\mathbb{Z}_8, \circ) \cong C_2 \times C_2$, the Klein 4-group. On the other hand, one observes that the element $[2]_8$ has order 4 in $(2\mathbb{Z}_8, +)$, thus $(2\mathbb{Z}_8, +) \cong C_4$, as claimed.

13.11. One possible example of such a ring Λ may be constructed as follows. Let $R = (2\mathbb{Z}, +, \cdot)$ be the ring of even integers under the usual addition and multiplication. Note that R is a commutative ring without zero divisors, but does not have an identity element, since $1 \notin R$. Moreover, let $\Lambda = M_2(R)$ be the set of (2×2)-matrices with entries in R. As usual, we define addition and multiplication of matrices $A, B \in M_2(R)$ via

$$\begin{pmatrix} a_{1,1} & a_{1,2} \\ a_{2,1} & a_{2,2} \end{pmatrix} + \begin{pmatrix} b_{1,1} & b_{1,2} \\ b_{2,1} & b_{2,2} \end{pmatrix} = \begin{pmatrix} a_{1,1}+b_{1,1} & a_{1,2}+b_{1,2} \\ a_{2,1}+b_{2,1} & a_{2,2}+b_{2,2} \end{pmatrix},$$

$$\begin{pmatrix} a_{1,1} & a_{1,2} \\ a_{2,1} & a_{2,2} \end{pmatrix} \begin{pmatrix} b_{1,1} & b_{1,2} \\ b_{2,1} & b_{2,2} \end{pmatrix} = \begin{pmatrix} a_{1,1}b_{1,1}+a_{1,2}b_{2,1} & a_{1,1}b_{1,2}+a_{1,2}b_{2,2} \\ a_{2,1}b_{1,1}+a_{2,2}b_{2,1} & a_{2,1}b_{1,2}+a_{2,2}b_{2,2} \end{pmatrix}.$$

It is straightforward to show that, with these operations of addition and multiplication, Λ forms a ring. However, the only candidate for an identity element would be the 2-dimensional identity matrix

$$\begin{pmatrix} 1 & 0 \\ 0 & 1 \end{pmatrix},$$

which does not lie in Λ, since $1 \notin R$. Also, Λ is not commutative; for instance,

$$\begin{pmatrix} 2 & 2 \\ 4 & 6 \end{pmatrix} \begin{pmatrix} 2 & 4 \\ 0 & 8 \end{pmatrix} = \begin{pmatrix} 4 & 24 \\ 8 & 64 \end{pmatrix} \neq \begin{pmatrix} 20 & 28 \\ 32 & 48 \end{pmatrix} = \begin{pmatrix} 2 & 4 \\ 0 & 8 \end{pmatrix} \begin{pmatrix} 2 & 2 \\ 4 & 6 \end{pmatrix}.$$

Moreover, Λ has zero divisors; for instance, we have

$$\begin{pmatrix} 0 & 2 \\ 0 & 4 \end{pmatrix} \begin{pmatrix} 2 & 8 \\ 0 & 0 \end{pmatrix} = \begin{pmatrix} 0 & 0 \\ 0 & 0 \end{pmatrix}, \text{ while } \begin{pmatrix} 0 & 2 \\ 0 & 4 \end{pmatrix}, \begin{pmatrix} 2 & 8 \\ 0 & 0 \end{pmatrix} \neq \begin{pmatrix} 0 & 0 \\ 0 & 0 \end{pmatrix}.$$

13.12. We shall discuss here the case of left ideals; the other two cases being similar, and left to the reader.

(i) Since $0 \in \Lambda_\nu$ for all $\nu \in N$, we have $0 \in \Lambda'$; in particular, $\Lambda' \neq \emptyset$. Next, let $\lambda_1', \lambda_2' \in \Lambda'$. Then $\lambda_1', \lambda_2' \in \Lambda_\nu$ for all $\nu \in N$, thus $\lambda_1' - \lambda_2' \in \Lambda_\nu$ for all ν, since each Λ_ν is a subring of Λ. Hence, $\lambda_1' - \lambda_2' \in \Lambda'$, so that Λ' is closed under taking differences. Next, let $\lambda' \in \Lambda'$ and $\lambda \in \Lambda$ be arbitrary elements. Then $\lambda' \in \Lambda_\nu$ for all ν, so $\lambda\lambda' \in \Lambda_\nu$ for all $\nu \in N$ since each Λ_ν is a left ideal of Λ by hypothesis, hence $\lambda\lambda' \in \Lambda'$, showing that Λ' is again a left ideal of Λ.

(ii) Set

$$\mathscr{S} := \{\Lambda' : \Lambda' \text{ a left ideal of } \Lambda, \text{ and } S \subseteq \Lambda'\}.$$

We have $\mathscr{S} \neq \emptyset$, since $\Lambda \in \mathscr{S}$. Let

$$\langle S \rangle_\ell := \bigcap_{\Lambda' \in \mathscr{S}} \Lambda'.$$

By Part (i), $\langle S \rangle_\ell$ is a left ideal of Λ, and we have $S \subseteq \langle S \rangle_\ell$ by definition of $\langle S \rangle_\ell$. Moreover, if Λ' is any left ideal of Λ containing S, then $\Lambda' \in \mathscr{S}$, thus $\langle S \rangle_\ell \leq \Lambda'$. Hence, $\langle S \rangle_\ell$ is the smallest left ideal of Λ containing the subset S, as required.

13.13. Suppose that $\lambda = \lambda_1 + \lambda_2$ and $\mu = \mu_1 + \mu_2$ are elements of $\Lambda_1 + \Lambda_2$, where $\lambda_i, \mu_i \in \Lambda_i$ for $i = 1, 2$. Then $\lambda_i - \mu_i \in \Lambda_i$, so

$$\lambda - \mu = (\lambda_1 + \lambda_2) - (\mu_1 + \mu_2) = (\lambda_1 - \mu_1) + (\lambda_2 - \mu_2) \in \Lambda_1 + \Lambda_2.$$

Also, for $\lambda = \lambda_1 + \lambda_2 \in \Lambda_1 + \Lambda_2$ and $\mu \in \Lambda$, we have

$$\mu\lambda = \mu(\lambda_1 + \lambda_2) = \mu\lambda_1 + \mu\lambda_2 \in \Lambda_1 + \Lambda_2.$$

Finally, since $\lambda_i = \lambda_i + 0 \in \Lambda_1 + \Lambda_2$ for $\lambda_i \in \Lambda_i$, we have $\Lambda_i \leq \Lambda_1 + \Lambda_2$; in particular, $\Lambda_1 + \Lambda_2 \neq \emptyset$, whence our claim.

13.14. Denote by L the right-hand side of (13.10). Since $\lambda \in \langle \lambda \rangle_\ell$ and $\langle \lambda \rangle_\ell$ is closed under addition, we have $n\lambda \in \langle \lambda \rangle_\ell$ for all integers n. Also, since $\langle \lambda \rangle_\ell$ is a left ideal, we have $\varepsilon\lambda \in \langle \lambda \rangle_\ell$ for all $\varepsilon \in \Lambda$. Hence, $L \subseteq \langle \lambda \rangle_\ell$. On the other hand, for $n, n_1, n_2 \in \mathbb{Z}$ and $\varepsilon, \varepsilon_1, \varepsilon_2, \delta \in \Lambda$,

$$(n_1\lambda + \varepsilon_1\lambda) - (n_2\lambda + \varepsilon_2\lambda) = (n_1 - n_2)\lambda + (\varepsilon_1 - \varepsilon_2)\lambda,$$
$$\delta(n\lambda + \varepsilon\lambda) = (n\delta + \delta\varepsilon)\lambda,$$

which, together with the fact that $\lambda = 1 \cdot \lambda + 0 \cdot \lambda \in L$, shows that L is a left ideal of Λ containing the element λ. Thus, by definition of $\langle \lambda \rangle_\ell$, we have $\langle \lambda \rangle_\ell \subseteq L$, whence the result.

13.15. (a) Let $I \leq \mathbb{Z}$ be an ideal. Then either $I = 0\mathbb{Z}$ is the zero ideal, or there exists an element n_0 of minimal positive value in I. If $n \in I$ is an arbitrary element, then division with remainder gives

$$n = qn_0 + r, \quad q, r \in \mathbb{Z}, 0 \leq r < n_0.$$

It follows that $r = n - qn_0 \in I$, thus necessarily $r = 0$ by minimality of n_0. Hence, $I = n_0\mathbb{Z}$ is a principal ideal.

(b) By Part (a), each ideal in $\mathbb{Z}$ is of the form $n\mathbb{Z}$ for some non-negative integer n. Moreover, we have $a\mathbb{Z} \leq b\mathbb{Z}$ if, and only if, $b \mid a$, and $a\mathbb{Z} = b\mathbb{Z}$ if, and only if, $a = b$. Hence, the maximal ideals of $\mathbb{Z}$ are precisely those ideals of the form $p\mathbb{Z}$ with some prime number p. Also, if $n \in \mathbb{Z}$ is such that $n^2 \in p\mathbb{Z}$, then $p \mid n^2$, implying $p \mid n$ by the fundamental theorem of arithmetic. Hence, $n \in p\mathbb{Z}$, and $p\mathbb{Z}$ satisfies Condition (13.7). By Corollary 13.11,

$$\mathfrak{J}(\mathbb{Z}) = \bigcap_{p \,\text{prime}} p\mathbb{Z} = \{0\},$$

as claimed. (To see the second equality, note that, for $n > 1$ and a prime q with $q \nmid n$, we have $n \notin q\mathbb{Z}$.)

13.16. Let n be an integer with $n > 1$, and let $\mathbb{Z}_n = \mathbb{Z}/n\mathbb{Z}$. The ideals in $\mathbb{Z}_n$ are of the form $m\mathbb{Z}/n\mathbb{Z}$, where m is a positive integer such that $m \mid n$; in particular,

$m\mathbb{Z}/n\mathbb{Z}$ is a maximal ideal of $\mathbb{Z}_n$ if, and only if, $m = p$ is a prime number dividing n. Also, if

$$(m+n\mathbb{Z})^2 = m^2 + n\mathbb{Z} \in p\mathbb{Z}/n\mathbb{Z},$$

then $m^2 \in p\mathbb{Z}$, implying $m \in p\mathbb{Z}$, so that $m + n\mathbb{Z} \in p\mathbb{Z}/n\mathbb{Z}$. Hence, the maximal ideals $p\mathbb{Z}/n\mathbb{Z}$ of $\mathbb{Z}_n$ satisfy Condition (13.7). By Corollary 13.11,

$$\mathfrak{J}(\mathbb{Z}_n) = \bigcap_{\substack{p\,\text{prime}\\ p|n}} p\mathbb{Z}/n\mathbb{Z} = \Big(\bigcap_{\substack{p\,\text{prime}\\ p|n}} p\mathbb{Z}\Big)/n\mathbb{Z} = p_1\dots p_r\mathbb{Z}/n\mathbb{Z},$$

where $p_1, p_2, \dots, p_r$ are the prime numbers dividing n. In particular, we have $\mathfrak{J}(\mathbb{Z}_n) = \{\overline{0}\}$ if, and only if, n is square-free. Also, $\mathfrak{J}(\mathbb{Z}_n) < \mathbb{Z}_n$ for $n \geq 2$.

14.1. By Lemma 14.4, the field $K = k(i)$ is algebraically closed, where i is a root of the polynomial x^2+1. As the field extension $k \leq K$ has degree 2, the elements 1 and i form a basis for K over k; thus, each element $c \in K$ can be uniquely written in the form $c = a + bi$, where $a, b \in k$. Define a map $^- : K \to K$ by $\overline{a+bi} = a - bi$. Then, as in the case of the field extension $\mathbb{R} \leq \mathbb{C}$, one sees that $^-$ is a field automorphism of K of order 2, such that $\overline{c} = c \iff c \in k$.

Now suppose that $p(x) \in k[x]$ is an irreducible polynomial of degree at least 2. Over K, the polynomial $p(x)$ decomposes as a product of linear factors with roots coming from $K \setminus k$. Let $c = a + bi \in K$ be such that $p(c) = 0$. Then $c \neq \overline{c}$, and $\overline{p(c)} = p(\overline{c}) = 0$. Hence, $p(x)$ must be divisible by

$$(x-c)(x-\overline{c}) = x^2 - (c+\overline{c})x + c\overline{c},$$

and we have

$$c + \overline{c} = (a+bi) + (a-bi) = 2a \in k$$

as well as

$$c\overline{c} = (a+bi)(a-bi) = a^2 - (bi)^2 = a^2 + b^2 \in k.$$

Since $p(x)$ is assumed to be irreducible, we must have $p(x) = \alpha(x-c)(x-\overline{c})$ for some $\alpha \in k$, whence our claim.

14.2. As in the proof of Lemma 14.8, we consider K as a well-ordered set

$$1 = a_0, a_1, \dots, a_\omega, \dots \tag{14.42}$$

and, for each ordinal involved in (14.42), we define fields k_ν, k_ν^* as follows: we set $k_0 = k_0^* = k$ and, if k_μ, k_μ^* are defined for $\mu < \nu$, we let

$$k_\nu^* := \bigcup_{\mu<\nu} k_\mu^*$$

and

$$k_\nu := \begin{cases} k_\nu^*(a_\nu) & \text{if } \alpha \notin \Sigma_\square(k_\nu^*(a_\nu)), \\ k_\nu^* & \text{otherwise.} \end{cases}$$

As before, Satz 2 in [75, §2] guarantees that all k_ν^* are fields, thus so are the k_ν, and that $\alpha \notin \Sigma_\square(k_\nu)$ for all ν occurring in (14.42), which implies that, in the field

$$\Omega := \bigcup_\nu k_\nu,$$

the element α is not a sum of squares. However, if $a = a_\nu \in K \setminus \Omega$, then $a \notin k_\nu$, so that, in $k_\nu^*(a)$, α is a sum of squares. A fortiori, we have $\alpha \in \Sigma_\square(\Omega(a))$, which shows that, in each proper algebraic extension Ω' of Ω, the element α is a sum of squares, as required.

14.3. Suppose first that $s > M$. Since $M \geq 1$ by definition, we have $s > 1$ thus, in particular

$$s^n \geq s^{n-1} \geq \cdots \geq s. \tag{14.43}$$

Also, for $1 \leq \nu \leq n$, we have

$$a_\nu s^{n-\nu} \geq -|a_\nu| s^{n-\nu} \geq -|a_\nu| s^{n-1}$$

by (14.43), which yields

$$\sum_{\nu=1}^{n} a_\nu s^{n-\nu} \geq -(|a_1| + \cdots + |a_n|) s^{n-1}.$$

Hence, for $s > M$,

$$f(s) = s^n + \sum_{\nu=1}^{n} a_\nu s^{n-\nu} \geq s^n - (|a_1| + \cdots + |a_n|) s^{n-1}.$$

However, by definition of M,

$$(|a_1| + \cdots + |a_n|) s^{n-1} \leq M s^{n-1} < s \cdot s^{n-1} = s^n,$$

implying $f(s) > s^n - s^n = 0$, as claimed.

Next, suppose that $s < -M$, and consider the quantity

$$(-1)^n f(s) = (-s)^n + \sum_{\nu=1}^{n} (-1)^\nu a_\nu (-s)^{n-\nu}.$$

As $-s > M \geq 1$, we have

$$(-s)^n \geq (-s)^{n-1} \geq \cdots \geq -s. \tag{14.44}$$

Combining (14.44) with the triangle inequality for the absolute value in K, we get

$$\left| \sum_{\nu=1}^{n} (-1)^\nu a_\nu (-s)^{n-\nu} \right| \leq \sum_{\nu=1}^{n} |a_\nu| (-s)^{n-\nu} \leq \sum_{\nu=1}^{n} |a_\nu| (-s)^{n-1}$$
$$= (|a_1| + \cdots + |a_n|)(-s)^{n-1}.$$

Since $s < -M$ by hypothesis, and $M \geq |a_1| + \cdots + |a_n|$ by definition of M,

$$(|a_1| + \cdots + |a_n|)(-s)^{n-1} \leq M(-s)^{n-1} < (-s) \cdot (-s)^{n-1} = (-s)^n,$$

so that

$$(-1)^n f(s) \geq (-s)^n - \left| \sum_{\nu=1}^{n} (-1)^\nu a_\nu (-s)^{n-\nu} \right| > (-s)^n - (-s)^n = 0,$$

again as desired.

REFERENCES

[1] J. Aczél, *Lectures on Functional Equations and Their Applications*, Academic Press, New York, 1966.

[2] S. A. Amitsur, A general theory of radicals. I: Radicals in complete lattices, *Amer. J. Math.* **74** (1952), 774–786.

[3] S. A. Amitsur, A general theory of radicals. II: Radicals in rings and bicategories, *Amer. J. Math.* **75** (1954), 100–125.

[4] S. A. Amitsur, A general theory of radicals. III: Applications, *Amer. J. Math.* **75** (1954), 126–136.

[5] M. F. Atiyah and I. G. Macdonald, *Introduction to Commutative Algebra*, Addison-Wesley Publ. Co., 1969.

[6] T. M. Apostol, *Mathematical Analysis*, Addison-Wesley Publ. Co., third printing 1969.

[7] J. R. Argand, Réflexions sur la nouvelle théorie des imaginaires, suivies d'une application à la démonstration d'un théorème d'analyse, *Annales de Math.* **5** (1814), 197–209.

[8] E. Artin, Kennzeichnung des Körpers der reellen algebraischen Zahlen, *Abh. Math. Sem. Univ. Hamburg* **3** (1924), 319–323.

[9] E. Artin, Über die Zerlegung definiter Funktionen in Quadrate, *Abh. Math. Sem. Univ. Hamburg* **5** (1927), 100–115.

[10] E. Artin and O. Schreier, Algebraische Konstruktion reeller Körper, *Abh. Math. Sem. Univ. Hamburg* **5** (1926), 85–99.

[11] R. Baer, Abelian groups that are direct summands of every containing abelian group, *Bull. Amer. Math. Soc.* **46** (1940), 800–806.

[12] G. Birkhoff, On the structure of abstract algebras, *Proc. Cambridge Phil. Soc.* **31** (1935), 433–454.

[13] G. Birkhoff, Subdirect unions in universal algebra, *Bull. Amer. Math. Soc.* **50** (1944), 764–768.

[14] G. Birkhoff, *Lattice Theory*, American Mathematical Society Colloquium Publications vol. XXV, third edition, American Mathematical Society, 1973.

[15] N. Bourbaki, *Algebra Vol. I*, Springer-Verlag, 2nd printing, 1989.

[16] G. Cantor, Über einen die trigonometrischen Reihen betreffenden Lehrsatz, *J. reine u. angew. Math.* **72** (1870), 130–138.

[17] G. Cantor, Beweis, daß eine für jeden reellen Wert von x durch eine trigonometrische Reihe gegebene Funktion $f(x)$ sich nur auf eine einzige Weise in dieser Form darstellen läßt, *J. reine u. angew. Math.* **72** (1870), 139–142.

[18] G. Cantor, Über die Ausdehnung eines Satzes aus der Theorie der trigonometrischen Reihen, *Math. Ann.* **5** (1872), 123–132.

[19] G. Cantor, Über eine Eigenschaft des Inbegriffes aller reellen algebraischen Zahlen, *J. reine u. angew. Math.* **77** (1874), 258–262.

[20] G. Cantor, Über eine elementare Frage der Mannigfaltigkeitslehre, *Jahresber. DMV* **1** (1891), 75–78.

[21] G. Cantor, Beiträge zur Begründung der transfiniten Mengenlehre, *Math. Ann.* **46** (1895), 481–512 and **49** (1897), 207–246.

[22] H. Cartan and S. Eilenberg, *Homological Algebra*, Princeton, 1956.

[23] P. M. Cohn, *Algebra Vol. 1*, second edition, John Wiley & Sons, 1993.

[24] J.-G. Darboux, Sur la composition des forces en statique, *Bull. Sci. Math.* **9** (1875), 281–288.

[25] J.-G. Darboux, Sur le theórème de la géometrie projective, *Math. Ann.* **17** (1880), 33–42.

[26] J. W. Dauben, *Georg Cantor – His Mathematics and Philosophy of the Infinite*, Harvard University Press, 1979.
[27] A. C. Davis, A characterisation of complete lattices, *Pacific J. Math.* **5** (1955), 311–319.
[28] R. P. Dilworth and A. M. Gleason, A generalized Cantor theorem, *Proc. Amer. Math. Soc.* **13** (1962), 704–705.
[29] Ph. Dwinger, *Introduction To Boolean Algebras*, Physica-Verlag, 1961.
[30] H. D. Ebbinghaus (et al.), *Zahlen*, second edition, Springer Verlag, 1992.
[31] A. A. Fraenkel, *Abstract Set Theory*, third revised edition, North-Holland Publishing Comnpany, 1966.
[32] P. Gabriel, Des catégories abéliennes, *Bull. Soc. Math. France* **90** (1962), 323–448.
[33] G. Grätzer, *Universal Algebra*, 2nd edition (with updates), Springer, 2008.
[34] P. R. Halmos, *Measure Theory*, D. van Nostrand Company, Inc., Princeton, ninth printing, 1964.
[35] G. K. W. Hamel, Eine Basis aller Zahlen und die unstetigen Lösungen der Funktionalgleichung $f(x+y) = f(x) + f(y)$, *Math. Ann.* **60** (1905), 459–462.
[36] H. Herrlich, *Axiom of Choice*, Lecture Notes in Math. Vol. 1876, Springer Verlag, 2006.
[37] D. Hilbert, Über invariante Eigenschaften spezieller binärer Formen, insbesondere der Kugelfunktionen, Doctoral dissertation, Königsberg, 1885.
[38] D. Hilbert, Mathematische Probleme, *Nachrichten von der Königlichen Gesellschaft der Wissenschaften zu Göttingen*, Math. Phys. Klasse Heft 3 (1900), 253–297.
[39] D. Hilbert, *Grundlagen der Geometrie*, Leipzig, 1922.
[40] D. Hilbert, Über ternäre definite Formen, *Acta Math.* **17** (1893), 169–197.
[41] A. Hajnal and A. Kertész, Some new algebraic equivalents of the axiom of choice, *Publ. Math. Debrecen* **19** (1973), 339–340.
[42] W. Hodges, Krull implies Zorn, *J. London Math. Soc. (2)* **19** (1979), 285–287.
[43] H. Herrlich and G. E. Strecker, *Category Theory*, Allyn and Bacon Inc., 1973.
[44] N. Jacobson, The radical and semi-simplicity for arbitrary rings, *Amer. J. Math.* **67** (1945), 695–707.
[45] N. Jacobson, *Structure of Rings*, Amer. Math. Soc. Colloq. Publ. XXXVII, 1956.
[46] T. J. Jech, *The Axiom of Choice*, North-Holland Publ. Co., 1973.
[47] F. Kasch, *Modules and Rings*, Academic Press, 1982.
[48] A. Kertész, A characterisation of the Jacobson radical, *Proc. Amer. Math. Soc.* **14** (1963), 595–597.
[49] A. Kertész, *Lectures on Artinian Rings*, Teubner-Verlag, 1968.
[50] A. Kertész, *Einführung in die Transfinite Algebra*, Birkhäuser Verlag, 1975.
[51] G. Köthe, Die Struktur der Ringe, deren Restklassenring nach dem Radikal vollständig reduzibel ist, *Math. Zeit.* **32** (1930), 161–186.
[52] C. Kuratowski, Une méthode d'elimination des nombres transfinis des raisonnements mathematiques, *Fund. Math.* **3** (1922), 76–108.
[53] K. Kuratowski and A. Mostowski, *Set Theory*, North Holland Publ. Co., 1968.
[54] E. Landau, Über die Darstellung definiter binärer Formen durch Quadrate, *Math. Ann.* **57** (1903), 53–64.
[55] E. Landau, Über die Darstellung definiter Funktionen durch Quadrate, *Math. Ann.* **62** (1906), 272–285.
[56] E. Landau, Über die Zerlegung total positiver Zahlen in Quadrate, Göttinger Nachr. 1919, 392–396.
[57] S. Lang, *Algebra*, Addison-Wesley Publ. Co., Inc., 1965.
[58] F. Levi, Arithmetische Gesetze im Gebiete diskreter Gruppen, *Rend. Palermo* **35** (1913), 225–236.
[59] J. C. Marica and S. J. Bryant, Unary algebras, *Pacific J. Math.* **10** (1960), 1347–1359.

[60] H. v. Mangoldt and K. Knopp, *Einfuehrung in die Hoehere Mathematik*, S. Hirzel Verlag, 1974.

[61] T. Nakayama, A remark on finitely generated modules, *Nagoya Math. Journal* **3** (1951), 139–140.

[62] E. Noether, Abstrakter Aufbau der Idealtheorie in Algebraischen Zahl- und Funktionenkörpern, *Math. Ann.* **96** (1927), 26–61.

[63] M. Novotný, Über Abbildungen von Mengen, *Pacific J. Math.* **13** (1963), 1359–1369.

[64] J. G. Oxley, *Matroid Theory*, Oxford University Press, 1992.

[65] B. Peirce, Linear associative algebra, *Amer. J. Math.* **4** (1881), 97–229.

[66] N. Popescu and L. Popescu, *Theory of Categories*, Sijthoff & Noordhoff International Publishers, 1979.

[67] R. Rado, Axiomatic treatment of rank in infinite sets, *Canad. J. Math.* **1** (1949), 337–343.

[68] C. Reid, *Hilbert*, Springer-Verlag, 1996.

[69] J. J. Rotman, *Advanced Modern Algebra*, Pearson Education, Inc., 2002.

[70] H. Schubert, *Categories*, Springer-Verlag, 1972.

[71] I. Schur, Neue Begründung der Theorie der Gruppencharaktere, *Sitzungsberichte der Preussischen Akademie der Wissenschaften*, 1905, Physikalisch-Mathematische Klasse, 406–432.

[72] D. S. Scott, Prime ideal theorems for rings, lattices and Boolean algebras, *Bull. Amer. Math. Soc.* **60** (1954), 390.

[73] J.-P. Serre, *Trees*, Springer-Verlag, 1980.

[74] A. Solian, *Theory of Modules*, John Wiley & Sons, 1977.

[75] E. Steinitz, *Algebraische Theorie der Körper* (R. Baer and H. Hasse, eds.), Chelsea Publ. Co., 1950.

[76] P. Suppes, *Axiomatic Set Theory*, van Nostrand Reinhold Company, 1960.

[77] F. Szász, Lösung eines Problems bezüglich einer Charakterisierung des Jacobsonschen Radikals, *Acta Math. Acad. Sci. Hung.* **18** (1967), 261–272.

[78] F. Szász, Äquivalenzrelation für die Charakterisierung des Jacobsonschen Radikals, *Acta Math. Acad. Sci. Hung.* **22** (1971), 85–86.

[79] A. Tarski, A lattice-theoretic fixpoint theorem and its applications, *Pacific J. Math.* **5** (1955), 285–309.

[80] G. Vitali, Sul problema della misura dei gruppi di punti di una retta, Bologna, 1905.

[81] B. L. van der Waerden, *Algebra I*, ninth edition, Springer-Verlag, 1993.

[82] M. W. Weaver, On the commutativity of a correspondence and a permutation, *Pacific J. Math.* **10** (1960), 105–111.

[83] H. Weber, Lehrbuch der Algebra Bd. I, F. Vieweg und Sohn, 1898.

[84] J. H. M. Wedderburn, On hypercomplex numbers, *Proc. LMS* Vol. s2-6 (1908), 77–118.

[85] H. Whitney, On the abstract properties of linear dependence, *Amer. J. Math.* **57** (1935), 509–533.

For EU product safety concerns, contact us at Calle de José Abascal, 56–1°, 28003 Madrid, Spain or eugpsr@cambridge.org.

www.ingramcontent.com/pod-product-compliance
Ingram Content Group UK Ltd.
Pitfield, Milton Keynes, MK11 3LW, UK
UKHW021808160426
469997UK00018B/239
9781009737876